Polyphosphazenes in Biomedicine, Engineering, and Pioneering Synthesis

ACS SYMPOSIUM SERIES **1298**

Polyphosphazenes in Biomedicine, Engineering, and Pioneering Synthesis

Alexander K. Andrianov, Editor
Institute for Bioscience and Biotechnology Research
University of Maryland
Rockville, Maryland

Harry R. Allcock, Editor
Department of Chemistry
The Pennsylvania State University
University Park, Pennsylvania

Sponsored by the
ACS Division of Polymeric Materials: Science and Engineering
and
ACS Division of Polymer Chemistry, Inc.

American Chemical Society, Washington, DC

Distributed in print by Oxford University Press

Library of Congress Cataloging-in-Publication Data

Names: Andrianov, Alexander K., editor. | Allcock, H. R., editor. | American Chemical Society. Division of Polymeric Materials: Science and Engineering. | American Chemical Society. Division of Polymer Chemistry.
Title: Polyphosphazenes in biomedicine, engineering, and pioneering synthesis / Alexander K. Andrianov, editor (Institute for Bioscience and Biotechnology Research, University of Maryland, Rockville, Maryland), Harry R. Allcock, editor (Department of Chemistry, The Pennsylvania State University, University Park, Pennsylvania) ; sponsored by the ACS Division of Polymeric Materials: Science and Engineering, and ACS Division of Polymer Chemistry.
Description: Washington, DC : American Chemical Society, [2018] | Series: ACS symposium series ; 1298 | Includes bibliographical references and index.
Identifiers: LCCN 2018033781 (print) | LCCN 2018036987 (ebook) | ISBN 9780841233607 (ebook) | ISBN 9780841233614
Subjects: LCSH: Polyphosphazenes. | Polyphosphazenes--Industrial applications.
Classification: LCC TP248.P675 (ebook) | LCC TP248.P675 P65 2018 (print) | DDC 668.9/2--dc23
LC record available at https://lccn.loc.gov/2018033781

Distributed in print by Oxford University Press

Foreword

The ACS Symposium Series was first published in 1974 to provide a mechanism for publishing symposia quickly in book form. The purpose of the series is to publish timely, comprehensive books developed from the ACS sponsored symposia based on current scientific research. Occasionally, books are developed from symposia sponsored by other organizations when the topic is of keen interest to the chemistry audience.

Before agreeing to publish a book, the proposed table of contents is reviewed for appropriate and comprehensive coverage and for interest to the audience. Some papers may be excluded to better focus the book; others may be added to provide comprehensiveness. When appropriate, overview or introductory chapters are added. Drafts of chapters are peer-reviewed prior to final acceptance or rejection, and manuscripts are prepared in camera-ready format.

As a rule, only original research papers and original review papers are included in the volumes. Verbatim reproductions of previous published papers are not accepted.

ACS Books Department

Contents

Fundamentals: Synthesis, Degradation Pathways, and Molecular Modeling

Indexes

Preface

The synthesis and utilization of polymer molecules with inorganic elements in the backbone has been an important objective since the discovery and commercialization of organosilicon ("silicone") polymers in the 1940s. The broader possibilities for inorganic–organic polymers are now evident from the widespread development of the polyphosphazene platform. The continuing expansion of this system at both the fundamental and applied levels provides ample evidence for the validity of the "inorganic polymer" concept.

The origins of polyphosphazenes can be traced back to the 1890s; however, the study of *organo*phosphazene polymers, which began in the 1960s, has expanded rapidly in recent years, until today hundreds of different macromolecules with widely diverse properties are now known. Many have been studied from a fundamental chemistry viewpoint, and a growing number have been optimized for practical applications, especially in biomedicine and advanced engineering. A major difference between classical polymer science and polyphosphazenes is the ease with which different organic, organometallic, or inorganic side groups can be utilized in the phosphazene system — an advantage that facilitates the development of unique structure–property relationships and practical uses. Thus, the recent developments of this system are part of a widening international interest in this field, which promises to further diversify the chemistry and applications of this system.

For this reason, a symposium on the subject of "*Polyphosphazenes in Biomedicine, Engineering & Pioneering Synthesis*" was held at a recent meeting of the American Chemical Society (ACS) in August 2017 in Washington, DC. The chapters in this book provide a summary of the international contributions reported at that meeting, the purpose of which was to bring together a broad range of topics, research investigators, and representatives from industry to discuss the current status of different aspects of this field.

Our deepest appreciation goes to the ACS Division of Polymeric Materials: Science and Engineering, the ACS Division of Polymer Chemistry, White Square Chemical, Inc., and CeloNova BioSciences, Inc. Without the generous support of these institutions, the symposium and this book could not have been possible. Their contributions have allowed us to take a glance into the present status and future developments in this growing field of polymer science and applications.

Alexander K. Andrianov
Institute for Bioscience and Biotechnology Research
University of Maryland
Rockville, Maryland 20850, United States

Harry R. Allcock
Evan Pugh Professor
Department of Chemistry
The Pennsylvania State University
University Park, Pennsylvania 16802, United States

Overview

Chapter 1

Synthesis, Structures, and Emerging Uses for Poly(organophosphazenes)

Harry R. Allcock*

Evan Pugh Professor of Chemistry, The Pennsylvania State University, University Park, Pennsylvania 16802, United States
***E-mail: hra1@psu.edu.**

Several different methods exist for the synthesis of poly(organophosphazenes), including ring-opening polymerization followed by macromolecular substitution, and several condensation methods. The macromolecular substitution alternative allows the linkage of a very wide range of different side groups to the inorganic backbone via reactions that permit the structure of the final polymer to be controlled by the electronic and steric characteristics of the side groups. The product polymers are elastomers, film- and fiber-formers, gels, membranes, and bioactive or biostable macromolecules that have many actual and potential biomedical or engineering uses. These opportunities are illustrated by a table of structure-property relationships which provides guidance for the design of polymers with predictable combinations of characteristics and potential uses.

Background

The first report of a polyphosphazene was by Henry Stokes in 1897 who found that heating the chlorophosphazene cyclic trimer (**1**), tetramer, or higher ring systems led to a striking increase in the viscosity of the melt and the formation of a rubbery elastomer (Scheme 1) (*1*). This polymer was later described as "inorganic rubber". Inorganic rubber bore many resemblances to crosslinked natural rubber and it would have been utilized technologically were it not for its sensitivity to atmospheric hydrolysis and its subsequent breakdown to ammonium chloride and phosphates, a change that destroyed all of its elastomeric properties. It was swelled by benzene, but did not dissolve in that or any other solvent.

My involvement with this field began during a postdoctoral year spent at Purdue University, where the U.S. Army was supporting a program to find alternatives to poly(organosiloxanes) (silicones) and classical organic polymers. Our first instincts were to follow the silicone synthesis example by first preparing organic-derivatives of cyclophosphazene rings $(NPR_2)_3$ (**3**) followed by thermolysis of these (Scheme 1). However, the small molecule organo*cyclo*phosphazenes did not polymerize, although some examples do undergo ring-expansion reactions. Thus, this initiative came to an end, and I moved on to work in other laboratories to learn something about classical organic polymer chemistry.

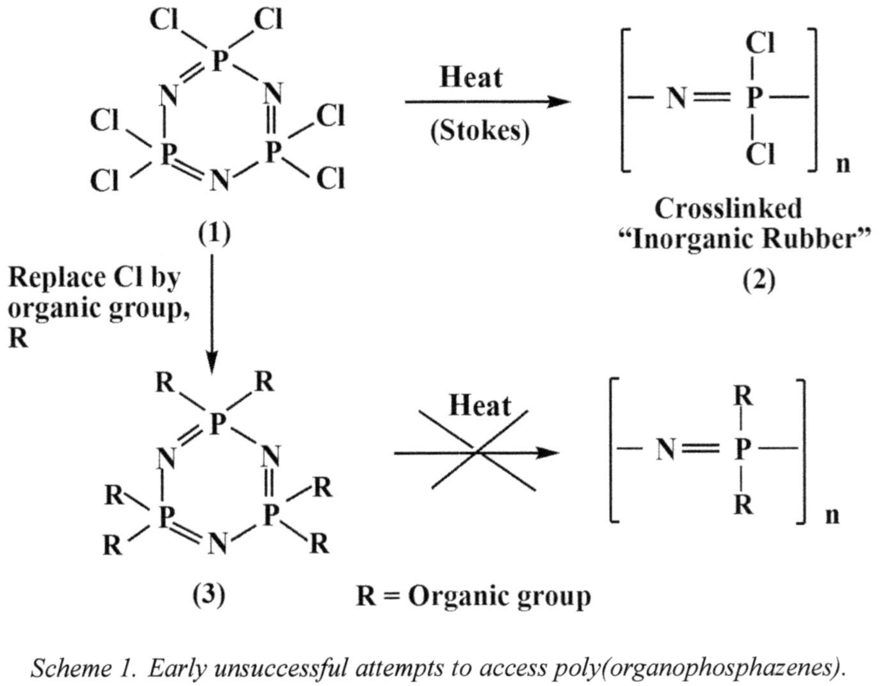

Scheme 1. Early unsuccessful attempts to access poly(organophosphazenes).

It was during the five years I spent in the American Cyanamid long-range research laboratory in Stamford, Connecticut that we began to understand why no progress had been possible with this system. It seemed likely that the crosslinking of inorganic rubber was due to presence of P-OH groups along the polymer chains caused by hydrolysis of P-Cl bonds in either the cyclic trimer or the high polymer, or probably in both. Subsequent reactions between the P-Cl and P-OH groups on separate chains would cause formation of the crosslinks and, because inorganic rubber was crosslinked and insoluble, direct replacement of all the chlorine atoms in the system by organic groups was not feasible.

Thus, together with my coworker, Robert Kugel, we established a protocol to purify the cyclic starting material, $(NPCl_2)_3$ (**1**), with rigorous exclusion of moisture to ensure that no P-OH groups could be formed. This compound was then heated at ~250°C in an evacuated sealed glass tube, and it polymerized to a clear, colorless rubbery material (**4**) that *was completely soluble in benzene or tetrahydrofuran.* Solutions of this polymer were then treated with the sodium salts of alcohols or phenols or with amines to replace all the chlorine atoms and yield stable, uncrosslinked polymers (Scheme 2). The first polymers we synthesized had ethoxy, methoxy, trifluoroethoxy, phenoxy, or methylamino side groups. They had very high molecular weights, typically corresponding to an average of 15,000 repeating units, and were stable in the atmosphere. In retrospect, it is remarkable that all ~30,000 chlorine atoms along each polymer chain can be replaced by organic groups, but these results illustrated the broad synthetic potential of the macromolecular substitution method. This work was published as a communication in the *Journal of the American Chemical Society* (*2*) and as full papers in *Inorganic Chemistry* (*3*, *4*), These polymers proved to be the forerunners of several hundred different macromolecules that have been produced by this method since that time both in our program at the Pennsylvania State University and in other laboratories worldwide (*5–7*).

Overview of Synthesis Methods

Macromolecular Substitution

The original synthesis process that depends on a ring-opening polymerization of hexachlorocyclotriphosphazene, $(NPCl_2)_3$, followed by replacement of the chlorine atoms by organic nucleophiles (*2–4*), has yielded the largest number of different polyphosphazenes. This process is summarized in Scheme 2.

The essential features of the macromolecular substitution method are as follows:

(1) The chlorine replacement process is a nucleophilic substitution reaction. In principle, almost any organic or inorganic nucleophile should be appropriate for this reaction. In practice, certain constraints exist, which are as follows.

(2) Bulky nucleophiles have difficulty replacing all the chlorine atoms along a polyphosphazene chain. Sometimes, only half of the chlorine atoms (often one chlorine per phosphorus) are replaced due to steric inhibition.

With the bulkiest nucleophiles, such as polycyclic aryloxides an even smaller fraction of chlorine atoms are replaced. However, the remaining P-Cl bonds are available for reactions with less-hindered or more reactive nucleophilic reagents. This is important because unreacted P-Cl bonds, even in a highly organic-substituted polymer, would remain as sites of hydrolytic instability.

(3) Strong nucleophiles, such as fluoroalkoxides or fluoro- or nitro-aryloxides provide the most facile reactions. Non-fluorinated nucleophiles such as alkoxides or aryloxides, require more forcing reaction conditions such as long reaction times or higher temperatures.

(4) Primary alkylamino-reagents (RNH_2) require special reaction conditions because they can lead to crosslinking or other side reactions due to the bi-functionality of the nucleophile. This can be corrected by careful control of the conditions or by the presence of bulky organic units. However, arylamines have a lower reactivity due to their steric characteristics. Some well-studied amino side group systems are shown in Chart 1.

(5) Mixed-substituent polymers are easy to produce using simultaneous or sequential exposure of poly(dichlorophosphazene) to two or more different nucleophiles. The ratios of the two side groups depend on some combination of reagent nucleophilicity and the relative amounts of the different reagents in the reaction mixture.

(6) Because the uncatalyzed ring-opening polymerization of $(NPCl_2)_3$ gives a broad molecular weight distribution, this is reflected in the polydispersity of the final polymer. This challenge can be avoided if the poly(dichlorophosphazene) is produced by a living cationic condensation polymerization route (see later).

(7) Reactions of poly(dichlorophosphazene) with organometallic reagents can be complicated because of coordination between the skeletal nitrogen atoms and the incoming reagents. Each skeletal nitrogen has a lone pair of electrons that can coordinate to transition metals in competition with the halogen-replacement process. This side reaction can be minimized by the presence of side group fluorine atoms which withdraw electrons from the skeletal nitrogen atoms. Thus, it follows that most of the successful reactions of phosphazene polymers with organometallic reagents require the presence of P-F side units rather than P-Cl units.

(8) Because of the factors described above, it is prudent for investigators to explore new macromolecular substitution processes by first carrying out small molecule model reactions using the more facile interactions of the cyclic oligomers $(NPCl_2)_{3\ \mathrm{or}\ 4}$ or $(NFF_2)_3$. The products from these reactions are more easily characterized by small molecule analytical methods such as NMR spectroscopy or X-ray crystallography.

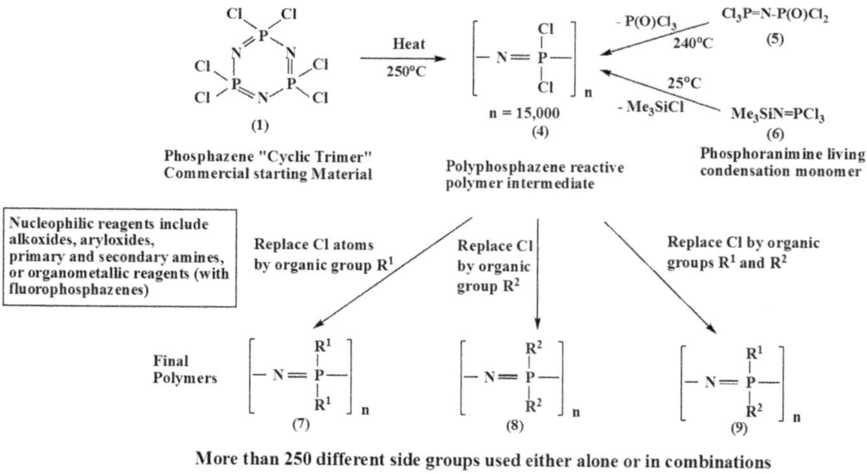

Scheme 2. The macromolecular substitution approach to poly(organophosphazene) synthesis.

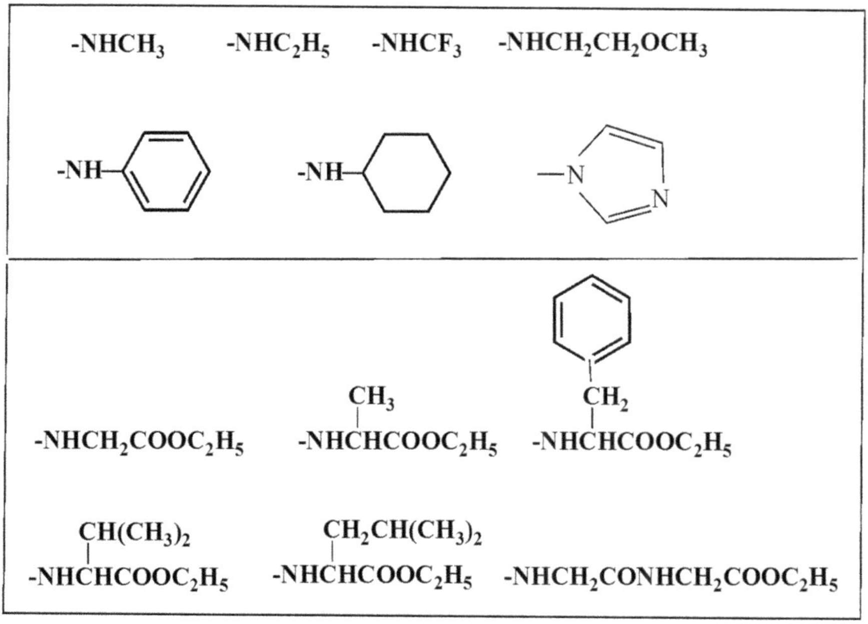

Chart 1. Typical amino side groups that have been linked to a polyphosphazene chain.

Condensation Routes to Polyphosphazenes

An important variation to the macromolecular substitution process is the use of alternative synthesis routes to poly(dichlorophosphazene). Two options exist. First, de Jeager in France invented a high temperature process in which $Cl_3P{=}N{-}POCl_2$ eliminates $POCl_3$ to yield a broad molecular weight distribution poly(dichlorophosphazene) (Equation 1) (*8*, *9*).

Second, a useful alternative became apparent from a collaborative effort between our program at Penn State and the Manners group at the University of Toronto (*10–12*). The method is summarized in Equation 2. This is a "living", room temperature polymerization of compound 6 catalyzed by small amounts of Lewis acids such as PCl_5. The polymer chain lengths are controlled precisely by the ratio of initiator to monomer (one initiator molecule is responsible for the growth of one chain). Moreover, after polymerization, the still-living chain end can be utilized to couple to the end of an organic polymer or to initiate polymerization of an organic monomer or to couple to another type of polymer. Thus, hybrid phosphazene-organic or phosphazene-silicone block copolymers are now accessible, a fact that allows the attributes of polyphosphazenes to be combined with the benefits of other types of polymers.

Equations 1-3. Condensation synthesis of polyphosphazenes.

$$Cl_3P{=}NPOCl_2 \xrightarrow[-\ POCl_3]{} (NPCl_2)_n \qquad (1)$$

$$Cl_3PNSi(CH_3)_3 \xrightarrow[-\ ClSi(CH_3)_3]{} (NPCl_2)_n \qquad (2)$$

$$(CH_3)_3SiN{=}P(R)_2\text{-}OR' \xrightarrow[-\ Si(CH_3)_3OR']{} (NPR_2)_n \qquad (3)$$

Third, Wisian-Neilson and Neilson in Texas developed an alternative to the post-polymerization substitution routes described above. In their process (reaction 3) (*13*, *14*), an organosilicon-substituted monomer, such as $(CH_3)_3SiN{=}PR_2OR'$, polymerizes to give an organic-substituted phosphazene polymer directly. A separate chapter in this book describes this method in detail. Related condensation methods have also been reported by Flindt and Rose (*15*) and by Montague and Matyjasewski (*16*)

Control of Structure-Property Relationships

Underlying Principles

The polyphosphazene system, and especially the macromolecular substitution route, provide the scientist with two valuable ways to control properties. First, the backbone of these polymers has special characteristics that cannot be found elsewhere. These polymers are "inorganic" in terms of the elements in the skeleton, but most of them are also "organic" based on the side group structure and chemistry. The inorganic aspects provide the underling thermo-oxidative stability, coordination ligand behavior, and have a unique influence on the side groups. On the other hand, the side groups protect the skeleton from hydrolysis or favor hydrolysis depending on the type of group, affect the torsional mobility of the backbone, and play a major role as the polymer molecules interact with each other and with the environment. It is the mutual influence of the skeleton and the side groups on each other that provides much of the intellectual and practical utility of polyphosphazenes.

Role of the Backbone

The phosphorus-nitrogen backbone in polyphosphazenes imposes its unique characteristics on all phosphazene polymers. These characteristics include resistance to combustion, extreme chain flexibility due to the low barrier to torsion of the skeletal bonds, and (depending on side groups) transparency throughout the visible and near-ultraviolet region of the spectrum. The backbone also becomes sensitive to hydrolysis to phosphate and ammonia if specific side groups are present, and this property is utilized for the development of bioerodible polymers. From a thermal stability viewpoint, the backbone resists oxidative decomposition at temperatures up to ~300°C, above which temperature depolymerization to cyclic oligomers is a common reaction. The backbone nitrogen atoms possess a lone pair of electrons that can bind transition metal complexes or (in some instances) protons and this provides access to catalytic, or chemotherapeutic polymers similar to cis-platinum (*17*), or to polymer-metal complexes with catalytic or electronic properties (*18*, *19*).

Influence of Different Side Groups

Different side groups linked to the phosphorus atoms yield different properties. They control solubility in different solvents, and materials characteristics that range from elastomers to film-formers, or fibers. The side groups also determine hydrophilicity, hydrophobicity, biostability, or bioerodibility. The choice of side groups also allows the polymers to be optimized for membranes, fibers, films, ultraviolet stability, bone regeneration matrices, drug delivery, bio-imaging devices, fire- and radiation-resistant materials. Some of these structure-property relationships are shown in Table 1.

Table 1. Different properties generated by the presence of different side groups linked to a polyphosphazene polymer chain

For Elastomer Formation OCH_3, OC_2H_5, OC_3H_7, OC_4H_9 $OCH_2CH_2OCH_3$ $OCH_2CH_2OCH_2CH_2OCH_3$ OCH_2CF_3 / $OCH_2(CF_2)_xCF_2H$ ** OC_6H_5 / $OC_6H_4CH_3$ ** OCH_2CF_3 / $CH_2Si(CH_3)_3$	For Fibers or Films OCH_2CF_3 OC_6H_5 OC_6H_4R Cycloalkanoxy	For Surface Hydrophobicity OCH_2CF_3 $OCH_2(CF_2)CF_2H$ OC_6H_6 $CH_2Si(CH_3)_3$
For High T_g's (above 37°C) NHC_6H_5 $OC_6H_4C_6H_5$ OC_6H_4COOH OC_6H_4COONa $OC_6H_4N{=}CHR$	For Solubility in Water $NHCH_3$ OCH_3 $OCH_2CH_2OCH_2CH_2OCH_3$ OC_6H_4COONa $OC_6H_4SO_3Na$ Glucosyl, Glyceryl	For Surface Hydrophilicity $OCH_2CH_2OCH_2CH_2OCH_3$ OC_6H_4COOH $OC_6H_4SO_3Na$ $OC_6H_4SO_3H$ $OC_6H_4NH_2$
For Bioerosion $NHCH_2COOC_2H_5$, etc Imidazolyl Glucosyl Glyceryl Ethoxy	For Biostability OCH_2CF_3 $OCH_2(CF_2)_nCF_2H$ OC_6H_5R $CH_2Si(CH_3)_3$ $OCH_2CH_2OCH_2CH_2OCH_3$	For Linkage of Bioactive Agents $OC_6H_4NH_2$ OC_6H_4COOH $OC_6H_4N^+R_3Br^-$ $OCH_2CH_2OCH_2CH_2NH_2$
For Solid Ionic Conductivity $OCH_2CH_2OCH_2CH_2OCH_3$ $OCH_2CH_2(OCH_2CH_2)_XOCH_3$ OC_6H_5 / $OC_6H_4SO_3H$ OC_6H_5 / $OC_6H_4P(O)(OR)_2OH$ OC_6H_5 / $OC_6H_4S(O_2)NHS(O_2)CF_3$	For High Refractive Index $OC_6H_4C_6H_5$ O-Naphthalene O-Anthracene Cl, Br, I-Aryloxy S-containing side groups	For Liquid Crystallinity $OC_6H_4N{=}NC_6H_4OCH_3$ $O(CH_2CH_2O)_XC_6H_4N{=}N$-Aryl $OC_6H_4C_6H_4OCH_2C(Me)HEt$ OCH_2CF_3
For Non-Linear Optical Behavior $O(CH_2CH_2O)_XC_6H_4N{=}N$-Aryl $O(CH_2CH_2O)_XC_6H_4C{=}C$-Aryl OCH_2CH_2-$N(Me)C_6H_4NO_2$	For Microsphere Formation and Hydrogels OC_6H_4COOH /Na^+, Ca^{++} $OC_6H_4SO_3H$ $OCH_2CH_2OCH_2CH_2OCH_3$ $OCH_2CH_2(OCH_2CH_2)_XOCH_3$	For Micelle Formation Hydrophobic phosphazene plus hydrophilic organic block copolymer Hydrophilic polyphosphazene plus hydrophobic organic block copolymer
For Electronic Conductivity Thiophene Pyrrole Ferrocene For Polymer Dyes Azo compounds Phthalocyanines etc.	For Pyrolysis to Ceramics Borazine side groups Amino side groups For Fire Resistance Nearly all Polyphosphazenes	For Coordination to Transition Metals $OC_6H_4PPh_2$ Ferrocene units Imidazolyl Pyridinyl Carboranyl $OCH_2CH_2OCH_2CH_2SCH_3$

** Commercialized

Recent work in our program has focused on the synthesis and characterization of polymers that have side groups with increased structural complexity in an attempt to expand the property combinations and address some important engineering and biomedical challenges. The following sections summarize some of the fundamental questions that are being considered, which may guide the future evolution of the polyphosphazene field.

Side Group Chemical and Dimensional Limits

What limits exist to the linkage of large or multi-functional side groups to the polyphosphazene skeleton? This is an important question because increasingly complex side groups hold the key to generating new property combinations, particularly in materials science and medical technology. For example, many chemotherapeutic drug molecules are both bulky and possess multiple reaction sites. Blocking of all but one reactive site on the nucleophile, linkage to the phosphazene backbone sometimes via a flexible spacer group, followed by de-protection of the pendent units is a technique that allows such groups to be linked to the polymer chain.

Bulky side groups present special challenges in macromolecuar substitution reactions.. Because these reactions are nucleophilic substitution processes they are sensitive to the steric bulk of the nucleophile. Steric hindrance in an incoming nucleophile may limit the number of bulky side groups that can be introduced along a polymer chain, but incomplete substitution can be compensated by replacement of the remaining halogen atoms by less hindered nucleophiles. The ratio of bulky groups to less hindered side units will almost certainly affect the physical properties of the polymer such as glass- or melting temperatures, chain torsional mobility, etc.

Thus, the influence of steric hindrance on the substitution process has been studied in our laboratory by examining reactions of poly(dichlorophosphazene) with a series of similar reagents that differ only in the size of the organic or organometallic nucleophile. An example is shown in Chart 2 where the sodium salts of several cyclic alcohols were allowed to react with poly(dichlorophosphazene) in tetrahydrofuran or higher boiling solvents (*20*). The results showed that complete chlorine replacement occurred within 24 hours for cyclobutanoxide or cyclopentanoxide reagents, but revealed clear evidence of increased steric hindrance with cyclohexanoxide, cycloheptanoxide, or cyclooctanoxide reagents. In these cases, the remaining chlorine atoms could readily be replaced by trifluoroethoxy groups. Moreover, separation of the bulky cyclohexanoxy units from the reactive center by the presence of one or two methylene spacer groups allowed 100% of the chlorine atoms to be replaced. This illustrates the techniques that need to be applied in general when sterically hindered side groups are to be linked to the polyphosphazene backbone. Polymers of this type are film-formers rather than elastomers (Figure 1d).

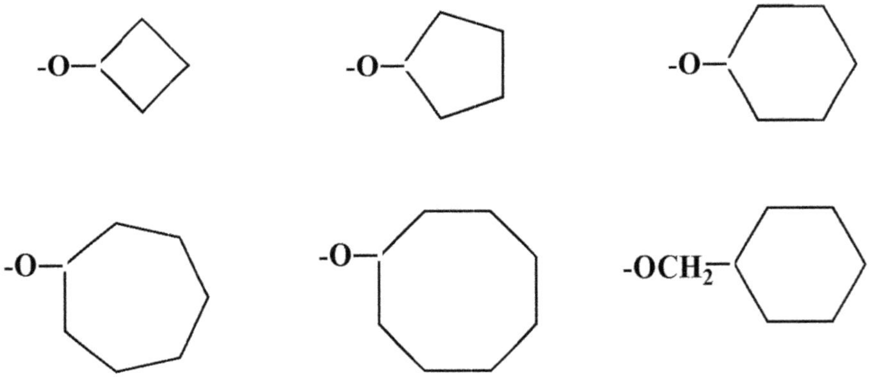

Chart 2. Cyclic organic side groups that facilitate an investigation of the influence of reagent steric hindrance on the macromolecular substitution process.

The use of flexible spacer groups between the backbone and the bulky side group has also been utilized with numerous nucleophiles to expand their reactivity with poly(dichlorophosphazene). Examples are shown in Chart 3. Sodium biphenylene oxide will replace all the 30,000 or so chlorine atoms along a poly(dichlorophosphazene) chain to give polymer 10, but it requires forcing reaction conditions such as high temperatures and long reaction times. However, the oxo-nucleophiles based on 11-17 replace ony a fraction of the chlorine atoms in the chloro-polymer even under forcing conditions. Moreover, full substitution is not possible even when a long spacer unit in present, as in 18, where the massive silsequioxane groups block the adjacent P-Cl bonds and prevent 100% substitution (*21*).

Because bulky side groups such as **10-15** in Chart 3 encounter serious steric hindrance during the chlorine replacement steps, complete halogen replacement must involve the introduction of less-hindered co-substituents such as $-OCH_2CF_3$ groups. This phenomenon can be an advantage in some circumstances. For instance, low percentages of **10-15** can convert a microcrystalline polyphosphazene, such as $[NP(OCH_2CF_3)_2]_n$ into a rubbery elastomer without the system being crosslinked (*22–24*). They serve as "inter-digitation" sites that interrupt the molecular symmetry and prevent crystallization, and the inter-digitation prevents the chains from sliding past each other when under tension (Figure 2). It seems likely that other bulky co-substituents might serve the same function.

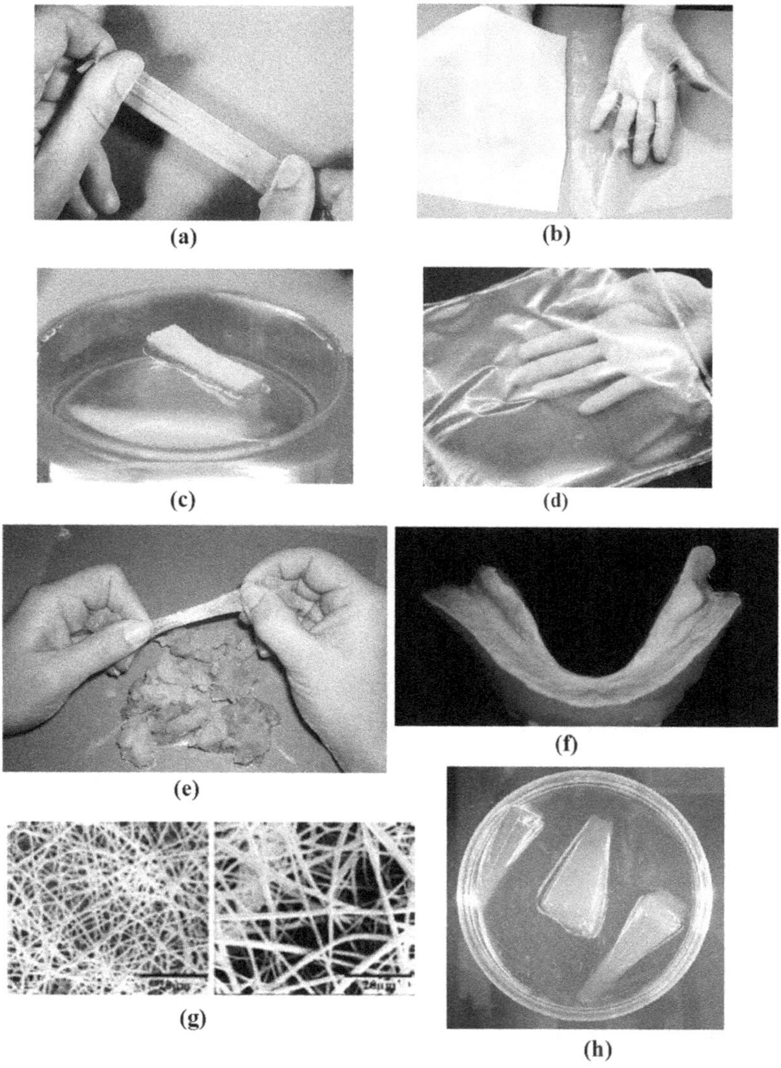

Figure 1. Examples of polyphosphazenes with different side groups and morphologies. (a) Solution-cast and stretch-oriented film of $[NP(OCH_2CF_3)_2]_n$. (b) Solution-cast films of (left) $[NP(OCH_2CF_3)_2]_n$, and (right) $[NP(OCH2(CF_2)_4CF_2H)_2]_n$. (c) An expanded foam of $[NP(OCH_2CF_3)_2]_n$ after absorbing super-critical carbon dioxide and exposure to atmospheric pressure. (d) A solution-cast film of a bis(cyclopentanoxy)phosphazene polymer. (e) PN-F elastomer derived from the side groups shown in illustration (b). (f) A denture prosthetic derived from PN-F elastomer (courtesy of L. Gettleman). (g) Electrospun nanofibers of a polyphosphazene with both ethyl glycinate and ciprofloxacin antibiotic side groups as an experimental controlled drug release system. (h) A radiation crosslinked hydrogel derived from the polymer MEEP – $[NP(OCH_2CH_2CH_2OCH_3)_2]_n$. pH-Retractable hydrogels are accessible from crosslinked derivatives of PCPP, $[NP(OC_6H_4COOH)_2]_n$.

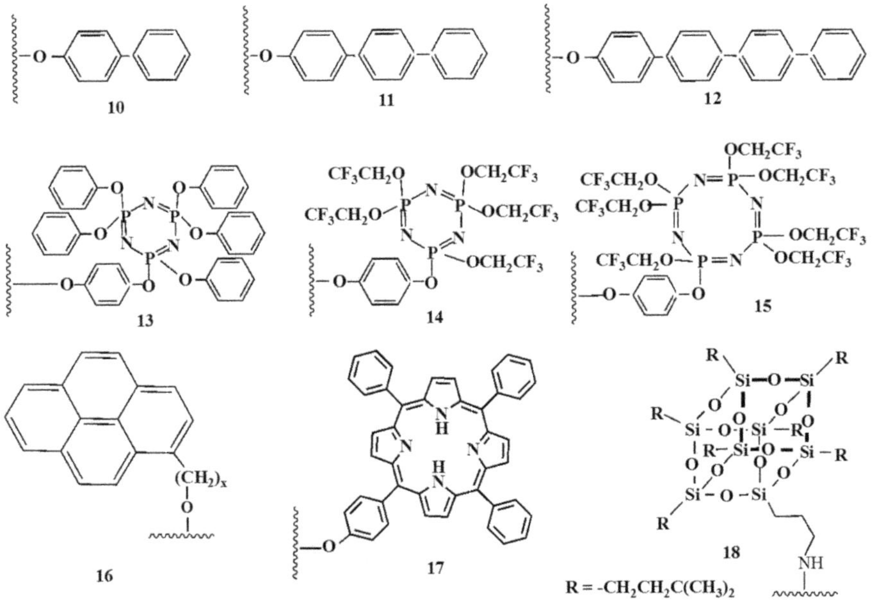

Chart 3. Bulky side groups that impart special properties to polyphosphazenes including elasticity when they are present in small quantities.

Another example of the need to link bulky organic side groups to a polyphosphazene chain is when colored films, or photonic or electronic materials are targeted for device applications. Thus, phosphazene polymers with bulky chromophore molecules, such as the example shown as **17** (Chart 2), present an opportunity to utilize the macromolecular substitution approach (*25*). Few of the numerous red, blue, or green dye molecule anions will replace all the chlorine atoms in poly(dichlorophosphazene), but the synthesis of polymers with only a small ratio of the dye plus transparent co-substituents will generate the required optical density. Numerous applications exist where polymers of this type have advantages over free dyes dissolved in glassy polymers or insoluble particulate inorganic dyes suspended in a glassy matrix, as in color photosensors. This work illustrates the value of polyphosphazenes as *carrier polymers* for side units that serve some non-traditional function, such as those described in the following sections.

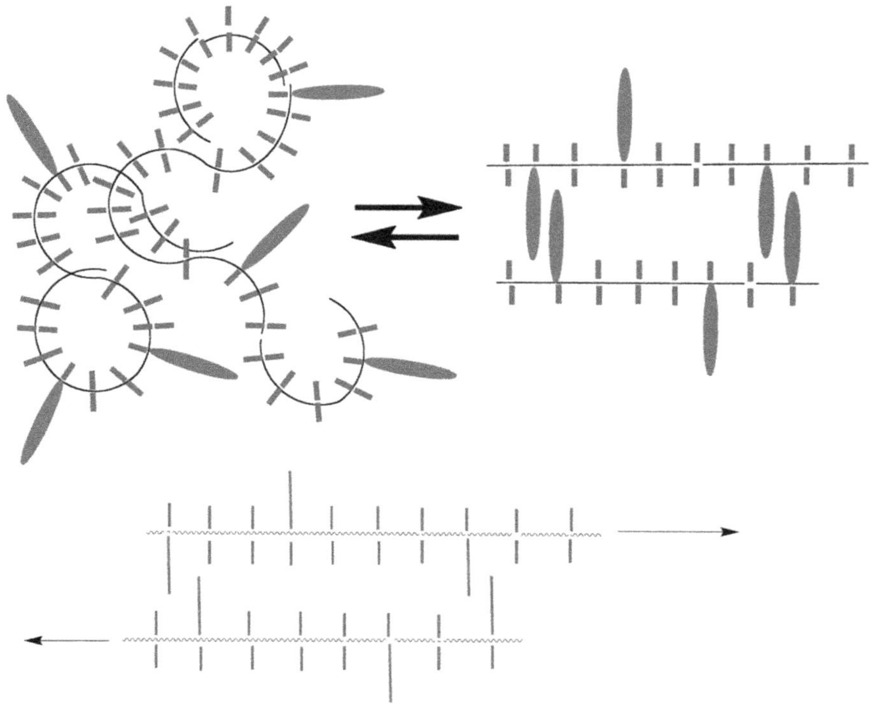

*Figure 2. Low percentages of bulky substituents, such as **11-15**, can convert a microcrystalline polyphosphazene into a rubbery elastomer without the system being crosslinked. They serve as "interdigitation" sites in the material that interrupt the molecular symmetry to avoid crystallization, and prevent the chains from sliding past each other when under tension. It seems likely that other bulky co-substituents might serve the same functions.*

Multifunctional Side Groups

A second challenge in the macromolecular substitution approach is to link multi-functional organic side groups to the polyphosphazene chain without crosslinking the macromolecules through the remaining functional groups. Numerous biologically interesting side groups such as sugars, steroids, and drug molecules fall into this category. A solution to this problem is to protect all but one of the reactive sites on the nucleophile before coupling the side group to the polymer, and then to de-protect the other active groups. This requires some synthetic delicacy in order to prevent unwanted decomposition elsewhere along the polymer chain. However, it has been accomplished for a number of biomedically useful systems (*26–32*).

Materials Science Aspects

Films, Fibers, Foams, and Coatings

Numerous polyphosphazenes with a wide range of different side groups can be readily fabricated into transparent or opalescent, fire-resistant films by solvent casting techniques (Figures 1b and d). Fibers are accessible by classical fiber precipitation and orientation techniques. Melt spinning is a technique that has not yet been exploited. Electrospinning to yield fiber mats for composite materials or medical applications has been studied. (Figure 1g) (*33*). Established examples of polymers for film- or fiber- fabrication include polyphosphazenes with side groups such as $-OCH_2CF_3$, $-OCH_2CCl_3$, $-OCH_2(CF_2)_xCF_2H$, $-O(CH_2)_xCH_3$, or $-OC_6H_5$. The above-mentioned expansion of the system to include film-forming polymers with cycloalkanoxy side groups (*20*) illustrates the increasing scope of the field (Figure 1d). Perhaps surprisingly, some of these polymers remain fire resistant in spite of the high carbon content.

An expansion of the polyphosphazene field into the technology of rigid, porous solids (foams) has also been demonstrated by the dissolution of $[NP(CH_2CF_3)_2]_n$ in supercritical carbon dioxide followed by rapid evaporation of this solvent (*34*). These materials float on water and are non-flammable (Figure 1c).

Elastomers

Elastomeric character in a polymer is associated with long polymer chain lengths, highly flexible backbones, an absence of crystallinity, and a limited concentration of reactive side groups that allow crosslinking to prevent the polymer chains from being drawn past each other when subjected to mechanical tension. Impact resistance is another valuable property of elastomers especially for uses in aircraft, automobiles, or medical devices. This combination of properties is found in all the classical organic elastomers (natural rubber, polyisobutylene, styrene-butadiene, silicone rubber, etc.). Also, as mentioned above, it is found in a number of poly(organophosphazenes). Several examples have been mentioned earlier. The first practical polyphosphazene elastomer (known as PN-F) is a polymer with two different fluoroalkoxy side groups on each phosphorus and a minor loading of o-allylphenoxy units as a crosslinker (*35*, *36*). Illustrations of this material and some of its uses are shown in Figures 1e and f. More recent examples synthesized in our program use non-fluorinated substituents such as linear alkoxy groups (*37*) or, as discussed, minor concentrations of bulky side groups such as **11-15** that prevent chain slippage by physical inter-digitation (Figure 2) (*22–24*). The synthesis of hybrid phosphazene-poly(organonosiloxane) (silicone) elastomers was mentioned earlier (*38–41*).

Elastomeric polyphosphazenes are of technological interest in a number of different applications, that include oil- and fire-resistant tubes and coatings, low-temperature flexible polymers, extreme impact-resistance materials, and biomedical elastomers for uses in cardiovascular devices or catheters. In many biomedical applications, they are alternatives to silicone elastomers or polyurethanes.

Liquid Crystalline (LC) Polymers

Many polymers possess a materials phase that lies between the glass transition temperature and the melting point. This phase contains polymer chains that are partly ordered and partly randomized. This is the liquid crystalline or mesogenic state. There may be only one liquid crystalline transition or there could be several at different temperatures below the true melting point. Several polyphosphazenes show liquid crystalline transitions (*42*). A number of these polymers behave in this way because they contain mesogenic side groups such as biphenyleneoxy or azobiphenylene side groups (*43, 44*) (Table 1). Others, with relatively simple side chains, such as CF_3CH_2O- groups, undergo several liquid crystalline transitions between the glass transition temperature at -66°C and the melting temperature at 242°C (*42*).

Polymer liquid crystallinity has its materials advantages and disadvantages. Such polymers can be utilized in liquid crystalline optical devices where the LC transition must be raised above ambient temperatures. However, it can also complicate the engineering behavior (strength, elasticity, fiber properties) as the temperature is raised.

Energy-Related Applications

The fire-retardant characteristics of most polyphosphazenes together with the lithium ion or proton conduction of specific derivatives makes them attractive candidates for use as non-flammable electrolytes in lithium ion batteries or for hydrogen- or methanol-driven fuel cells. For lithium battery uses, polyphosphazenes with oligo-ethyleneoxy side chains are excellent lithium ion conductors (Table 1), especially when plasticized by small amount (~5-10%) of propylene carbonate (*45, 46*). They also resist combustion which is a key requirement for aerospace and automobile applications. Specific polyphosphazenes have also been examined for use in primary seawater battery applications (*47*). Proton conductive fuel cell membranes derived from aryloxy phosphazenes with sulfonic acid or other acidic functional groups have shown promise in preliminary experiments (*48, 49*).

Membranes

The ease of tuning polyphosphazene properties by side group variations makes them good candidates for a variety of membrane applications. Three types of phosphazene membranes have been investigated – solid membranes for liquid or gas separations (*50*), and gel-type membranes for biological applications. Polymers with aryloxy side group were investigated for liquid separations (*50*). Recently. polymers with methoxyethoxyethoxy (MEEP) side groups have been studied for gas separations - particularly for CO_2 sequestration purposes to limit CO_2 buildup in the atmosphere (*51*).

Optical, Photonic, and Electronic Materials

Optical properties can be anticipated for a number of different polyphosphazenes. First, the backbone has a high electron density which provides a basis for high refractive indices. Second, the window for visible light transmission by the backbone extends across the whole visible spectrum and into the near ultraviolet. Third, the refractive index can be changed via the introduction of different side groups. For example, aryloxy side groups add an incremental increase to the refractive index (*52*, *53*). Fourth, colored filters are accessible by the linkage of organic dyes to the backbone (*25*). Fifth, the choice of side groups that favor films, elastomers, or fibers is a valuable property for many optical and photonic applications. Finally, linkage of photo- or electro-responsive side groups opens opportunities to form materials that respond to light intensity and electric current (*52*). An example of a photochromic system is the linkage of spiropyran side groups to the polymer for potential use in sunglasses (*54*).

Combination of Polyphosphazenes with other Polymers and Materials

The formation of composites between polyphosphazenes and classical organic polymers or biopolymers has been studied for many years. Thus, the combination of polyphosphazenes with other polymers offers broad opportunities to expand properties. Both block copolymers and polymer blends have been explored. Block copolymers of polyphosphazenes and polysiloxanes or polyphoshazenes and organic polymers synthesized via the living cationic polymerization route (Reaction 2) have provided a productive route to these hybrids.

A second type of hybrid system is one in which polyphosphazenes become intimately associated with other materials such as layered solids or carbon nanotubes. A recent example is our demonstration that single wall carbon nanotubes form strong associations with, and are solubilized by, a variety of polyphosphazenes that bear alkoxy or aryloxy side groups (*55*). The mechanism of this interaction is still being established, but one explanation is that each polyphosphazene molecule wraps itself around a nanotube assisted by interactions between the pi-electrons of the nanotubes and those of the phosphazene skeleton and/or the carbon-hydrogen bonds of the phosphazene side groups. Other interactions of polyphosphazenes with tubular or layered structures can be anticipated in the future, and this is a prime target for future work.

Biomedical Applications

Reasons for Biomedical Interest

Of all the aspects of polyphosphazene science, their utilization in biology and medicine may be one of their most important immediate contributions to science and technology. Starting from the early days of polyphosphazene research, the biological aspects have been the subject of keen interest. This began with the synthesis of polyphosphazene-platinum coordination adducts as candidates for anti-tumor agents (*56*), and continued through the development of polymers with steroidal side groups (*57*), microspheres for mammalian cell and vaccine delivery vehicles (*58*), bioerodible polymers for bone regeneration and other applications (*59–65*), and coatings for gold nanospheres for bioimaging and drug delivery procedures (*66*). The current interest in polyphosphazenes as coatings for biomedical devices, such as heart pump or other vascular prostheses (*67*), drug delivery from hydrogels, or use of elastomers in dental devices (*68*), are other examples of this trend. Several other chapters in this book provide a detailed illustration of the importance of this aspect of the field. One of the most interesting aspects of the macromolecular substitution process is the ease with which biologically useful side groups can be linked to the preformed backbone. Thus, once again the polyphosphazene is a *carrier polymer* that also has the capacity to bioerode and release the bioactive agent (*69*).

Polymers that hydrolyze to non-toxic products have many uses in experimental biomedicine. These uses range from controlled drug delivery devices to matrices for the regeneration of soft tissues or living bone. The simplest example is the ethoxy-substituted elastomeric polymer $[NP(OCH_2CH_3)_2]_n$ which hydrolyzes slowly to ethanol, phosphate, and ammonia, products that can be metabolized or excreted rapidly (*70*). The most widely explored system of this type includes polyphosphazenes with amino acid ethyl ester side groups (*59*, *60*), which hydrolyze to phosphate, ammonia, amino acid and ethanol. Moreover, the rate of hydrolysis depends of the type of amino acid and the ester function. The utility of this system for bone regeneration has been demonstrated in a large number of experiments and publications from the Laurencin group at the University of Connecticut (*61–65*). Another chapter in this book describes this application in more detail.

Hydrogels

A wide range of properties and potential uses are accessible through polyphosphazene hydrogels (*71–75*). A hydrogel is usually a lightly crosslinked, hydrophilic polymer that would normally be completely soluble in water, but is prevented from dissolving by crosslinks. Typically, a hydrogel is 90% or more water, but the gel has shape and form. It may be biostable or sensitive to hydrolytic decomposition. A typical polyphosphazene hydrogel polymer bears water-solubilizing alkyl ether side groups such as $-OCH_2CH_2OCH_2CH_2OCH_3$ units that are crosslinked by exposure to X-rays or gamma rays. Hydrogels are excellent carriers for solutions of drug molecules from which they can be extruded by contraction of the gel induced by temperature changes. Hydrogels may also be fabricated into artificial muscles and other devices that respond to pH changes or to additional crosslinking caused by the presence of transition metal ions in the structure. Expansion and contraction of the gel can occur via the response of acidic or basic co-substituents on the polymer, or by changes to the temperature and the effect that has on the hydrogen bonding between the polymer and water.

One of the oldest and most widely used hydrogel precursors is poly(carboxyphenoxy phosphazene) (PCPP) (*76*) (Scheme 3) mentioned in several sections of this book. Because it is a polyelectrolyte, it can exist as a water-soluble form (sodium or potassium salt) or as an ionically crosslinked system (Ca^{++} salt), and this has led to its use for cell or vaccine encapsulation and delivery. This topic also is covered in detail in separate chapters of this book.

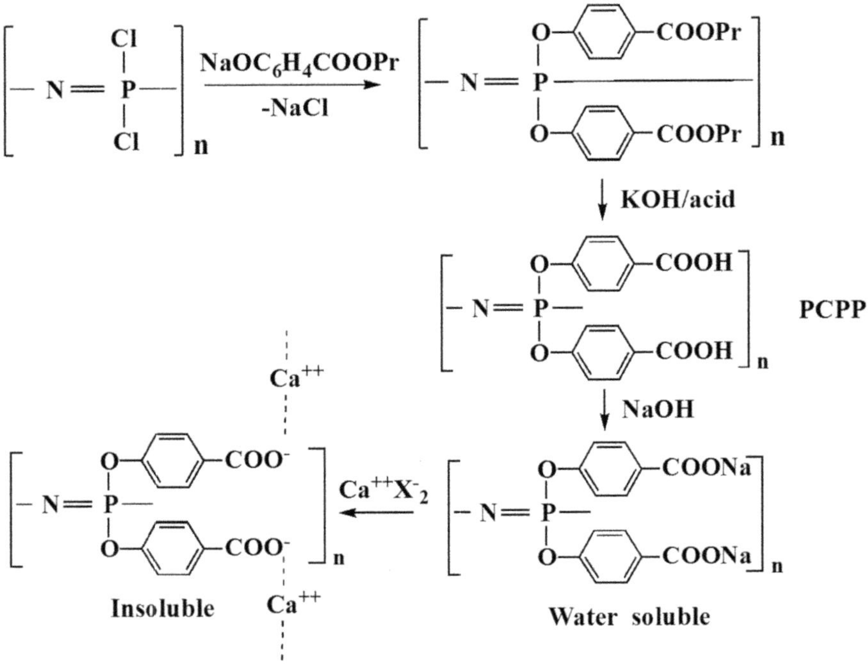

Scheme 3. Synthesis of PCPP

Vesicles, Microspheres, and Nanosphere Coatings

The development of polyphosphazene technology for the encapsulation of mammalian cells in vesicles, as a starting point to the construction of artificial liver or pancreas devices, was one of the earliest experimental biomedical uses for polyphosphazene chemistry (*58*). The polymer PCPP (Scheme 3) was specifically developed for this application. Subsequently this polymer was utilized for the oral delivery of anti-influenza vaccines. This development is described in detail by Andrianov in a separate chapter of this book. This same system is also the basis of variable transmissive membranes, first developed by Allcock and Kwon, that respond to changes in pH, ion strength, and other environmental changes, and which are possible precursors to devices for the controlled release of drugs in response to changes in body physiology (*76*). In this sense, the use of PCPP to coat gold nanospheres for medical imaging or drug delivery is the most recent illustration of the biomedical opportunities that exist for polyphosphazenes (*66*).

Final Comments

The synthesis of long chain polymers from organic small molecules was one of the triumphs of 20th century chemistry. However, the pace of discovery of new organic polymers has slowed in recent years as most of the readily accessible macromolecules have been synthesized, examined, and commercialized. Moreover, many needed property combinations have remained inaccessible due to the limitations imposed by the chemistry of carbon. A hope for the future is that many of the other 99 elements in the Periodic Table may allow access to these important property combinations. The chemistry of phosphorus, and the polyphosphazenes in particular, provides an example of how explorations outside the realm of hydrocarbon-based polymers provides an answer to this challenge. An extraordinary range of different side groups has been linked to a polyphosphazene backbone – alkoxy-, aryloxy, alkyl ether, fluoroalkoxy, amino acid esters, drug molecules, transition metal groups, and many more examples, and these polymers display a wide range of characteristics that are different from those of their all-organic counterparts. Thus, the polyphosphazene system offers not only an example of the benefits that result from the utilization of non-carbon elements in polymers, but also provides new polymers with precisely those property combinations that are lacking in conventional synthetic macromolecules.

The continued expansion of the polyphosphazene field is evident from the range of topics presented at this conference and summarized in this book. Moreover, an increasing number of research groups have entered the field in recent years. In terms of the number of different polymers now known, polyphosphazenes have emerged as the largest field of hybrid inorganic-backbone macromolecules and, indeed, as one of the largest and most diverse areas in polymer chemistry. The fact that many of these polymers are accessible via macromolecular substitution from only one precursor polymer – poly(dichlorophosphazene) - provides an enormous simplification both for future synthetic work and understanding structure-property relationships. Moreover, a number of other inorganic intermediates can be visualized, and this should expand the scope of this and other related systems even further.

References

1. Stokes, H. N. On the Chloronitrides of Phosphorus (II). *Am. Chem. J.* **1897**, *19*, 782–785.
2. Allcock, H. R.; Kugel, R. L. Synthesis of High Polymeric Alkoxy- and Aryloxyphosphonitriles. *J. Am. Chem. Soc.* **1965**, *87*, 4216–4217.
3. Allcock, H. R.; Kugel, R. L.; Valan, K. J. High Molecular Weight Poly[alkoxy- and aryloxyphosphazenes]. *Inorg. Chem.* **1966**, *5*, 1709–1715.
4. Allcock, H. R.; Kugel, R. L. High Molecular Weight Poly(diaminophosphazenes). *Inorg. Chem.* **1966**, *5*, 1716–1718.
5. Allcock, H. R. *Chemistry and Applications of Polyphosphazenes*; John Wiley & Sons, Hoboken, NJ, 2003.
6. *Polyphosphazenes for Biomedical Applications*; Andrianov, Ed.; John Wiley & Sons, Hoboken, NJ, 2009.

7. *Synthesis and Characterization of Cyclophosphazenes*, Gleria, M.; De Jaeger, Eds.; Nova Science Publishers, New York, 2004.
8. De Jaeger, R.; Potin, P. Poly(dichlorophosphazene) from P-Trichloro-N-Dichlorophosphoryl Mono-phosphazenene $Cl_3P{=}N{-}POCl_2$. In *Synthesis and Characterization of Cyclophosphazenes*; Gleria, M.; De Jaeger, R. Eds.; Nova Science Publishers, NY. 2004; pp 25−48.
9. Elayan, S. S.; Peterson, E. S.; Allen, C. W. Synthesis of Poly(dichlorophosphazene) by the Melt Phase Polymerization of a P-Trichloro-N-(dichlorophosphoryl)monophosphazene. *J. Inorg. Organomet. Polym Mater.* **2017**, *27*, S119.
10. Honeyman, C. H.; Manners, I.; Morrissey, C. T.; Allcock, H. R. Ambient Temperature Synthesis of Poly(dichlorophosphazene) with Molecular Weight Control. *J. Am. Chem. Soc.* **1995**, *117*, 7035–7036.
11. Allcock, H. R.; Crane, C. A.; Morrissey, C. T.; Nelson, J. M.; Reeves, S. D.; Honeyman, C. H.; Manners, I. "Living" Cationic Polymerization of Phosphoranimines as an Ambient Temperature Route to Polyphosphazenes with Controlled Molecular Weights. *Macromolecules* **1996**, *29*, 7740–7747.
12. Allcock, H. R.; Nelson, J. M.; Reeves, S. D.; Honeyman, C. H.; Manners, I. Ambient Temperature Direct Synthesis of Poly(organophosphazenes) via the Living Cationic Polymerization of Organo-substituted Phosphoranimines. *Macromolecules* **1997**, *30*, 50–56.
13. Neilson, R. H.; Wisian-Neilson, P. Poly(alky/arylphosphazenes) and their Precursors. *Chem. Rev.* **1988**, *88*, 541–562.
14. Wisian-Neilson, P. Polyphosphazenes from Condensation Polymerization. In *Biomedical Applications of Polyphosphazenes*; Andrianov, A. K. Ed.; John Wiley & Sons, Hoboken, NJ, 2009; pp 155−167.
15. Flindt, E. P.; Rose, H. Trivalent-pentavalente Phosphosverbindindungen/ Phosphazene. IV. *Z. Anorg. Allg. Chem.* **1977**, 428–204.
16. Montague, R. A.; Matyjaszewski, K. Synthesis of Poly[bis(trifluoroethoxy)] Under Mild Conditions Using a Fluoride Initiator. *J. Am. Chem. Soc.* **1990**, *112*, 6721.
17. Allcock, H. R.; Allen, R. W.; O'Brien, J. P. Crystal and Molecular Structure of a Platinum-Cyclophosphazene Complex-cis-Dichloro[octa(methylamino)cyclotetraphosphazene]. *J. Am. Chem. Soc.* **1977**, *99*, 3984–3987.
18. Davidson, R. J.; Ainscough, E. W.; Brodie, A. M.; Jameson, G. B.; Waterland, M. R.; Allcock, H. R.; Hindenlang, M. D.; Moubaraki, B.; Murray, K.; Gordon, K.; Horvath, R.; Jameson, G. N. L. Towards an Iron(11) Spin Crossover Grated Polyphosphazenene Polymer. *Inorg. Chem.* **2012**, *51*, 8307–8316.
19. Davidson, R. J.; Ainscough, E. W.; Brodie, A. M.; Waterland, M. R.; Allcock, H. R.; Hindenlang, M. D.; Jameson, G. N. L. Avoiding Crosslinking in Iron-Polyphosphazene Metallo-Polymers. *Inorg. Chem. Commun.* **2015**, *51*, 1–3.
20. Tian, Z.; Hess, A.; Fellin, C. R.; Nulwala, H.; Allcock, H. R. Phosphazene High Polymers and Models with Cyclic Aliphatic Side Structure-Property Relationships. *Macromolecules* **2015**, *48*, 4301–4311.

21. Tong, C.; Tian, Z.; Li, Z.; Modzelewski, T.; Allcock, H. R. Synthesis and Characterization of Trifluoroethoxy Polyphosphazenes Containing POSS Side Groups. *Macromolecules* **2016**, *49*, 1313–1320.
22. Modzelewski, T.; Wilts, E.; Allcock, H. R. Elastomeric Polyphosphazenes with Phenoxy-cyclotriphosphazene Side Groups. *Macromolecules* **2015**, *48*, 7543–7549.
23. Modzelewski, T.; Wonderling, N. M.; Allcock, H. R. Polyphosphazene Elastomers Containing Interdigitated Oligo-p-phenyleneoxy Side Groups: Synthesis, Mechanical Properties and X-ray Scattering Studies. *Macromolecules* **2015**, *48*, 4882–4890.
24. Li, Z.; Chen, C.; Tian, Z.; Modzelewski, T.; Allcock, H. R. Polyphosphazenes with Cyclotetraphosphazene Side Groups: Synthesis and Elastomeric Properties. *J. Inorg. Organomet. Mater. Polym.* **2016**, *26*, 667–674.
25. Li, Z.; Allcock, H. R. Polyphosphazenes with Immobilized Dyes as Potential Color Filter Materials. *Appl. Mater. Interfaces* **2015**, *1*, 13518–13523.
26. Nichol, J. L.; Allcock, H. R. Polyphosphazenes with Amino Acid Citronellol Ester Side Groups for Biomedical Applications. *Eur. Polym. J.* **2015**, *62*, 214–221.
27. Nichol, J. L.; Morozowich, N. L.; Decker, T. E.; Allcock, H. R. Crosslinkable Citronellol-Containing Polyphosphazenes and their Biomedical Potential. *J. Polym. Sci. A* **2014**, *52*, 2258–2265.
28. Tian, Z.; Chen, C.; Allcock, H. R. Injectable and Biodegradable Supramolecular Hydrogels by Inclusion Complexation between Poly(organophosphazene) and a a-Cyclodextrin. *Macromolecules* **2013**, *45*, 2715–2724.
29. Tian, Z.; Zhang, Y.; Liu, X.; Chen, C.; Guiltinan, M. J.; Allcock, H. R. Biodegradable Polyphosphazenes Containing Antibiotics: Synthesis, Characterization and Hydrolytic Release Behavior. *Polym. Chem.* **2013**, *4*, 1826–1835.
30. Allcock, H. R.; Morozowich, N. L. Bioerodible Polyphosphazenes and their Medical Potential. *R.S.C. Polym. Chem.* **2012**, *3*, 578–590.
31. Allcock, H. R.; Scopelianos, A. G. Synthesis of Sugar-Coated Cyclic and Polymeric Phosphazenes, and their Oxidation, Reduction, and Acetylation Reactions. *Macromolecules* **1983**, *16*, 715–719.
32. Allcock, H. R.; Fuller, T. J.; Matsumura, K. Reactions of Steroid Salts with Hexachlorocyclotri-phosphazene. *J. Org. Chem.* **1981**, *46*, 13–22.
33. Allcock, H. R.; Steely, L. B.; Singh, A. Hydrophobic and Superhydrophobic Surfaces from Polyphosphazenes. *Polym. Int.* **2006**, *55*, 621–625.
34. Steely, L. B.; Li, Q.; Badding, J. V.; Allcock, H. R. Foam Formation with Fluorinated Polyphosphazenes by Liquid CO_2 Processing. *Polym. Sci. Eng.* **2008**, *48*, 683–686.
35. Rose, S. H. J. Synthesis of Phosphonitrilic Fluoroelastomers. *J. Polym. Sci., Ser. B* **1968**, *6*, 837–839.
36. *Technical Marketing Data*; Firestone Tire Co. and Ethyl Corp.
37. Chen, C.; Li, Z.; Allcock, H. R. (unpublished work).
38. Tong, C.; Wilts, E.; Allcock, H. R. (in press).

39. Allcock, H. R.; Prange, R. Properties of Poly(phosphazene-siloxane) Block Copolymers Synthesized via Telechelic Polyphosphazenes and Polysiloxane. *Macromolecules* **2001**, *34*, 6858–6865.
40. Prange, R.; Reeves, S. D.; Allcock, H. R. Polyphosphazene-Polystyrene Copolymers: Block and Graft Copolymers from Polyphosphazene and Polystyrene Macromonomers. *Macromolecules* **2000**, *33*, 5763–5765.
41. Allcock, H. R.; Kuharcik, S. E. Hybrid Phosphazene-Organosilicon Polymers, Part 11. High Polymers with Materials Synthesis and Properties. *J. Inorg. Organ. Polym.* **1996**, *6*, 1–41.
42. Kojima, M.; Magill, J. H. Spherulitic Crystallization in Poly[bis(trifluoroethoxy)phosphazene] BPFP. *Polymer* **1985**, *26*, 1971–1978.
43. Kim, C.; Allcock, H. R. A Liquid Crystalline Poly(organophosphazene). *Macromolecules* **1987**, *20*, 1726–1727.
44. Klingenberg, E. H.; Allcock, H. R. The Synthesis of Liquid Crystal Phosphazenes Containing Chiral Mesogens. *Macromolecules* **2005**, *28*, 4351–4360.
45. Allcock, H. R.; O'Connor, S. M. J.; Olmeijer, D.; Napierala, M. E.; Cameron, C. G. Polyphosphazenes Bearing Branched and Linear Oligoethyleneoxy Side Groups as Solid Solvents. *Macromolecules* **1996**, *23*, 7544–7552.
46. Allcock, H. R.; O'Connor, S. M. J.; Olmeijer, D. L; Napierala, M. E.; Cameron, C. G. Polyphosphazenes Bearing Branched and Linear Oligoethyleneoxy Side Groups as Solid Solvents for Ionic Conduction. *Macromolecules* **1996**, *23*, 7544–7552.
47. Welna, D. T.; Stone, D. A.; Allcock, H. R. Lithium Ion Conductive Polymers as Prospective Membranes in Lithium-Seawater Batteries. *Chem. Mater.* **2006**, *18*, 4486–4492.
48. Hofmann, M. A.; Ambler, C.; Maher, A. E.; Chalkova, E.; Zhou, X. Y.; Lvov, S. N.; Allcock, H. R. Synthesis of Polyphosphazenes with Sulfonimide Side Groups. *Macromolecules* **2002**, *35*, 6490–6493.
49. Allcock, H. R.; Hofmann, M. A.; Ambler, C. M.; Lvov, S. N.; Zhou, X. Y.; Chalkova, E.; Weston, J. Phenyl Phosphonic Acid Functionalize Poly[aryloxyphosphazenes] as Proton Conducting Membranes for Direct Methanol Fuel Cells. *J. Membr. Sci.* **2002**, *202*, 47–54.
50. Peterson, E. S.; Stone, M. L.; Cummings, D. G.; McCaffrey, R. R. Separations of Hazardous Organics from Gas and Liquid Feedstreams Using Phosphazene Polymer Membranes. *Sep. Sci. Technol.* **1993**, *28*, 271–278.
51. Chen, C.; Ren, Y.; Li, X.; Nulwala, H.; Allcock, H. R. (unpublished work) 2018.
52. Allcock, H. R.; Mang, M. N.; Dembek, A. A.; Wynne, K. J. Poly(organophosphazenes) with Phenylphenoxy and Related Bulky Side Groups. Synthesis, Thermal Transition Behavior, and Optical Properties. *Macromolecules* **1989**, *22*, 4179–4190.
53. Dembek, A. A.; Kim, C.; Allcock, H. R.; Devine, R. L. S.; Steier, W. H.; Spangler, C. W. A Second-Order Nonlinear Optical Poly(organophosphazene). *Chem. Mater.* **1990**, *2*, 97–99.

54. Allcock, H. R.; Kim, C. Photochromic Polyphosphazenes with Spiropyran Units. *Macromolecules* **1991**, *24*, 2846–2851.
55. Ren, Y.; Allcock, H. R. *Macromolecules*submitted.
56. Allcock, H. R.; Allen, R. W.; O'Brien, J. P. Synthesis of Platinum Derivatives of Polymeric and Cyclic Phosphazenes. *J. Am. Chem. Soc.* **1977**, *99*, 3984–3987.
57. Allcock, H. R.; Fuller, T. J.; Matsumura, K. Reactions of Steroid Salts with Hexachlorcyclotriphosphazene. *J. Org. Chem.* **1981**, *46*, 13–22.
58. Cohen, S.; Bano, M. C.; Visscher, K. B.; Chow, M.; Allcock, H. R.; Langer, R. An Ionically-Crosslinkable Polyphosphazene: A Novel Polymer for Microencapsulation. *J. Am. Chem. Soc.* **1990**, *112*, 7832–7833.
59. Allcock, H. R.; Morozowich, N. L. Bioerodible Polyphosphazenes and their Medical Potential. *R.S.C. Polym. Chem.* **2012**, *3*, 578–590.
60. Allcock, H. R.; Fuller, T. J.; Mack, D. P.; Matusumura, K.; Smeltz, K. M. Synthesis of Poly[(amino acid alkyl ester)phosphazenes]. *Macromolecules* **1977**, *10*, 824–830.
61. Deng, M.; Nair, L. S.; Nukavarapu, S. P.; Jiang, T.; Kanner, W.; Li, X.; Kumber, S. G.; Weikel, A. L.; Krogman, N. R.; Allcock, H. R.; Laurencin, C. T. Di-Peptide Based Polyphosphazene and Polyester Blends for Bone Tissue Engineering. *In Vitro* and *In Vivo* Biodegradability and Biocompatibility Studies. *Biomaterials* **2010**, *31* (18), 4898–4908.
62. Krogman, N. R.; Weikel, A. L.; Nguyen, K. A.; Nukavarapu, S. P.; Nair, L. S.; Laurencin, C. T.; Allcock, H. R. Hydrogen Bonding in Blends of Polyesters with Di-Peptide-Containing Polyphosphazenes. *J. Appl. Polym. Sci.* **2010**, *115*, 431–437.
63. Deng, M.; Nair, L. S.; Nukavarapu, S.; Kumbar, S. G.; Jiang, T.; Weikel, A. L.; Krogman, N. R.; Allcock, H. R.; Laurencin, C. T. Biodegradable Di-Peptide-Based Polyphosphazene Blends for Regenerative Medicine: Analysis, Characterization, and Implications of a Unique Polymer Matrix Erosion on Morphology and in situ Porous Structure. *Adv. Funct. Mater.* **2010**, *20*, 2794–2806.
64. Weikel, A. L.; Owens, S. G.; Fushimi, T.; Allcock, H. R. Synthesis and Characterization of Methionine- and Cysteine-Substituted Phosphazenes. *Macromolecules* **2010**, *43*, 5205–5210.
65. Deng, M.; Kumbar, S. G.; Nair, L. S.; Weikel, A. L.; Allcock, H. R.; Laurencin, C. T. Biomimetic Structures: Biological Implications of Dipeptide-substituted Polyphosphazene-Polyester Blend Nanofiber Matrices for Load-Bearing Bone Regeneration. *Adv. Funct. Mater.* **2011**, *21*, 2641–2651.
66. Cheheltani, R.; Ezzibdeh, R. M.; Chhour, P.; Chandrika, K.; Jurcova, M.; Blundell, C.; Litt, H. I.; Ferrari, V. A.; Allcock, H. R.; Sehgal, C. M.; Cormode, D. P. Tunable Biodegradable Gold Nanoparticles as Contrast Agents for Computed Tomography and Photoacustical Imaging. *Biomaterials* **2016**, *102*, 87–97.
67. Xu, L.-C.; Li, Z.; Tian, Z.; Chen, C.; Allcock, H. R.; Siedlecki, C. A. A New Textured Polyphosphazene Biomaterial with Improved Blood Coagulation and Microbial Infection Responses. *Acta Biomater.* **2018**, *67*, 87–98.

68. Gettleman, L.; Gebert, P. H. U.S. Patent 04,661, 065, 1987.
69. Allcock, H. R.; Fuller, T. J. Phosphazene High Polymers with Steroidal Side Groups. *Macromolecules* **1980**, *13*, 1338–1345.
70. Nichol, J. L; Hotham, I. T.; Allcock, H. R. Ethoxyphosphazene Polymers and their Hydrolytic Behavior. *Polym. Degrad. Stab.* **2014**, *109*, 92–96.
71. Allcock, H. R.; Phelps, M. V. B.; Barrett, E. W.; Pishko, M. V.; Koh, W.-G. Ultraviolet Photolithograpic Development for Polyphosphazene Hydrogel Microstructures for Potential Use in Microarray Biosensors. *Chem. Mater.* **2006**, *18*, 609–613.
72. Barrett, E. W.; Phelps, M. V. B.; Silva, R. J.; Gaumond, R. P.; Allcock, H. R. Patterning Poly(organophosphazenes) for Selective Cell Adhesion Applications. *Biomacromolecules* **2005**, *6*, 1689–1698.
73. Barrett, E. W.; Phelps, M. V. B.; Allcock, H. R.; Silva, R. J.; Gaumond, R. P. Microcontact Printing of Poly(organophosphazenes): Potential Applications for Selective Cell Adhesion. *PMSE Prepr.(ACS Div. Polym. Mater. Sci. Eng.)* **2004**, *91*, 633–634.
74. Phelps, M. V. B.; Barrett, E. W.; Allcock, H. R. Cytotoxicity and Cell Encapsulation Studies of Poly[bis(methoxyethoxyethoxy)phosphazene] and Its Hydrogels. *Adv. Mater.* **2005**, *17*, 859–862.
75. Fei, S.-T.; Phelps, M. V. B.; Wang, Y.; Barrett, E.; Gandhi, F.; Allcock, H. R. A Redox Responsive Gel Based on Ionic Crosslinking. *Soft Matter* **2006**, *2*, 397–401.
76. Allcock, H. R.; Kwon, S. An ionically-Crosslinkable Polyphosphazene: Poly[di(carboxylatophenoxy-phosphazene)] and Its Hydrogels and Membranes. *Macromolecules* **1989**, *22*, 75–79.

Chapter 2

Self-Assembling Ionic Polyphosphazenes and Their Biomedical Applications

Alexander K. Andrianov*

Institute for Bioscience and Biotechnology Research, University of Maryland, 9600 Gudelsky Dr., Rockville, Maryland 20850, United States
***E-mail: aandrianov@umd.edu.**

Ionic polyphosphazenes represent a distinct class of polyelectrolytes with unique structural characteristics and the ability to undergo hydrolytic degradation under physiological conditions. Their potential applications in life sciences span from immunostimulation and vaccine delivery to biodegradable nanoparticulate encapsulation systems based on ionotropic hydrogels. Many biologically relevant properties of these macromolecules stem from their ability to interact with biological targets on molecular and cellular levels. The present review summarizes current knowledge on the self-assembly behavior of ionic polyphosphazenes in aqueous solutions and discusses potential role of supramolecular systems in the development of new biomedical applications of these versatile macromolecules.

Introduction

Polyphosphazene chemistry has long been a source of inspiration to researchers in their quest for novel custom-designed polymers for life sciences applications (*1*, *2*). These synthetic macromolecules, which are comprised of phosphorus-nitrogen backbone and organic side groups, provide undprecedented structural diversity due to the unique macromolecular substitution synthetic pathway (*1*). Water-soluble polyphosphazenes, in particular, have drawn considerable attention as multifunctional drug and vaccine delivery vehicles (*3*, *4*). The foremost reason for a continuous interest in these hybrid organic-inorganic macromolecules lies in their ability to undergo hydrolytic degradation in aqueous solutions (*5*). Such processes typically result in the release of physiologically "benign by design" degradation products and are well controlled through the selection of appropriate side groups, linkers, and even formulation components (*5–8*). However, the fact that some applications of polyphosphazenes in life sciences are based on their ability to spontaneously self-assemble with proteins and other important biological targets is frequently overlooked. Nevertheless, it is the ability of ionic polyphosphazenes to interact with biological entities that led to the development of a novel class of immunoadjuvants and their advancement to clinical trials (*9*). More evidence continues to emerge pointing to the importance of such interactions with complex biological systems, both on molecular and cellular levels (*10–12*). The present review is an attempt to summarize current knowledge on the self-assembly behavior of ionic polyphosphazenes and discuss the potential significance of these processes for expanding applications of polyphosphazenes in life sciences.

Polyphosphazenes as Polyelectrolytes

Although intrinsic hydrolytic degradability remains a key feature of ionic polyphosphazenes (*6*, *13*), the following factors also play an important role in differentiating these macromolecules from conventional polyelectrolytes. Polyphosphazenes are characterized by high skeletal flexibility - the inherent barrier to torsion of phosphorus-nitrogen bonds appears to be in the same region as the barrier for poly(dimethylsiloxane) (*1*). Their structural diversity relies on organic chemistry methods rather than polymerization processes (*1*). The structure of the monomer unit typically includes two or more ionic or ionizable groups. These charges are not distributed evenly along the backbone, but rather form "pairs" along the chain, which makes it difficult to apply commonly used parameters, such as linear charge density, and directly compare them with conventional polyelectrolytes. Nevertheless, a formal approach based on the Manning theory (*14*, *15*) using a skeletal bond length of 1.6 Å (*1*) and Bjerrum length of 7.15 Å, yields an extremely high value of apparent linear charge density – 4.47. The above structural characteristics may render ionic polyphosphazenes some of the most unusual polyelectrolytes known to date. From the application standpoint, this can be of critical importance as interactions of polyions with proteins and biologically relevant surfaces are typically governed by polymer linear charge density and intrinsic chain stiffness (flexibility) (*16–18*). However, the role of the polyphosphazene backbone's exceptional flexibility and the potentially "high linear density" of functional groups is yet to be elucidated.

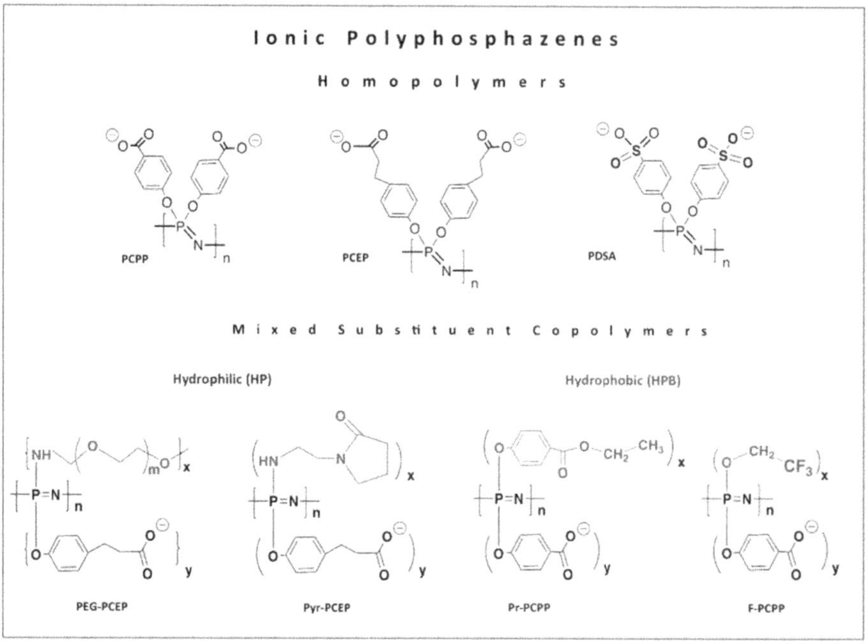

Figure 1. Structures of ionic polyphosphazenes.

Most of the research on ionic polyphosphazenes has been carried out with negatively charged polymers containing carboxylic or sulfonic acid groups. For the purpose of the present chapter, which attempts to link their self-assembly behavior with biological applications, polyphosphazene polyelectrolytes can be provisionally divided into three groups (Figure 1). Perhaps the most investigated family of polyphosphazene polyions consists of homopolymers, for which the term "high ionic group content" polyphosphazenes can be tentatively used. This group includes the first ionic polyphosphazene synthesized, PCPP (*19*), and its close relative, PCEP (*20*), which have been primarily investigated as immunoadjuvants and vaccine delivery vehicles (*9*). Sulfonated polyphosphazenes, including poly[diphenoxyphosphazenedisulfonic acid] (PDSA) (*21*, *22*), drew much attention primarily as proton conductive polymers (*23*, *24*), but have not yet been studied in context of biomedical applications. The second group of polyelectrolytes includes mixed substituent copolymers containing acidic groups and hydrophilic neutral chains - "Hydrophilic (HP)" polyelectrolytes (Figure 1). Copolymers of PCEP containing pyrrolidone side groups (Pyr-PCEP) were synthesized with the purpose of controlling solubility and degradation profiles of polyacids and in an effort to modulate interactions of ionic homopolymers with biological targets, such as cells (*25*, *26*). Polyelectrolytes containing relatively long PEG chains (PEG-PCEP) were designed as delivery vehicles for proteins, which can potentially eliminate or reduce interactions with biological targets. They will be discussed in a separate chapter of this book (A. Martinez et al., Chapter 6). Finally, a group of hydrophobically modified polyelectrolytes (HPB, Figure 1) such as polyacids containing fluorinated (F-PCPP) (*27*) and propyl paraben (Pr-PCPP) side groups (*28*) were also introduced, primarily for creating hydrophobic microspheres or coatings through the use of layer-by-layer nano-assembly processes. Cationic polyphosphazenes have also been synthesized and investigated mainly for gene delivery applications (*29–39*), however this lies outside the scope of the present review.

Polyphosphazenes with High Content of Ionic Groups as Immunoadjuvants

Self-Assembly with Proteins

The discovery of the potent immunoadjuvant properties of ionic polyphosphazenes in the early nineties (*40–43*) prompted more thorough physico-chemical investigation of PCPP adjuvanted vaccine formulations. Although, at the time no physico-chemical evidence of polymer association with vaccine antigen was established, the possibility of spontaneous self-assembly was always contemplated on the basis of the following considerations. It was observed that the immunoadjuvant activity of PCPP is significantly higher than that of similar conventional ionic polymers, such as alginic, poly(acrylic), and poly(methacrylic acids) (*41*, *44*). The immunoadjuvant effect of most ionic polymers can be greatly enhanced once they are covalently linked to the antigen (*45*). However, the very fact of such association between PCPP and vaccine antigens remained unproven.

Experimental support for the spontaneous self-assembly of PCPP with proteins was first generated for a model protein – bovine serum albumin, BSA. Mixing of polymer and protein solutions resulted in a spontaneous formation of multimeric complexes with a a relatively large number number of protein molecules (140) associated with a polyphosphazene chain (*46*). Further investigations were conducted using asymmetric flow field flow fractionation (AF4) (*10*), dynamic light scattering (DLS), analytical ultracentrifugation (*47*), high-sensitivity differential scanning, and isothermal titration calorimetry methods (*48*). In particular, the AF4 study investigated interactions of PCPP and PCEP with sixteen different proteins, including some of the most common antigens, at near physiological conditions in a phosphate buffered saline, PBS (pH 7.4) (*10*). The binding isotherm for PCPP and avidin, a model protein with high isoelectric point (pI = 10.5), revealed the apparent dissociation constant of approximately 2×10^{-7}. Although it was not possible to make an accurate comparison of the above apparent dissociation constant with values for other biological systems due to the simplified nature of the approximation, it is important to note that the micromolar range of dissociation constants is typical for binding of signaling protein to a biological target (*49*). Overall, it was concluded that the isoelectric point of the protein and its electrostatic potential surface played an important role in the spontaneous self-assembly of both ionic polymers with proteins. This was anticipated on the basis of previously reported observations on the electrostatic nature of interactions between polyelectrolytes and proteins (*50*).

In a somewhat unexpected finding, the glycosylated proteins showed much higher affinity to PCPP than their non-glycosylated counterparts. For example, the dissociation constant for a complex of PCPP with deglycosylated avidin was approximately two orders of magnitude higher than the one reported for the same protein containing glucans (*10*). This suggests that in addition to electrostatic interactions, hydrogen bonds can contribute to the formation of complexes. It may be worth mentioning that the formation of hydrogen bond complexes between PCPP and poly(ethylene oxide) at physiological conditions was reported and attributed to the presence of non-ionized carboxylic acid groups in PCPP at neutral pH (*51*). Another notable deviation of the strictly electrostatic model of interactions with ionic polyphosphazenes is BSA – a protein in which hydrophobic cleft is a major determinant in interactions of with other proteins and polyelectrolytes (*52–54*). The superior binding of this protein to PCPP and PCEP, when compared to other proteins with similar isoelectric points, suggests the importance of hydrophobic interactions in self-assembly of ionic polyphosphazenes (*10, 47*). It was also reported that the formation of PCPP-lysozyme complexes proceeds through a cooperative mechanism due to interactions of the neighboring bound protein molecules (*48*). Furthermore, contribution to the binding mechanism comes from cooperative van der Waals bonds between dehydrated apolar surface groups of the protein caused by the protein-polymer and protein-protein interactions (*48*).

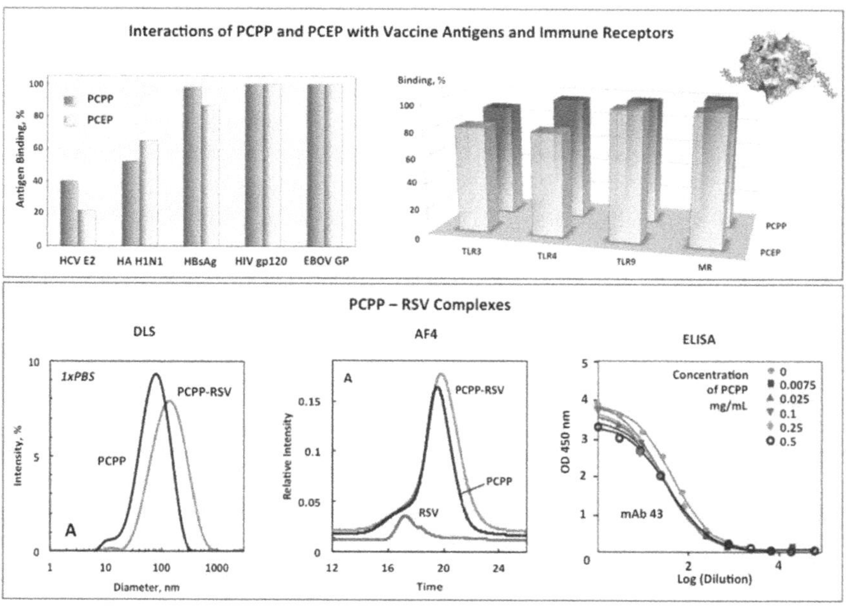

Figure 2. Binding of various vaccine antigens and soluble immune receptors by PCPP and PCEP as evaluated by AF4 method (top) and characterization of PCPP-RSV complexes by DLS, AF4 and ELISA (bottom). Adapted with permission from ref. (10). Copyright 2016 American Chemical Society, and ref. (55). Copyright 2017 American Chemical Society.

Interestingly, although proteins show various avidity to ionic polyphosphazenes, all tested vaccine antigens, such as Hepatitis B and C, influenza (H1N1), HIV (gp120), Ebola (EBOV GP) (*10*), and Respiratory Syncytial Virus sF Subunit Vaccine (RSV) (*55*) displayed strong interactions with polyphosphazenes (Figure 2). Based on the apparently higher avidity of polyphosphazenes to vaccine antigens when compared to some other proteins tested, as well as on the strong immunopotentiation effects observed in vivo, it is tempting to speculate that antigen-polyphosphazene complexes remain stable in the presence of serum proteins. However, more research is needed to prove this.

Protein-polymer complexation can in certain cases lead to an increase in the size of the complex compared to the polyion alone or even potentially cause some undesirable protein aggregation (*10*). Therefore, evaluation of biologically relevant protein functionality has been an important part of studies on the self-assembly of ionic polyphosphazenes with vaccine antigens. Investigation of circular dichroism (CD) spectra of PCPP complexes with RSV revealed no major changes in protein conformation (*55*). Moreover, the antigenicity of PCPP-adjuvanted RSV sF formulations was evaluated in vitro using ELISA assays with antibodies targeting three different epitopes (*55*). The effect of polyphosphazene was found to be practically negligible, which suggests that despite the formation of the complex, the antigen remains largely accessible to antibodies (Figure 2) (*55*). Finally, all PCPP adjuvanted formulations demonstrated in vivo immunoadjuvant performance superior to the RSV protein alone (*55*). The effect of PCPP addition on H5N1 influenza vaccine was also studied using single radial immunodiffusion immunoprecipitation (SRID) method, which showed no reduction in antigenicity (*56*). Moreover, in thermal stability experiments the PCPP adjuvanted H5N1 formulations demonstrated up to 3.5 fold increase in antigen half-life, which demonstrates the antigen stabilizing effect of PCPP (*56*). This is also consistent with findings on the improved thermal stability of horseradish peroxidase (HRP) in the presence of PCPP (*57*). It can be therefore concluded that spontaneous self-assembly of ionic polyphosphazenes with proteins may also present an interest as a convenient approach to stabilizing vaccines and improving their shelf life.

Interactions with Cells

Virtually all adjuvant systems developed to date are focused on two main mechanisms - specific immune activation (intrinsic immunoadjuvant effect) and delivery-depot effect (*58*). Protein binding properties of ionic polyphosphazenes discussed above may imply that delivery of vaccine antigens constitutes their primary role in enhancing immune responses. The intrinsic immunoadjuvant activity generally results from interactions of molecular adjuvants with antigen-presenting cells (*58*). Therefore, it was of interest to investigate activity of ionic polyphosphazenes on a cellular level.

The ability of synthetic polyelectrolytes to interact with membrane proteins of immunocompetent cells has been suggested as one of the important characteristics explaining their immunoadjuvant activity (*45*). Nevertheless the relevant studies on this subject remain scarce. To that end, Toll-Like Receptors (TLRs) –

membrane proteins representing some of the most essential types of immune receptors and functioning as primary sensors of the innate immune system (*59–61*) - are an important model for such investigations. The ability of PCPP and PCEP to interact with soluble TLRs (TLR3, −4, −9) in solution at near physiological conditions (PBS, pH 7.4) was explored using AF4 method (*10*). Both polyphosphazenes demonstrated strong avidity to soluble receptors (Figure 2), which may suggest a possibility of direct activation of immune cells by PCPP and PCEP through the TLR signaling pathway, either on the external cell surface (TLR4) or endosomal (TLR3 and −9) levels. Similar results were observed for the mannose receptor, another type of pattern-recognition immune receptor (Figure 2) (*10*). The observed strong interactions of ionic polyphosphazenes with membrane proteins, such as TLRs, which are typically characterized with regions of high lipophilic character and significant content of carbohydrates, can be potentially explained with the high affinity of polyphosphazenes to glycosylated and amphipathic proteins discussed above.

This molecular level interaction study with membrane proteins in solution was further extended to evaluate the stimulatory effect of PCPP and PCEP in cellular assays with engineered HEK293 cells, which overexpress human TLR genes - specifically TLR2, −3, −4, −5, −7, −8, and −9 (*10*). Although it was clear that both PCPP and PCEP were able to stimulate cellular responses, the latter does not appear specific for most TLRs as the TLR^{-} negative control cell line data also show some activation. The strongest responses for TLR overexpressing cells with minimal non-specific stimulation were observed in the case of TLR 8 and TLR 9 for PCPP and the same receptors, plus TLR 3, for PCEP (*10*). Interestingly, all of the above receptors are typically associated with nucleic acid agonists, which bear some formal structural similarities to polyphosphazenes.

Dendritic cells (DCs) are the most potent antigen-presenting cells and the ability of immunoadjuvants to induce their activation and maturation is of fundamental importance (*62*). The effect of PCPP on maturation, activation and antigen presentation by human adult and newborn dendritic cells (DCs) was studied in vitro (*11*). PCPP treatment induced DC activation as evaluated by upregulation of co-stimulatory molecules and production of cytokines. Moreover, when formulated with HIV group-specific antigen, it induced maturation of DCs and release of mixed Th1/Th2 cytokine responses, promoting both cellular and potentially humoral responses to the formulated antigen (*11*). It was concluded that the PCPP vaccine formulation had intrinsic adjuvant activity, could facilitate effective delivery of antigen to DCs, and may be advantageous for induction of beneficial T cell-mediated immunity (*11*).

The above findings on interactions of polyphosphazenes with cells correlate well with results on intramuscular injection of mice with PCEP, which induced significant recruitment of neutrophils, macrophages, monocytes, DCs, and lymphocytes at the site of injection as well as in the draining lymph nodes (*12*). Flow cytometric analysis showed that the majority of the recruited immune cells took up and/or were associated with PCEP at the injection site (*12*). Furthermore, in vivo, PCEP induced time-dependent changes in the gene expression of many "adjuvant core response genes" including cytokines, chemokines, innate immune receptors, interferon induced genes, adhesion molecules, and antigen-presentation genes (*63*).

As discussed above, it appears that PCPP and PCEP are capable of displaying multiple functions as immunoadjuvants. These macromolecules spontaneously self-assemble with antigenic proteins into non-covalent complexes, interact with cellular receptors, and stimulate and induce maturation of the most important antigen-presenting DCs. Obviously, ionic groups are critical in establishing all types of these interactions and it is very important to monitor relevant characteristics of the complexes to achieve optimal results. It was observed that the immunoadjuvant activity of hydrophobically modified PCPP (Pr-PCPP) rises almost linearly as the content of carboxylic acid in the polymer increases (*28*). Therefore, it is not surprising that complete saturation of the complex with protein molecules can lead to some loss of immunoadjuvant activity, emphasizing the fact that the availability of carboxylic acid groups and extended conformation of the complex is important for its interactions with immunocompetent cells and achieving optimal immune response (*46*).

A striking difference in the behavior of PCPP and PCEP was observed for their potential endosomolytic activity (*10*). Studies were conducted using red blood cells (RBC) as endosomal membrane models in the pH range of 6.0-7.5 (*64*, *65*). The ability of vaccine carriers to increase the amount of antigen that escapes from endosomes into the cytoplasm was previously connected with the enhancement of cross-presentation of antigens by DCs, which plays a central role in the induction of efficient immune responses, especially CD8+ T-cell responses (*66–68*). Contrary to PCPP, which did not show any membrane disruptive activity in the above pH range, PCEP was found to be disruptive to membranes within the pH environment of early endosomes (pH 6.0 - 6.9) (*69*). Membranolytic properties of some polyacids are typically realized through the pH triggered conformational changes and formation of hydrophobic aggregates during acidification in an early endosomal environment (*64*). It can be hypothesized that more hydrophobic structure of PCEP, as compared to PCPP, may play critical role in interactions with cellular membranes. The finding of pH dependent membrane disruptive activity of PCEP can provide new insights for better understanding of the differences in immunoadjuvant activities of these polyphosphazenes.

Applications of Polyphosphazenes in Vaccines

Spontaneous self-assembly of ionic polyphosphazenes with antigenic proteins resulting in the formation of soluble non-covalent complexes has important implications for their biological applications (*9*). After extensive investigations with bacterial and viral antigens in multiple animal models (*9*, *70*), PCPP has been advanced into several clinical studies demonstrating both an immunoadjuvant effect and a good safety profile (*71–74*). Newer generation polyphosphazene adjuvants, which include PCEP have been synthesized and shown promising results in animal studies (*20*, *75*). Polyphosphazenes have been conformed into microparticle formulations, which displayed potential in parenteral, oral, and intranasal delivery (*76–79*), and used in combinations with other adjuvants (*80*, *81*). Finally, physical and mechanical properties of PCPP as a polymeric material were exploited to micro-fabricate PCPP microneedles for intradermal delivery of vaccines, which demonstrated superior performance compared to parenteral injections (*82–84*).

"Hydrophilically Modified" Mixed Substituent Ionic Polyphosphazenes

Findings on the ability of polyphosphazene polyelectrolytes to interact both on the molecular level with proteins and on the cellular level to induce activation of immunocompetent cells raised an important question on whether their activity on the cellular level can be suppressed or modulated to extend the technology to the delivery of protein therapeutics. Macromolecular drugs are an increasingly important class of drugs (*85*), however their applications are severely limited due to their short half-life in vivo (*86*), undesirable antigenicity (*86*), and low uptake by targeted cells (*87*) such as cancer cells. Various approaches have been developed to address the challenge and thus far PEGylation – creation of a steric shell by covalent modification of proteins with poly(ethylene glycol), PEG, appears to be most successful commercially (*88*, *89*). Attachment of highly hydrated and flexible PEG chains increases the size of the protein, thereby preventing its elimination through glomerular filtration, and renders it invisible for the immune system, resulting in reduced clearance by phagocytes of the reticuloendothelial system (*90–92*). Therefore, the underlying concept for the development of polyphosphazene carriers for protein drug delivery was to maintain the ionic content sufficient for enabling non-covalent binding with the protein payload and to introduce neutral hydrophilic side groups, which would create steric shield around the protein and reduce its immunogenicity. It was also desirable to maintain the pH-dependent membrane disruptive feature of PCEP, which would facilitate endosomal escape of the protein resulting in the cytosolic delivery of the protein – a stimuli-responsive "smart polymer" feature (*93–95*). From the application standpoint, the development of an alternative "non-covalent PEGylation" approach is extremely attractive as it may result in a simplified manufacturing in which PEGylation is achieved on the formulation level by simple mixing of solutions, and significant reduction in production costs, which are associated with the need to purify protein from PEGylation reaction by-products (*96*). The non-covalent PEGylation approach can also be potentially extended to protein molecules for which covalent PEGylation is currently challenging.

Two different types of polyphosphazenes have been synthesized and explored for the purpose of therapeutic protein delivery. They are either polyacids containing grafted PEG side chains (described in details in a separate chapter of this book – Martinez et al., Chapter 6) or mixed substituent copolymers with pyrrolidone side groups (Pyr-PCPP) (*25*, *26*), which were designed with the expectation that neutral groups can reduce unwanted interactions with immunocompetent cells. All of these polymers demonstrated hydrolytic degradation at near physiological conditions (PBS, pH 7.4, 37 °C) and accelerated conditions (PBS, pH 7.4, 55 °C) (*25*). Nevertheless the rate of degradation was dramatically lower at ambient temperature and 4 °C, which should allow for an adequate shelf life of these biodegradable polymers (*25*).

Self-Assembly with Proteins

The ability of hydrophilic ionic copolymers to self-assemble with proteins at near physiological conditions (PBS, pH 7.4) was investigated for polyphosphazenes containing pyrrolidone and phenyl propionic acid side groups and avidin (*25*). All mixed substituent polyphosphazenes were able to bind avidin as shown by AF4 method. Formation of the complex, however, did not interfere with the ability of this protein to bind its low molecular weight substrate, biotin, as both unbound and complexed avidin showed the same affinity to it (*25*). Importantly, maximum protein loading increased with the content of carboxylic acid groups in the polymer (Figure 3) indicating that the majority of ionic groups in such complexes may be consumed in interactions with protein and are potentially "hidden" from the immune system (*25*). This assumption is supported with the observed reduction in antibody binding to the protein complexed with PEGylated polyphosphazene (PEG-PCPP), which is discussed in a separate chapter of this book (A. Martinez, et al., Chapter 6).

Cellular Uptake of Polymer–Protein Complexes

The potential of Pyr-PCPP copolymers to facilitate interaction of a model protein, FITC-labeled avidin, with cells was examined in vitro using oral adenosquamous carcinoma Cal27 cells (*25*). It was found that copolymers drastically (up to 50 fold) enhanced association of protein with cells, although better performance was achieved for copolymer with higher content of acidic groups. Furthermore, the technique of additional staining of surface accumulated protein molecules (*97*) was applied (Figure 3), which allowed differentiation between cell-internalized (green) and cell-adsorbed protein (yellow). Based on this calculated percentage of internalization, Pyr-PCPP copolymers improved the uptake of protein cargo by cells up to 21-fold (*25*). It was concluded that polyphosphazene polyacids facilitated cell-surface interaction followed by time-dependent, vesicular mediated, and saturable internalization of a model protein cargo into cancer cells, demonstrating potential for intracellular delivery (*25*).

Biocompatibility of these polymers with blood components was assessed in the hemolysis assay (*98*), which demonstrated lack of cellular toxicity at neutral pH. However, acidification of the solution below pH 6 triggered membrane lysis, a behavior that is characteristic to hydrophobically modified polyacids (*64*, *95*). Generally, acidic conditions trigger coil to globule conformational changes, which in turn cause adsorption of the polymer to the outer leaflet of the bilayer, and subsequently membrane expansion and disruption (*64*). Since the threshold of polymer membranolytic activity generally corresponds to endosomal pH (pH 5.0-6.5) (*69*) and correlation between hemolytic efficiency and endosomal disruption had been previously established (*99*), such polyphosphazene polyacids may present an interest as endosomolytic carriers facilitating delivery into the cytoplasm (*64, 95, 100, 101*).

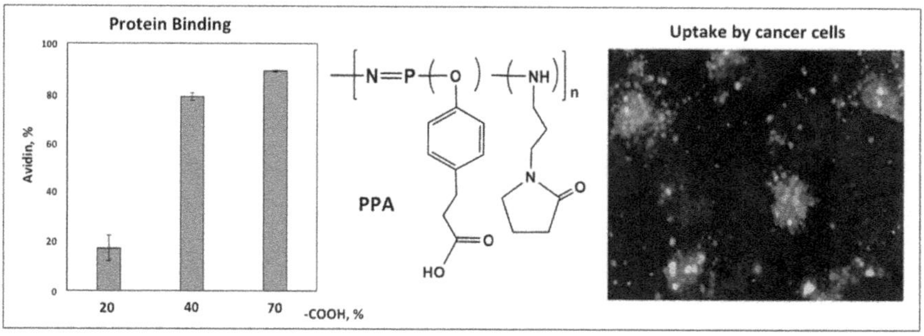

Figure 3. Hydrophilic Pyr-PCEP mixed substituent polyphosphazenes: avidin binding as a function of acid group content in the polymer (left) and uptake of avidin-PPA complexes by cancer cells (right). Adapted with permission from ref. (25). Copyright 2017 American Chemical Society.

Hydrophobically Modified Mixed Substituent Ionic Polyphosphazenes

Hydrophobically modified ionic polyphosphazenes – copolymers of PCPP containing propyl paraben side groups (Pr-PCPP in Figure 1), were initially synthesized to confirm critical importance of carboxylic acid groups for the immunoadjuvant activity of PCPP (*28*). Similarly to PCPP, aqueous solutions of these polymers displayed phase separation upon addition of acid, which, as expected, appeared to be more pronounced for copolymers with higher content of hydrophobe (*28*). Surprisingly, it was observed that Pr-PCPP copolymers showed little (polymers with low hydrophobe content) or no sensitivity (polymers with high hydrophobe content) to solutions of sodium chloride, which under the same conditions, caused precipitation of PCPP (*28*). This emphasized the importance of steric factors in condensation of sodium counterions with PCPP and suggested that Pr-PCPP materials can be advantageous for applications when such interactions need to be reduced (*28*).

Perhaps the most captivating class of hydrophobically modified ionic polyphosphazenes synthesized to date is a hybrid system containing trifluoroethoxy groups – the main side groups of fluorinated elastomers (*102*) - and carboxylic acid side groups of PCPP (F-PCPP in Figure 1) (*27*). Hydrophobic and superhydrophobic coatings of fluorinated polyphosphazenes (*103*) are recognized for their outstanding biocompatibility (*104*) and are important constituents of clinically validated injectable microspheres (*105, 106*) and medical devices (*107, 108*). Somewhat unexpectedly, F-PCPP copolymers displayed solubility in water with the content of fluoroethoxy- groups of up to 60% (mol.) and were soluble in water-ethanol mixture even when the hydrophobe content reached 97% (mol.) (*27*). This resulted in the development of aqueous based methods for the production of fluorinated nano- and microparticulates (*27*) and hydrophobic coatings through layer-by-layer deposition technique in aqueous solutions, which are discussed in a separate chapter of this book (S.A. Sukhishvili et al., Chapter 5).

Nanoparticulate Delivery Vehicles Based on Ionic Polyphosphazenes

Formation of ionotropic hydrogels with excellent mechanical properties has long been one of the prime features of PCPP, differentiating it from many other synthetic polyacids (*19*, *109*). Similarly to solutions of the well-known natural biocompatible polymer, alginic acid (*110*), aqueous solutions of PCPP easily cross-link under mild conditions in the presence of calcium salts (*19*, *109*). Initially, PCPP hydrogels have been studied as biomaterials for cell encapsulation (*109*, *111*), matrices for protein release (*13*, *112*), mucosal delivery of vaccines (*77*, *78*, *113*), and encapsulation materials for biomedical imaging applications (*114*, *115*). Calcium induced cross-linking of PCPP was also employed for forming a composite with hydroxyapatite under physiological conditions as self-setting cement for bone replacement applications (*116*, *117*).

One of the major challenges in the development of hydrogel based polyphosphazene technology was meeting the needs of various biomedical applications, which required precise control of hydrogel size and architecture, frequently on the nanoscale level. Earlier methods for producing PCPP hydrogel particulates were limited to various types of spray nozzles (*78*, *109*, *118*), which resulted in sizes in the millimeter to micrometer range and were not feasible for manufacturing due to their frequent clogging, scale-up issues, and concerns related to the containment of biologics. The development of nozzle-free methods followed an interesting discovery of unusual ionic selectivity of PCPP. It was found that PCPP undergoes phase separation in the presence of sodium chloride solutions of medium concentration, but remains in solution in the presence of other monovalent ions, such as lithium and potassium, as well as at low and high concentration of sodium chloride (*28*, *119*). This finding led to a development of two-step microparticle fabrication method, in which an aqueous solution of PCPP first undergoes coacervation induced by addition of sodium chloride and then growing coacervate microdroplets are cross-linked with calcium chloride to yield particles in the micrometer size range (*76*). The method allowed for control of microparticle sizes and was easy to scale-up (*76*). A single-step production of PCPP micro- and nanoparticulates was then suggested utilizing ion complexation of polyphosphazenes with physiologically benign amines, such as spermine and spermidine (*120*). Substitution of calcium ions with spermine not only led to the extension of the technology to the nanoscale level, but also improved stability of particulates to phosphate ions and resulted in a process which was easier to control (*120*). Moreover, the method was extended to polyphosphazenes, which were not suitable for the original coacervation process due to a better stability in the presence of sodium ions, such as PCEP (*20*), and various copolymers of PCPP (*26*, *27*). One of the most important features of ionically cross-linked systems is their high encapsulation efficiency for many proteins (*76*, *120*), which first self-assemble with polyphosphazenes into complexes and then undergo the micro- or nano-encapsulation process. New advances in the technology have been reported, which include the use of microfluidic mixing techniques for improved control of size and polydispersity (*121*). It was also demonstrated that both surface and core loading of nanocrystals and proteins in ionically cross-linked PCPP nanoparticles could be selectively achieved, which opens new opportunities in diagnostics and theranostics applications (*122*).

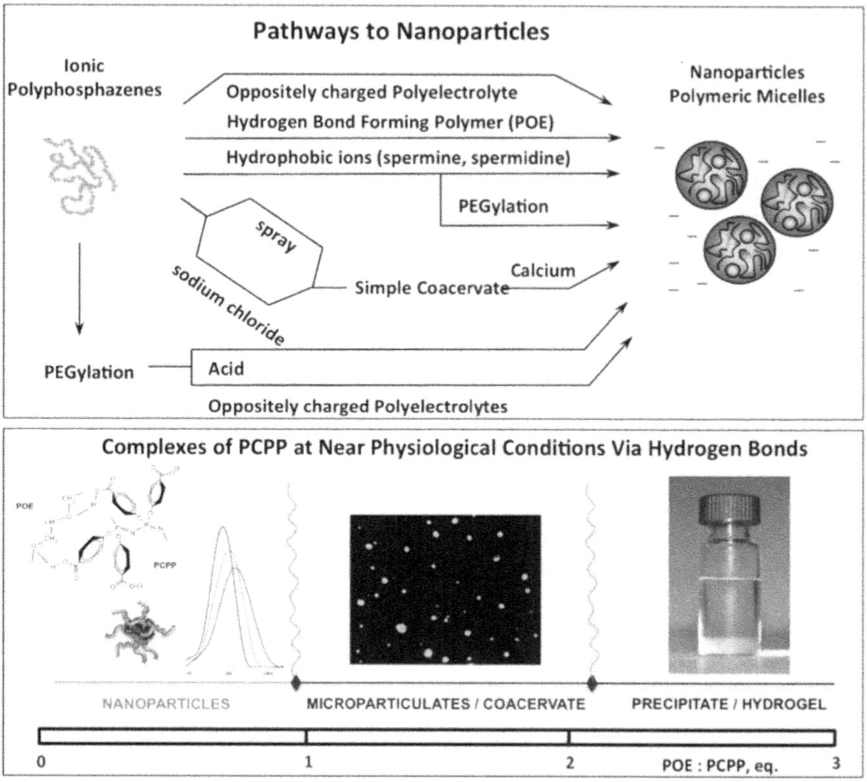

Figure 4. Various pathways to nanoparticle formulations using ionic polyphosphazenes (top) and complexes of PCPP formed at near physiological conditions with poly(oxy ethylene) (POE) via hydrogen bonds (bottom). Adapted with permission from ref. (51). Copyright 2016 The Authors. Published by Elsevier Ltd. This is an open access article under the CCBY-NC-ND license (http://creativecommons.org/licenses/by-nc-nd/4.0/).

Although by far the largest part of research on polyphosphazene hydrogel nanoparticulates was carried out using PCPP, it maybe worth mentioning that stability of ionotropic PCPP gels under physiological conditions is yet to be explored in more detail. Stabilization approaches have been suggested through PEGylation of nanoparticles using the formation of polyion complexes between PCPP and a block-copolymer of PEG and poly(L-lysine) (*121*) or adding PCPP with PEG side groups (*123*). Ionic polymers containing hydrophilic pyrrolidone groups and phenyl propionic acid groups are also an interesting class of sterically stabilized molecules that can easily form spermine cross-linked nanoparticles (*25*). Interestingly, they can also spontaneously self-assemble under acidic conditions in polymeric micelles with very narrow polydispersity (*25*). This behavior can be explained by poor solubility of ionic functionalities at low pH and may suggest presence of some blocky structure in these randomly substituted polymers (*25*).

A newly established alternative pathway to ionic hydrogels involves polymer cross-linking through hydrogen bonds, which in the case of polyphosphazene polyelectrolytes can be achieved at a neutral pH (Figure 4) (*51*). It is well established that the formation of interpolymer complexes depends on the degree of ionization of poly(carboxylic acid) and thus on environmental pH. Typically, such complexes are only formed in weakly or strongly acidic media and dissociate upon increase in pH, which limits their in vitro and in vivo utility (*124–126*). The ability of PCPP to form hydrogels at near physiological conditions may be explained by incomplete dissociation of PCPP at this pH and the ionic strength resulting in a significant content of non-ionized carboxylic acid groups (*6*, *28*). The method allows preparation of particulates on the micro- and nanoscale level and can be also useful for making hydrogel biomaterials under mild conditions in situ (Figure 4) (*51*).

Conclusions

Water-soluble ionic polyphosphazenes constitute an interesting class of polyelectrolytes with peculiar structural characteristics and the ability to undergo hydrolytic degradation in aqueous solutions. Polyphosphazene polyacids, such as PCPP and PCEP, have already demonstrated considerable potential as vaccine delivery vehicles and immunoadjuvants, which stems from their ability to spontaneously self-assemble with vaccine antigens and stimulate immunocompetent cells. However, it can be envisioned that while "protein loading capacity" of polyphosphazene carriers can be maintained through appropriate ionic content, the nature and extent of interactions with components of the immune system can be modulated via the synthesis of mixed substituent polymers containing hydrophilic "steric hindrance" side groups, such as PEG, providing the basis for extending polyphosphazene applications to drug delivery. Furthermore, hydrophobically modified ionic polyphosphazenes, especially fluorinated polyelectrolytes, can potentially offer new opportunities in constructing hydrophobic and superhydrophobic surfaces from aqueous solutions. Finally, broad environmental sensitivity of these polyions along with their ability to form hydrogels through ionic interactions and hydrogen bonds may facilitate the development of highly versatile methods for encapsulation of drugs and imaging agents into polyphosphazene nanoparticulate systems.

References

1. Allcock, H. R. *Chemistry and Applications of Polyphosphazenes*; Wiley: Hoboken, NJ, 2002; p 725.
2. Andrianov, A. K. *Polyphosphazenes for Biomedical Applications*; John Wiley & Sons: Hoboken, NJ, 2009; p 457.
3. Andrianov, A. K. Water-soluble polyphosphazenes for biomedical applications. *J. Inorg. Organomet. Polym. Mater.* **2006**, *16*, 397–406.
4. Teasdale, I.; Brüggemann, O. Polyphosphazenes: Multifunctional, biodegradable vehicles for drug and gene delivery. *Polymers* **2013**, *5*, 161–187.
5. Allcock, H. R.; Morozowich, N. L. Bioerodible polyphosphazenes and their medical potential. *Polym. Chem.* **2012**, *3*, 578–590.
6. DeCollibus, D. P.; Marin, A.; Andrianov, A. K. Effect of environmental factors on hydrolytic degradation of water-soluble polyphosphazene polyelectrolyte in aqueous solutions. *Biomacromolecules* **2010**, 487–492.
7. Andrianov, A. K.; Marin, A. Degradation of polyaminophosphazenes: Effects of hydrolytic environment and polymer processing. *Biomacromolecules* **2006**, *7*, 1581–1586.
8. Wilfert, S.; Iturmendi, A.; Schoefberger, W.; Kryeziu, K.; Heffeter, P.; Berger, W.; Brüggemann, O.; Teasdale, I. Water-soluble, biocompatible polyphosphazenes with controllable and pH-promoted degradation behavior. *J. Polym. Sci., Part A: Polym. Chem.* **2014**, *52*, 287–294.
9. Andrianov, A. K., Polyphosphazene vaccine delivery vehicles: State of development and perspectives. In *Polyphosphazenes for biomedical applications*; Andrianov, A. K., Ed.; Wiley: Hoboken, NJ, 2009; pp 47−63.
10. Andrianov, A. K.; Marin, A.; Fuerst, T. R. Molecular-level interactions of polyphosphazene immunoadjuvants and their potential role in antigen presentation and cell stimulation. *Biomacromolecules* **2016**, *17*, 3732–3742.
11. Palmer, C. D.; Ninković, J.; Prokopowicz, Z. M.; Mancuso, C. J.; Marin, A.; Andrianov, A. K.; Dowling, D. J.; Levy, O. The effect of stable macromolecular complexes of ionic polyphosphazene on HIV Gag antigen and on activation of human dendritic cells and presentation to T-cells. *Biomaterials* **2014**, *35*, 8876–8886.
12. Awate, S.; Wilson, H. L.; Singh, B.; Babiuk, L. A.; Mutwiri, G. The adjuvant PCEP induces recruitment of myeloid and lymphoid cells at the injection site and draining lymph node. *Vaccine* **2014**, *32*, 2420–2427.
13. Andrianov, A. K.; Payne, L. G.; Visscher, K. B.; Allcock, H. R.; Langer, R. Hydrolytic degradation of ionically cross-linked polyphosphazene microspheres. *J. Appl. Polym. Sci.* **1994**, *53*, 1573–1578.
14. Manning, G. S. Counterion condensation theory constructed from different models. *Phys. A (Amsterdam, Neth.)* **1996**, *231*, 236–253.
15. Cini, N.; Ball, V. Polyphosphates as inorganic polyelectrolytes interacting with oppositely charged ions, polymers and deposited on surfaces: fundamentals and applications. *Adv. Colloid Interface Sci.* **2014**, *209*, 84–97.

16. Mattison, K. W.; Dubin, P. L.; Brittain, I. J. Complex formation between bovine serum albumin and strong polyelectrolytes: Effect of polymer charge density. *J. Phys. Chem. B* **1998**, *102*, 3830–3836.
17. Muthukumar, M. Pattern recognition by polyelectrolytes. *J. Chem. Phys.* **1995**, *103*, 4723–4731.
18. Brynda, M.; Chodanowski, P.; Stoll, S. Polyelectrolyte–particle complex formation. Polyelectrolyte linear charge density and ionic concentration effects. Monte Carlo simulations. *Colloid Polymer Sci.* **2002**, *280*, 789–797.
19. Allcock, H. R.; Kwon, S. An ionically cross-linkable polyphosphazene: Poly[bis(carboxylatophenoxy)phosphazene] and its hydrogels and membranes. *Macromolecules* **1989**, *22*, 75–79.
20. Andrianov, A. K.; Marin, A.; Chen, J. Synthesis, properties, and biological activity of poly[di(sodium carboxylatoethylphenoxy)phosphazene]. *Biomacromolecules* **2006**, *7*, 394–399.
21. Andrianov, A. K.; Marin, A.; Chen, J.; Sargent, J.; Corbett, N. Novel route to sulfonated polyphosphazenes: Single-step synthesis using "noncovalent protection" of sulfonic acid functionality. *Macromolecules* **2004**, *37*, 4075–4080.
22. Allcock, H. R.; Fitzpatrick, R. J.; Salvati, L. Sulfonation of (aryloxy)-and (arylamino) phosphazenes: small-molecule compounds, polymers, and surfaces. *Chem. Mater.* **1991**, *3*, 1120–1132.
23. Guo, Q.; N. Pintauro, P.; Tang, H.; O'Connor, S. Sulfonated and crosslinked polyphosphazene-based proton-exchange membranes. *J. Membr. Sci.* **1999**, *154*, 175–181.
24. Fedkin, M. V.; Zhou, X.; Hofmann, M. A.; Chalkova, E.; Weston, J. A.; Allcock, H. R.; Lvov, S. N. Evaluation of methanol crossover in proton-conducting polyphosphazene membranes. *Mater. Lett.* **2002**, *52*, 192–196.
25. Martinez, A. P.; Qamar, B.; Fuerst, T. R.; Muro, S.; Andrianov, A. K. Biodegradable "smart" polyphosphazenes with intrinsic multifunctionality as intracellular protein delivery vehicles. *Biomacromolecules* **2017**, *18*, 2000–2011.
26. Andrianov, A. K.; Marin, A.; Peterson, P. Water-soluble biodegradable polyphosphazenes containing N-ethylpyrrolidone groups. *Macromolecules* **2005**, *38*, 7972–7976.
27. Andrianov, A. K.; Marin, A.; Peterson, P.; Chen, J. Fluorinated polyphosphazene polyelectrolytes. *J. Appl. Polym. Sci.* **2007**, *103*, 53–58.
28. Andrianov, A. K.; Svirkin, Y. Y.; LeGolvan, M. P. Synthesis and biologically relevant properties of polyphosphazene polyacids. *Biomacromolecules* **2004**, *5*, 1999–2006.
29. Kwon, S.-K. Synthesis of water-soluble aminoaryloxy-methylamino cosubstituted polyphosphazenes as carrier species for biologically active agents. *Bull. Korean Chem. Soc.* **2001**, *22*, 1243–1247.
30. de Wolf, H.; de Raad, M.; Snel, C.; van Steenbergen, M.; Fens, M.; Storm, G.; Hennink, W. Biodegradable poly(2-dimethylamino ethylamino)phosphazene for in vivo gene delivery to tumor cells. Effect of polymer molecular weight. *Pharm. Res.* **2007**, *24*, 1572–1580.

31. de Wolf, H. K.; Luten, J.; Snel, C. J.; Oussoren, C.; Hennink, W. E.; Storm, G. In vivo tumor transfection mediated by polyplexes based on biodegradable poly(DMAEA)-phosphazene. *J. Controlled Release* **2005**, *109*, 275–287.
32. Yang, Y.; Xu, Z.; Chen, S.; Gao, Y.; Gu, W.; Chen, L.; Pei, Y.; Li, Y. Histidylated cationic polyorganophosphazene/DNA self-assembled nanoparticles for gene delivery. *Int. J. Pharm.* **2008**, *353*, 277–282.
33. Yang, Y.; Xu, Z.; Jiang, J.; Gao, Y.; Gu, W.; Chen, L.; Tang, X.; Li, Y. Poly(imidazole/DMAEA)phosphazene/DNA self-assembled nanoparticles for gene delivery: Synthesis and in vitro transfection. *J. Controlled Release* **2008**, *127*, 273–279.
34. Yang, Y.; Zhang, Z.; Chen, L.; Gu, W.; Li, Y. Galactosylated poly(2-(2-aminoethyoxy)ethoxy)phosphazene/DNA complex nanoparticles: In vitro and in vivo evaluation for gene delivery. *Biomacromolecules* **2010**, *11*, 927–933.
35. Yang, Y.; Zhang, Z.; Chen, L.; Gu, W.; Li, Y. Urocanic acid improves transfection efficiency of polyphosphazene with primary amino groups for gene delivery. *Bioconjugate Chem.* **2010**, *21*, 419–426.
36. Luten, J.; van Nostrum, C. F.; De Smedt, S. C.; Hennink, W. E. Biodegradable polymers as non-viral carriers for plasmid DNA delivery. *J. Controlled Release* **2008**, *126*, 97–110.
37. Luten, J.; van Steenbergen, M. J.; Lok, M. C.; de Graaff, A. M.; van Nostrum, C. F.; Talsma, H.; Hennink, W. E. Degradable PEG-folate coated poly(DMAEA-co-BA)phosphazene-based polyplexes exhibit receptor-specific gene expression. *Eur. J. Pharm. Sci.* **2008**, *33*, 241–251.
38. Luten, J.; Van Steenis, J. H.; Van Someren, R.; Kemmink, J.; Schuurmans-Nieuwenbroek, N. M. E.; Koning, G. A.; Crommelin, D. J. A.; Van Nostrum, C. F.; Hennink, W. E. Water-soluble biodegradable cationic polyphosphazenes for gene delivery. *J. Controlled Release* **2003**, *89*, 483–497.
39. Allcock, H. R.; McIntosh, M. B.; Klingenberg, E. H.; Napierala, M. E. Functionalized polyphosphazenes: Polymers with pendent tertiary trialkylamino groups. *Macromolecules* **1998**, *31*, 5255–5263.
40. Andrianov, A. K.; Jenkins, S. A.; Payne, L. G.; Roberts, B. E. *Phosphazene Polyelectrolytes as Immunoadjuvants*. U.S. Patent 5,494,673, 1996.
41. Payne, L. G.; Jenkins, S. A.; Andrianov, A.; Roberts, B. E. Water-soluble phosphazene polymers for parenteral and mucosal vaccine delivery. *Pharm. Biotechnol.* **1995**, *6*, 473–493.
42. Payne, L. G.; Jenkins, S. A.; Woods, A. L.; Grund, E. M.; Geribo, W. E.; Loebelenz, J. R.; Andrianov, A. K.; Roberts, B. E. Poly[di(carboxylatophenoxy)phosphazene] (PCPP) is a potent immunoadjuvant for an influenza vaccine. *Vaccine* **1998**, *16*, 92–98.
43. Andrianov, A. K.; Sargent, J. R.; Sule, S. S.; Le Golvan, M. P.; Woods, A. L.; Jenkins, S. A.; Payne, L. G. Synthesis, physico-chemical properties and immunoadjuvant activity of water-soluble phosphazene polyacids. *J. Bioact. Compatible Polym.* **1998**, *13*, 243–256.

44. Payne, L. G.; Van Nest, G.; Barchfeld, G. L.; Siber, G. R.; Gupta, R. K.; Jenkins, S. A. PCPP as a parenteral adjuvant for diverse antigens. *Dev Biol Stand* **1998**, *92*, 79–87.
45. Kabanov, V. A. From synthetic polyelectrolytes to polymer-subunit vaccines. *Pure Appl. Chem.* **2004**, *76*, 1659–1677.
46. Andrianov, A. K.; Marin, A.; Roberts, B. E. Polyphosphazene polyelectrolytes: A link between the formation of noncovalent complexes with antigenic proteins and immunostimulating activity. *Biomacromolecules* **2005**, *6*, 1375–1379.
47. Burova, T. V.; Grinberg, N. V.; Dubovik, A. S.; Grinberg, V. Y. Conformational stability of bovine serum albumin in complexes with poly[di(carboxylatophenoxy)phosphazene]. *Polym. Sci., Ser. A* **2015**, *57*, 761–772.
48. Burova, T. V.; Grinberg, N. V.; Dubovik, A. S.; Olenichenko, E. A.; Orlov, V. N.; Grinberg, V. Y. Interpolyelectrolyte complexes of lysozyme with short poly[di(carboxylatophenoxy)phosphazene]. Binding energetics and protein conformational stability. *Polymer* **2017**, *108*, 97–104.
49. Kuriyan, J.; Konforti, B.; Wemmer, D. *The Molecules of Life: Physical and Chemical Principles*; Garland Science: New York and London, 2012; p 1008.
50. Kayitmazer, A. B.; Seeman, D.; Minsky, B. B.; Dubin, P. L.; Xu, Y. Protein–polyelectrolyte interactions. *Soft Matter* **2013**, *9*, 2553–2583.
51. Andrianov, A. K.; Marin, A.; Fuerst, T. R. Self-assembly of polyphosphazene immunoadjuvant with poly(ethylene oxide) enables advanced nanoscale delivery modalities and regulated pH-dependent cellular membrane activity. *Heliyon* **2016**, *2*Article e00102.
52. Chang, B.; Mahoney, R. Enzyme thermostabilization by bovine serum albumin and other proteins: evidence for hydrophobic interactions. *Biotechnol. Appl. Biochem.* **1995**, *22*, 203–214.
53. Li, Y.; Mattison, K. W.; Dubin, P. L.; Havel, H. A.; Edwards, S. L. Light scattering studies of the binding of bovine serum albumin to a cationic polyelectrolyte. *Biopolymers* **1996**, *38*, 527–533.
54. Silva, R. A.; Urzúa, M. D.; Petri, D. F. S.; Dubin, P. L. Protein adsorption onto polyelectrolyte layers: Effects of protein hydrophobicity and charge anisotropy. *Langmuir* **2010**, *26*, 14032–14038.
55. Cayatte, C.; Marin, A.; Rajani, G. M.; Schneider-Ohrum, K.; Snell Bennett, A.; Marshall, J. D.; Andrianov, A. K. PCPP-adjuvanted respiratory syncytial virus (RSV) sF subunit vaccine: Self-assembled supramolecular complexes enable enhanced immunogenicity and protection. *Mol. Pharmaceutics* **2017**, *14*, 2285–2293.
56. Andrianov, A. K.; Decollibus, D. P.; Marin, A.; Webb, A.; Griffin, Y.; Webby, R. J. PCPP-formulated H5N1 influenza vaccine displays improved stability and dose-sparing effect in lethal challenge studies. *J. Pharm. Sci.* **2011**, *100*, 1436–1443.
57. Marin, A.; DeCollibus, D. P.; Andrianov, A. K. Protein stabilization in aqueous solutions of polyphosphazene polyelectrolyte and non-ionic surfactants. *Biomacromolecules* **2010**, *11*, 2268–2273.

58. Ott, G.; Van Nest, G., Development of Vaccine Adjuvants: A Historical Perspective. In *Vaccine Adjuvants and Delivery Systems*; Singh, M., Ed.; John Wiley & Sons: Hoboken, NJ, 2007; pp 1–31.
59. O'Neill, L. A. J.; Golenbock, D.; Bowie, A. G. The history of Toll-like receptors redefining innate immunity. *Nat. Rev. Immunol.* **2013**, *13*, 453–460.
60. Sasai, M.; Yamamoto, M. Pathogen recognition receptors: Ligands and signaling pathways by Toll-like receptors. *Int. Rev. Immunol.* **2013**, *32*, 116–133.
61. Broz, P.; Monack, D. M. Newly described pattern recognition receptors team up against intracellular pathogens. *Nat. Rev. Immunol.* **2013**, *13*, 551–565.
62. Singh, M. *Vaccine Adjuvants and Delivery Systems*; Wiley-Interscience: Hoboken, NJ, 2007; p 449.
63. Awate, S.; Wilson, H. L.; Lai, K.; Babiuk, L. A.; Mutwiri, G. Activation of adjuvant core response genes by the novel adjuvant PCEP. *Mol. Immunol.* **2012**, *51*, 292–303.
64. Yessine, M.-A.; Leroux, J.-C. Membrane-destabilizing polyanions: interaction with lipid bilayers and endosomal escape of biomacromolecules. *Adv. Drug Delivery Rev.* **2004**, *56*, 999–1021.
65. Yessine, M.-A.; Lafleur, M.; Meier, C.; Petereit, H.-U.; Leroux, J.-C. Characterization of the membrane-destabilizing properties of different pH-sensitive methacrylic acid copolymers. *Biochim. Biophys. Acta, Biomembr.* **2003**, *1613*, 28–38.
66. Shen, H.; Ackerman, A. L.; Cody, V.; Giodini, A.; Hinson, E. R.; Cresswell, P.; Edelson, R. L.; Saltzman, W. M.; Hanlon, D. J. Enhanced and prolonged cross-presentation following endosomal escape of exogenous antigens encapsulated in biodegradable nanoparticles. *Immunology* **2006**, *117*, 78–88.
67. Stier, E. M.; Mandal, M.; Lee, K.-D. Differential cytosolic delivery and presentation of antigen by listeriolysin O-liposomes to macrophages and dendritic cells. *Mol. Pharmaceutics* **2005**, *2*, 74–82.
68. Lin, M.-L.; Zhan, Y.; Villadangos, J. A.; Lew, A. M. The cell biology of cross-presentation and the role of dendritic cell subsets. *Immunol. Cell Biol.* **2008**, *86*, 353–362.
69. Schmaljohann, D. Thermo-and pH-responsive polymers in drug delivery. *Adv. Drug Delivery Rev.* **2006**, *58*, 1655–1670.
70. Eng, N. F.; Garlapati, S.; Gerdts, V.; Potter, A.; Babiuk, L. A.; Mutwiri, G. K. The potential of polyphosphazenes for delivery of vaccine antigens and immunotherapeutic agents. *Curr. Drug Del.* **2010**, *7*, 13–20.
71. Bouveret Le Cam, N. N.; Ronco, J.; Francon, A.; Blondeau, C.; Fanget, B. Adjuvants for influenza vaccine. *Res. Immunol.* **1998**, *149*, 19–23.
72. Kim, J. H.; Kirsch, E. A.; Gilliam, B.; Michael, N. L.; VanCott, T. C.; Ratto-Kim, S.; Cox, J.; Nielsen, R.; Robb, M. L.; Caudrelier, P.; El Habib, R.; McNeil, J. In *A Phase I, Open Label, Dose Ranging Trial of The Pasteur Merieux Connaught (PMC) Oligomeric HIV-1 Gp160mn/LAI-2 Vaccine In HIV Seronegative Adults*, Abstracts of the 37th Annual Meeting of the Infectious Diseases Society of America, Philadelphia, PA, 1999; p 1028.

73. Thongcharoen, P.; Suriyanon, V.; Paris, R. M.; Khamboonruang, C.; de Souza, M. S.; Ratto-Kim, S.; Karnasuta, C.; Polonis, V. R.; Baglyos, L.; El Habib, R. A phase 1/2 comparative vaccine trial of the safety and immunogenicity of a CRF01_AE (Subtype E) candidate vaccine: ALVAC-HIV (vCP1521) prime with oligomeric gp160 (92TH023/LAI-DID) or bivalent gp120 (CM235/SF2) boost. *JAIDS J. Acquired Immune Defic. Syndromes* **2007**, *46*, 48.
74. Ison, M. G.; Mills, J.; Openshaw, P.; Zambon, M.; Osterhaus, A.; Hayden, F. Current research on respiratory viral infections: Fourth International Symposium. *Antiviral Res.* **2002**, *55*, 227–278.
75. Mutwiri, G.; Benjamin, P.; Soita, H.; Townsend, H.; Yost, R.; Roberts, B.; Andrianov, A. K.; Babiuk, L. A. Poly[di(sodium carboxylatoethylphenoxy)phosphazene] (PCEP) is a potent enhancer of mixed Th1/Th2 immune responses in mice immunized with influenza virus antigens. *Vaccine* **2007**, *25*, 1204–1213.
76. Andrianov, A. K.; Chen, J.; Sule, S. S. In *Ionically Cross-Linked Polyphosphazene Microspheres*; American Chemical Society, Polymer Preprints, Division of Polymer Chemistry; American Chemical Society: Anaheim, CA, 1999; pp 355–356.
77. Andrianov, A. K.; Payne, L. G. Polymeric carriers for oral uptake of microparticulates. *Adv. Drug Delivery Rev.* **1998**, *34*, 155–170.
78. Payne, L. G.; Jenkins, S. A.; Andrianov, A. K.; Roberts, B. E. Water-Soluble Phosphazene Polymers for Parenteral and Mucosal Vaccine Delivery. In *Vaccine Design*; Powell, M. F.; Newman, M. J., Eds.; Springer: New York, 1995; pp 473–493.
79. Garlapati, S.; Garg, R.; Brownlie, R.; Latimer, L.; Simko, E.; Hancock, R.; Babiuk, L.; Gerdts, V.; Potter, A. Enhanced immune responses and protection by vaccination with respiratory syncytial virus fusion protein formulated with CpG oligodeoxynucleotide and innate defense regulator peptide in polyphosphazene microparticles. *Vaccine* **2012**, *30*, 5206–5214.
80. Kovacs-Nolan, J.; Latimer, L.; Landi, A.; Jenssen, H.; Hancock, R. E. W.; Babiuk, L. A.; van Drunen Littel-van den Hurk, S. The novel adjuvant combination of CpG ODN, indolicidin and polyphosphazene induces potent antibody- and cell-mediated immune responses in mice. *Vaccine* **2009**, *27*, 2055–2064.
81. Kovacs-Nolan, J.; Mapletoft, J.; Lawman, Z.; Babiuk, L. Formulation of bovine respiratory syncytial virus fusion protein with CpG oligodeoxynucleotide, cationic host defence peptide and polyphosphazene enhances humoral and cellular responses and induces a protective type 1 immune response in mice. *J. Gen. Virol.* **2009**, *90*, 1892–1905.
82. Andrianov, A. K.; DeCollibus, D. P.; Gillis, H. A.; Kha, H. H.; Marin, A.; Prausnitz, M. R.; Babiuk, L. A.; Townsend, H.; Mutwiri, G. Poly[di(carboxylatophenoxy)phosphazene] is a potent adjuvant for intradermal immunization. *Proc. Natl. Acad. Sci. U.S.A.* **2009**, *106*, 18936–18941.

83. Andrianov, A.; Marin, A.; DeCollibus, D. Microneedles with intrinsic immunoadjuvant properties: Microfabrication, protein stability, and modulated release. *Pharm. Res.* **2011**, *28*, 58–65.
84. Andrianov, A. K.; Mutwiri, G. Intradermal immunization using coated microneedles containing an immunoadjuvant. *Vaccine* **2012**, *30*, 4355–4360.
85. Vaishya, R.; Khurana, V.; Patel, S.; Mitra, A. K. Long-term delivery of protein therapeutics. *Expert Opin. Drug Delivery* **2015**, *12*, 415–440.
86. Bruno, B. J.; Miller, G. D.; Lim, C. S. Basics and recent advances in peptide and protein drug delivery. *Ther. Del.* **2013**, *4*, 1443–1467.
87. Torchilin, V. Intracellular delivery of protein and peptide therapeutics. *Drug Discovery Today: Technol.* **2008**, *5*, e95–e103.
88. Turecek, P. L.; Bossard, M. J.; Schoetens, F.; Ivens, I. A. PEGylation of biopharmaceuticals: A review of chemistry and nonclinical safety information of approved drugs. *J. Pharm. Sci.* **2016**, *105*, 460–475.
89. Veronese, F. M.; Pasut, G. PEGylation, successful approach to drug delivery. *Drug Discovery Today* **2005**, *10*, 1451–1458.
90. Caliceti, P.; Veronese, F. M. Pharmacokinetic and biodistribution properties of poly(ethylene glycol)–protein conjugates. *Adv. Drug Delivery Rev.* **2003**, *55*, 1261–1277.
91. Tabrizi, M. A.; Tseng, C.-M. L.; Roskos, L. K. Elimination mechanisms of therapeutic monoclonal antibodies. *Drug Discovery Today* **2006**, *11*, 81–88.
92. Duncan, R. The dawning era of polymer therapeutics. *Nat. Rev. Drug Discovery* **2003**, *2*, 347–360.
93. Torchilin, V. P. Multifunctional, stimuli-sensitive nanoparticulate systems for drug delivery. *Nat. Rev. Drug Discovery* **2014**, *13*, 813–827.
94. Hoffman, A. S. Stimuli-responsive polymers: Biomedical applications and challenges for clinical translation. *Adv. Drug Delivery Rev.* **2013**, *65*, 10–16.
95. El-Sayed, M. E.; Hoffman, A. S.; Stayton, P. S. Smart polymeric carriers for enhanced intracellular delivery of therapeutic macromolecules. *Expert Opin. Biol. Ther.* **2005**, *5*, 23–32.
96. Fee, C. J.; Van Alstine, J. M. PEG-proteins: Reaction engineering and separation issues. *Chem. Eng. Sci.* **2006**, *61*, 924–939.
97. Muro, S.; Muzykantov, V. R.; Murciano, J.-C., Characterization of Endothelial Internalization and Targeting of Antibody—Enzyme Conjugates in Cell Cultures and in Laboratory Animals. In *Bioconjugation Protocols: Strategies and Methods*; Niemeyer, C. M., Ed.; Humana Press: Totowa, NJ, 2004; pp 21–36.
98. Dobrovolskaia, M. A.; Clogston, J. D.; Neun, B. W.; Hall, J. B.; Patri, A. K.; McNeil, S. E. Method for Analysis of Nanoparticle Hemolytic Properties in Vitro. *Nano Lett.* **2008**, *8*, 2180–2187.
99. Plank, C.; Oberhauser, B.; Mechtler, K.; Koch, C.; Wagner, E. The influence of endosome-disruptive peptides on gene transfer using synthetic virus-like gene transfer systems. *J. Biol. Chem.* **1994**, *269*, 12918–12924.
100. Muro, S. Challenges in design and characterization of ligand-targeted drug delivery systems. *J. Controlled Release* **2012**, *164*, 125–137.

101. Muro, S. New biotechnological and nanomedicine strategies for treatment of lysosomal storage disorders. *Wiley Interdiscip. Rev.: Nanomed. Nanobiotechnol.* **2010**, *2*, 189–204.
102. Allcock, H. R. Polyphosphazene elastomers, gels, and other soft materials. *Soft Matter* **2012**, *8*, 7521–7532.
103. Allcock, H. R.; Steely, L. B.; Singh, A. Hydrophobic and superhydrophobic surfaces from polyphosphazenes. *Polym. Int.* **2006**, *55*, 621–625.
104. Welle, A.; Grunze, M.; Tur, D. Plasma protein adsorption and platelet adhesion on poly[bis(trifluoroethoxy)phosphazene] and reference material surfaces. *J. Colloid Interface Sci.* **1998**, *197*, 263–274.
105. Smeets, A. J.; Nijenhuis, R. J.; van Rooij, W. J.; Lampmann, L. E. H.; Boekkooi, P. F.; Vervest, H. A. M.; De Vries, J.; Lohle, P. N. M. Embolization of uterine leiomyomas with polyzene F–coated hydrogel microspheres: Initial experience. *J. Vasc. Interv. Radiol.* **2010**, *21*, 1830–1834.
106. Stampfl, U.; Radeleff, B.; Sommer, C.; Stampfl, S.; Dahlke, A.; Bellemann, N.; Kauczor, H.-U.; Richter, G. M. Midterm results of uterine artery embolization using narrow-size calibrated embozene microspheres. *Cardiovasc. Intervent. Radiol.* **2011**, *34*, 295–305.
107. Styllou, P.; Silber, S. A case report of the new Polyzene™-F COBRA PzF™ Nanocoated Coronary Stent System (NCS): Addressing an unmet clinical need. *Cardiovasc. Revasc. Med.* **2016**, *17*, 209–211.
108. Cutlip, D. E.; Garratt, K. N.; Novack, V.; Barakat, M.; Meraj, P.; Maillard, L.; Erglis, A.; Jauhar, R.; Popma, J. J.; Stoler, R.; Silber, S. 9-Month clinical and angiographic outcomes of the COBRA polyzene-F nanocoated coronary stent system. *Cardiovasc. Interventions* **2017**, *10*, 160–167.
109. Cohen, S.; Bano, M. C.; Visscher, K. B.; Chow, M.; Allcock, H. R.; Langer, R. Ionically crosslinkable polyphosphazene: A novel polymer for microencapsulation. *J. Am. Chem. Soc.* **1990**, *112*, 7832–7833.
110. Lee, K. Y.; Mooney, D. J. Alginate: Properties and biomedical applications. *Prog. Polym. Sci.* **2012**, *37*, 106–126.
111. Bano, M. C.; Cohen, S.; Visscher, K. B.; Allcock, H. R.; Langer, R. A novel synthetic method for hybridoma cell encapsulation. *Nat. Biotechnol.* **1991**, *9*, 468–471.
112. Andrianov, A. K.; Cohen, S.; Visscher, K. B.; Payne, L. G.; Allcock, H. R.; Langer, R. Controlled release using ionotropic polyphosphazene hydrogels. *J. Controlled Release* **1993**, *27*, 69–77.
113. Andrianov, A. K.; Jenkins, S. A.; Payne, L. G.; Roberts, B. E. *Hydrogel Microencapsulated Vaccines*. U.S. Patent 5,529,777, 1996.
114. Cohen, S.; Andrianov, A. K.; Wheatley, M.; Allcock, H. R.; Langer, R. S. *Gas-Filled Polymeric Microbubbles for Ultrasound Imaging*. U.S. Patent 5,487,390, 1996.
115. Cohen, S.; Andrianov, A. K.; Wheatley, M.; Allcock, H. R.; Langer, R. S. *Polymeric Microparticles Containing Agents for Imaging*. U.S. Patent 5,562,099, 1996.
116. Greish, Y. E.; Bender, J. D.; Lakshmi, S.; Brown, P. W.; Allcock, H. R.; Laurencin, C. T. Formation of hydroxyapatite–polyphosphazene polymer

composites at physiologic temperature. *J. Biomed. Mater. Res., Part A* **2006**, *77A*, 416–425.

117. Greish, Y.; Bender, J.; Lakshmi, S.; Brown, P.; Allcock, H.; Laurencin, C. Composite formation from hydroxyapatite with sodium and potassium salts of polyphosphazene. *J. Mater. Sci. Mater. Med.* **2005**, *16*, 613–620.
118. Andrianov, A. K.; Langer, R.; Payne, L. G.; Roberts, B. E.; Jenkins, S. A.; Allcock, H. R. Polyphosphazene ionotropic gel microcapsules for controlled drug delivery. *Proc. Controlled Release Soc.* **1993**, 26–27.
119. Andrianov, A. K.; LeGolvan, M. P.; Svirkin, Y.; Sule, S. S. *Purification of Polyphosphazene Polyacids*. U.S. Patent 5,842,471, 1998.
120. Andrianov, A. K.; Chen, J. Polyphosphazene microspheres: Preparation by ionic complexation of phosphazene polyacids with spermine. *J. Appl. Polym. Sci.* **2006**, *101*, 414–419.
121. Cheheltani, R.; Ezzibdeh, R. M.; Chhour, P.; Pulaparthi, K.; Kim, J.; Jurcova, M.; Hsu, J. C.; Blundell, C.; Litt, H. I.; Ferrari, V. A.; Allcock, H. R.; Sehgal, C. M.; Cormode, D. P. Tunable, biodegradable gold nanoparticles as contrast agents for computed tomography and photoacoustic imaging. *Biomaterials* **2016**, *102*, 87–97.
122. Chhour, P.; Gallo, N.; Cheheltani, R.; Williams, D.; Al-Zaki, A.; Paik, T.; Nichol, J. L.; Tian, Z.; Naha, P. C.; Witschey, W. R.; Allcock, H. R.; Murray, C. B.; Tsourkas, A.; Cormode, D. P. Nanodisco Balls: Control over surface versus core loading of diagnostically active nanocrystals into polymer nanoparticles. *ACS Nano* **2014**, *8*, 9143–9153.
123. Andrianov, A. K. University of Maryland, unpublished results.
124. Khutoryanskiy, V. V. Hydrogen-bonded interpolymer complexes as materials for pharmaceutical applications. *Int. J. Pharm.* **2007**, *334*, 15–26.
125. Such, G. K.; Johnston, A. P. R.; Caruso, F. Engineered hydrogen-bonded polymer multilayers: From assembly to biomedical applications. *Chem. Soc. Rev.* **2011**, *40*, 19–29.
126. Kharlampieva, E.; Kozlovskaya, V.; Sukhishvili, S. A. Layer-by-layer hydrogen-bonded polymer films: From fundamentals to applications. *Adv. Mater.* **2009**, *21*, 3053–3065.

Biomaterials

Chapter 3

Polyphosphazene-Based Biomaterials for Regenerative Engineering

Kenneth S. Ogueri[1,2,3] and Cato T. Laurencin*,[1,2,3,4,5,6]

[1]Department of Materials Science and Engineering, University of Connecticut, Storrs, Connecticut 06269, United States
[2]Institute for Regenerative Engineering, University of Connecticut Health Center, Farmington, Connecticut 06030, United States
[3]Raymond and Beverly Sackler Center for Biomedical, Biological, Physical and Engineering Sciences, University of Connecticut Health Center, Farmington, Connecticut 06030, United States
[4]Department of Orthopaedic Surgery, University of Connecticut Health Center, Farmington, Connecticut 06030, United States
[5]Department of Biomedical Engineering, University of Connecticut, Storrs, Connecticut 06269, USA
[6]Department of Chemical and Biomolecular Engineering, University of Connecticut, Storrs, Connecticut 06269, United States
*E-mail: Laurencin@uchc.edu.

The emergence of regenerative engineering provides an innovative approach to tackling the challenging issues of tissue loss or failure. Regenerative engineering presents a tool that integrates the fields of advanced materials science, stem cell science, physics, developmental biology and clinical translation for the common goal of regenerating complex tissues and biological systems such as a knee or a whole limb. Biomaterials play an essential role in the success of this new approach. An ideal biomaterial for regenerative engineering should be biocompatible, have desired initial mechanical properties, should degrade in a rate that is appropriate for tissue regeneration, have resorbable degradation products, can present interconnected porous structures, be osteoconductive and allow for neovascularization. However, so far an ideal biomaterial that meets all these criteria listed above has not been developed. Polyphosphazene polymers offer an important and unique platform for the design and synthesis of novel biodegradable polymeric biomaterials with efficient control over degradation rates, mechanical properties, In vitro osteocompatibility, and in vivo biocompatibility. Their synthetic flexibility presents a library of biomaterials with a wide range of degradation rates and mechanical properties that can suit a variety of tissue regeneration needs. This chapter gives an overview of the design considerations and unique features of these biodegradable polyphosphazene-based biomaterials and their suitability for regenerative engineering applications.

Introduction

There is increasing interest in the use of polymeric biomaterials for biomedical applications such as tissue regeneration, controlled drug delivery, and immunomodulation, etc. (*1–6*) The importance of these biomaterials in the medical field is in part due to the significant increase in their degree of sophistication. Tremendous success has been recorded in the biomedical applications of biomaterials (*5*, *6*). However enormous challenges still exist in both the basic and translational aspects of the biomaterial design (*5*). The basic element entails the chemistry of the biomaterials which can be modulated to give customized materials properties (*5*). Materials with a wide range of properties can be designed through optimization and conferment of different chemical functionalities to meet the specific requirements for tissue regeneration (*2*). The applications of biomaterials in clinical setting constitute numerous unidentified and unaddressed issues based on previous experimentations (*7*, *8*). These problems are mainly related to the material-tissue interactions which are dependent on the chemical, physical and biological properties of the biomaterials (*5*). Materials interactions can be considered in two ways. 'Material dynamics' describes how the material affects the surrounding tissues and 'material kinetics' describes how the tissue in the microenvironment affects material properties (*9*).

To better address the many issues in biomaterial design and expedite progress, a collaborative approach termed 'regenerative engineering' has been put forward by Dr. Laurencin and co-workers (*10*, *11*). This new and innovative approach combines the efforts of researchers with expertise in chemistry, biology, materials, engineering and clinical practice to ultimately ensuring enhanced material performance and positive clinical results (*11*).

For the design of biodegradable polymeric materials, there are essential criteria that must be put into consideration. The material must (1) elicit minimal to mild tissue responses when in contact with living systems (2) have degradation time that corresponds to the regenerative process (3) have good initial mechanical properties in terms of structural integrity and dimensional stability (4) have metabolizable, and excretable degradation products and (5) have excellent processibility which could be helpful in product applications (*4*, *11*, *12*).

Biodegradable polymers can be utilized in tissue regeneration as porous 3D scaffolds for the migration, proliferation and subsequent development of biological tissues (*4*, *11*, *12*). These biological tissues possess complex structures that are made up of unique cell compositions, chemistry, and mechanical properties (*13*, *14*). Therefore, optimization of the material properties during design and synthesis allow for the attainment of improved material performance (*2*, *4*, *5*, *12*, *15*). The conferment of various moieties on the polymers is utilized in tailoring their degradation rates, environmental sensitivities, and mechanical properties (*5*, *16–18*). Materials with desired properties must be selected to meet specific functional requirements. The design flexibility of polymeric materials has led to the investigation of a wide range of polymer-based biomaterials for biomedical applications with novel materials continually being developed to meet new challenges (*4*, *5*, *16–18*).

Polyphosphazenes are among the few inorganic-organic hybrid polymers that have been thoroughly investigated as potential biomaterials for bone tissue regeneration (*2*, *19*). They are made up of a backbone of alternating phosphorus and nitrogen atoms and with each phosphorus atom bearing two organic side groups (*2*, *10*, *19*). In tissue regeneration, the side groups are hydrolytically active and selected based on their ability to sensitize the polymers to hydrolysis, which allow them to break down into non-toxic small molecules that can be metabolized or excreted from the body (*10*, *20*). Polyphosphazenes with amino acid esters and peptide esters are the most abundant class of these biodegradable polymers (*21*). Because amino acid esters and peptide esters are susceptible to hydrolysis, this allows water molecules to attack the polyphosphazene backbone and to break it down into ammonia, phosphates and corresponding side groups (*10*, *21*). These degradation products constitute a natural buffer as ammonia compound is basic and phosphate is amphoteric (*10*, *21*). Amphoteric compounds are compounds that can behave as an acid in the presence of a base and vice versa (*22*). These unique buffering effects have aroused enormous research interest in polyphosphazene blends with other clinically relevant polyesters such as poly(lactic-co-glycolic acid)(PLAGA), polylactic acid(PLA), polyglycolic acid (PGA), and polycaprolactone(PCL) (*10*, *21*, *23*, *24*). The acidic degradation products of these polyesters can be neutralized or stabilized by ammonia and phosphates (ammonium phosphate) from polyphosphazene degradation (*10*).

This chapter will focus on the design of biodegradable polyphosphazene-based biomaterials and their applications in regenerative engineering.

Synthesis of Polyphosphazenes

The most widely used route to poly (organo) phosphazenes is via a two-step reaction process starting from the commercially available cyclic trimer hexachlorocyclotriphosphazene (HCCTP) (*2*, *10*, *19*). The first step involves the synthesis of linear poly (dichlorophosphazene) (PDCP) which can be achieved via the controlled thermal ring-opening polymerization (ROP) of cyclic trimer at 250 °C under vacuum (*2*). There are many techniques available for the synthesis of PDCP (Figure 1), but the ROP is commonly used for bioerodible polyphosphazenes (*10*). Using a Lewis acid catalyst such as anhydrous $AlCl_3$ or BCl_3OPPh_3, ROP can also be carried out at a much lower temperature of

220°C (*25*). Thermal ring-opening polymerization of HCCTP usually yields high molecular weight PDCP (*10*). PDCP can also be produced using other routes such as the thermal or anionic polymerization of phosphoranimines, the thermal condensation of phosphorus azides, and the direct synthesis from PCl_5 and NH_4Cl (Figure 1) (*2, 10*). Regardless of the importance of the methods above, there exist several associated limitations such as the need for high temperature, lack of molecular weight control, the lower monomer conversion, and by-product generation. An alternative method which circumvents these limitations is via the living cationic polymerization of a phosphoranimine, catalyzed by phosphorus pentachloride in chloroform at room temperature (*2, 26, 27*). This route gives rise to shorter polymer chains but with living chain ends and a low polydispersity (*2, 10, 26*). Poly (dichlorophosphazene) is an air and moisture sensitive organic intermediate with labile P-Cl bonds which can readily have the attached chlorine atoms replaced with organic nucleophiles to obtain stable polymeric species (*10, 21*). This step of replacing the chlorine with different organic nucleophiles is called macromolecular substitution (Figure 2) (*21*). Macromolecular substitution provides the platform for the incorporation of varying side groups, which ultimately determines the polymer properties (*26*). A number of biodegradable polyphosphazenes, especially those whose side groups are hydrolytically susceptible (such as amino acid ester side group) have been extensively investigated as matrices for bone regeneration (*21*). Macromolecular substitution can be done using similar or dissimilar organic species where co-substitution of the side groups is carried out sequentially, allowing the initial incorporation of the bulkier group and subsequent reaction of the smaller or less bulky groups in solution phase (*10, 21*). Material properties such as degradation rates and physical properties can be fine-tuned using modification of types and ratios of the substituent side chains (*10, 20, 21*). This is crucial to the biomaterial designs aimed to meet the ever-changing needs of tissue regeneration. In general, macromolecular substitutions have utilized over 250 different side groups for the replacement of the labile chlorine species of polydichlorophosphazene (*10*).

Polyphosphazene Block Copolymers

In as much as the mixed-substituent macromolecular substitution gives rise to polyphosphazene copolymers, the nucleophilic side groups are randomly placed along the polyphosphazene backbone (*2*). Thus, random polyphosphazene copolymers are obtained from nucleophilic macromolecular substitution. The stereoregularity of block copolymers has proven to be very useful in numerous applications, and hence polyphosphazene block copolymers have been investigated for use in some applications including tissue regenerations (*21*). The living nature of the cationic polymerization routes allows for the incorporation of the second monomer to the active chain ends (*10, 21*). The macromolecular substitutions are subsequently done to yield a polymer with two segmental blocks that are made up of polyphosphazenes and the second polymer component. This process has resulted in the design of polyphosphazene-based materials with unique properties (*21*).

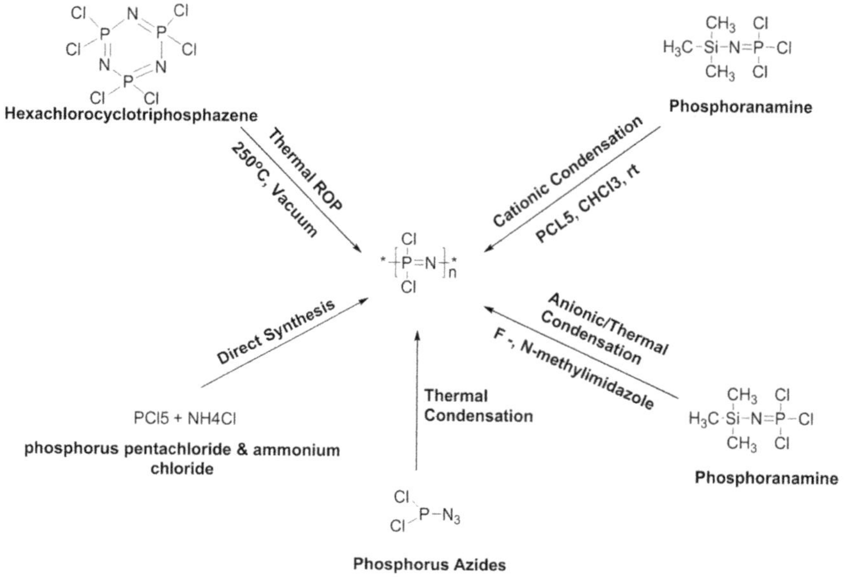

Figure 1. Different routes to the synthesis of poly (dichloro) phosphazene.

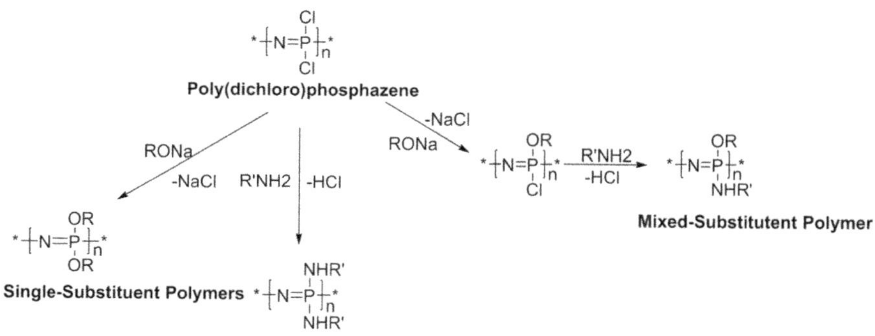

Figure 2. Single-substituent (simultaneous replacement of chlorine atoms) and mixed-substituent (sequential substitution of chlorine atoms) polymers obtained from macromolecular substitution.

Importance of Chlorine Replacement

One of the significant challenges in the synthesis of polyphosphazenes is ensuring that all the chlorine atoms in poly (dichlorophosphazene) are replaced by the intended organic side groups (*28*). ^{31}Phosphorus NMR is usually used to monitor the halogen substitution where the type of side groups influences the sensitivity of the phosphorous chemical shifts attached (*10*, *27*, *28*). For instance, the sharp chemical shift of P-Cl units in the polymer can be found around -18ppm, and that for the P-amino units can be located around 0ppm and it's usually broad (*10*, *21*). Any unreacted chlorine atoms could cause hydrolytic instability and may result in a slow release of hydrochloric acid and formation of P–OH units (*21*).

For tissue regeneration applications, when the polymer is placed on living tissue, the P-Cl may react with surrounding water to form P–OH units and hydrochloric acids (Figure 3) (*10*, *21*). These two chemicals can be injurious to the cells and tissues. To ensure a complete substitution of the chlorine atoms, in addition to monitoring the disappearance of peaks characteristic of P-Cl units with ^{31}P NMR, a stoichiometric excess of the organic nucleophile and extended reaction time are typically employed (*21*). The replacement of chlorine atoms can be influenced by specific factors and which include an appropriate solvent, reaction temperature and time, nature of nucleophile (nucleophilicity, steric characteristics), and solubility of side products (such as NaCl, Et_3NHCl) (*20*, *21*, *28*).

$$\text{P-Cl} + H_2O \rightleftharpoons \text{P-OH} + HCl$$

Labile Bond **Hydrochloric Acid**

Figure 3. Mechanism showing the hydrolytic sensitivity of the P–Cl bond. P-Cl in poly(dichloro)phosphazene reacts with water to give P-OH units and hydrochloric acids.

Design Strategy for Polyphosphazene Biomaterials

Incorporation of hydrolytically active organic side groups (such as amino acid esters, imidazolyl, glucosyl, glyceryl, and glycolate or lactate ester) can result in a class of polyphosphazenes that are biodegradable (*2*, *10*, *13*, *21*). When hydrolytically active side groups are attached to the polyphosphazene backbone, it can sensitize the backbone causing the polymers to break down into phosphate, ammonia, and corresponding side groups (*2*, *10*, *21*, *28*, *29*). In designing polyphosphazene biomaterials for tissue regeneration, the chemistry of the side groups is of paramount importance.

Side Group Chemistry

The introduction of different side groups with varying proportions in polyphosphazene synthesis can have tremendous control on properties of the polymer (*10*, *20*, *21*). Macromolecular substitution may involve mono-substitution using a single substituent group or co-substitution with two or more substituent groups (*10*). The degradation rates and mechanical properties can be fine-tuned via the optimization of the side group chemistry which entails the nature and ratios of the organic components (*2*, *10*). For instance, a hydrolytically active group such as amino acid ester will induce hydrolysis within the polymer backbone, whereas the presences of hydrophobic groups such as phenylphenoxy side groups inhibit hydrolysis (*30*). For this reason, macromolecular co-substitution of poly (dichlorophosphazene) with a hydrolytically active group and a hydrophobic group will allow the modulation of the degradation pattern to match specific tissue types and disease states (*30–32*).

Similarly, steric shielding of the side groups can be used to modulate degradation rates as well as the mechanical properties. The backbone can be shielded from hydrolytic attack by bulky groups, and amino acid ester with a bulky group at its alpha position hydrolyzes slower than the one with just hydrogen at its alpha position (*30*).

Also, Deng et al. (*23*, *24*) showed that the mechanical properties of polyphosphazene-based biomaterials correlate directly with the bulkiness of the side groups in a mixed-substituent polymer. In the study, the mechanical properties of a biodegradable polyphosphazene were enhanced by incorporating a bulky phenylphenoxy co-substituent at a specific ratio into the backbone. However, a higher proportion of phenylphenoxy may compromise the miscibility of the polymer with polyesters (*23*, *30*). So the ratio of the two side groups (phenylphenoxy and amino acid ester) will determine the hydrolytic breakdown and mechanical properties of this polymer, with an increase in the composition of aryloxy units resulting in polymers with the least sensitivity to water (*23*, *24*).

The type of linkage can play a role in the degradation rates of the polymer (*21*). For example, the resistance to hydrolysis is greater for polymers that bond to the organic side groups through Oxygen than the one with organic side groups linked through Nitrogen (*2*, *21*). This is not always true for all polyphosphazenes. Polyphosphazenes with aryloxy and alkoxy tend to possess high resistance to hydrolysis due to the presence of oxygen linkage (*2*, *21*, *32*).

Amino acid esters are used instead of amino acids to avert side reactions that may lead to crosslinking (*32*). The presence of esters can also affect hydrolysis rates and so as the length of an alkyl group. Intramolecular hydrogen bonding, inter molecular hydrogen bonding and crosslinking can have effects on the hydrolytic sensitivity of the polymers (*2*, *21*). The permeation of water into the polymer backbone can be hindered by extensive hydrogen bonding between the polymer chains and thus decreases hydrolysis. Deng et al. (*23*) have shown that a miscible blend with strong hydrogen bonding can undergo hydrolysis at a much lower rate than its components (polyphosphazenes and poly (lactic-co-glycolic acid)). The same is also applicable in the case of crosslinking as the presence of crosslinks reduce hydrolysis and hence degradation rates (*1*).

Degradation and Erosion Mechanism

For biodegradable polymers used in the biomedical field, it is essential to be aware of the possible toxicity of degradation intermediates or end products. In the course of degradation, the organic side groups are usually the first to be attacked by water molecules, and there are variations of polyphosphazene degradation mechanisms with different side groups (*10*, *20*, *29*). Figure 4 shows the three different pathways for polyphosphazene degradations where the amino acid ester is the side group which will sensitize the phosphazene backbone to hydrolysis. In the first mechanism, the ester linkage of the amino acid ester is cleaved first releasing alcohol. The resulting carboxylic moiety then attacks the phosphazene chain. This attack induces a breakdown of the secondary amine and phosphorous-nitrogen bonds (*21*). In the second mechanism, the ester functional group attacks the polymer backbone causing a collapse of the secondary amine and phosphazene bonds. This ultimately releases amino acid ester substituted phosphate which would eventually undergo further hydrolysis to produce amino acid, alcohol and ammonium phosphates (*10*, *21*). For the third mechanism, the cleavage occurs first on the secondary amine bond that links the amino acid ester to the phosphazene backbone. This produces a P-OH unit, and the proton attached to its oxygen then migrates to the nitrogen, sensitizing the polymer backbone to hydrolysis into ammonia and phosphate (*10*, *20*, *21*, *28*). The amino acid esters produced during the secondary amine breakdown will undergo further hydrolysis to release amino acid and alcohol (*10*, *20*, *21*). The common characteristics of the three mechanisms are the degradation of amino acid ester substituted polyphosphazene into non-toxic degradation products that are composed of ammonium phosphate and the corresponding side groups (*5*, *29*). Ammonium phosphate is an amphoteric compound which acts as a natural buffer in the local microenvironment both in vitro and in vivo (*33*, *34*). For a mixed substituent polyphosphazene, cleavage occurs first on the most hydrolytically active group followed by the least hydrolytically active or bulky groups such as p-phenylphenoxy, p-methylphenoxy, or tyrosine (*21*, *29*, *35*). Degradation study was carried out on the following polymers: methyl, ethyl, tertbutyl, and benzyl esters of glycine-, alanine-, valine-, and phenylalanine-substituted polyphosphazenes. Molecular weight changes of these polymers were determined using a gel permeation chromatography analysis. It was found that the hydrolytic sensitivity was inversely proportional to the size of the ester group, where an increase in ester size will result in a decrease in the hydrolytic sensitivity (*10*, *21*). Similarly, hydrolytic sensitivity will be higher for polymers with a smaller group (such as hydrogen) linked to the α-carbon atom of the amino acid ester side group (*30*). Whereas, polymers with bulky groups (such as phenyl) at the α-carbon of the amino component will show the least sensitivity to hydrolysis. Furthermore, Weikel et al. (*31*) examined the influence of pH on the hydrolytic sensitivity of dipeptide-based polyphosphazenes. It was found that the polymers experienced more rapid degradation at pH 4.0 and pH 7.0 than in basic media (pH 10) (*36*). This is as a result of the fast rate of hydrolysis of amino acid esters in an acidic and neural environment.

The unique buffering effect of the degradation products of polyphosphazene backbone makes it a promising candidate material for tissue regeneration (*23*, *24*, *37*).

Polyphosphazene Blends

Polymer blending offers a practical platform for combining beneficial features of two or more polymers into a single material with unique properties (*10*, *23*, *24*, *38*). The properties of the blend materials can be fine-tuned by merely adjusting the blend compositions (*10*). Inspired by the lingering issue of polyesters' acidic degradation products and quest to develop biomaterials with neutral degradation products and bioactivity, Laurencin and co-workers exploited the unique buffering effects of amino acid ester substituted polyphosphazene to design blend system with polyphosphazene and polyesters (*2, 21, 31*). Polyphosphazene-polyester blends are attractive materials for bone tissue regeneration applications because it combines the tunability and the pH-buffering effect of the polyphosphazene hydrolysis products with the high strength of polyesters (*2, 21, 23, 24, 31*). It was demonstrated that miscibility of the blend could be enhanced by using peptide group as co-substituent. The enhanced miscibility could be as a result of greater number of hydrogen bonding sites present in peptides than in amino acid ester (Figure 5) (*23, 24*).

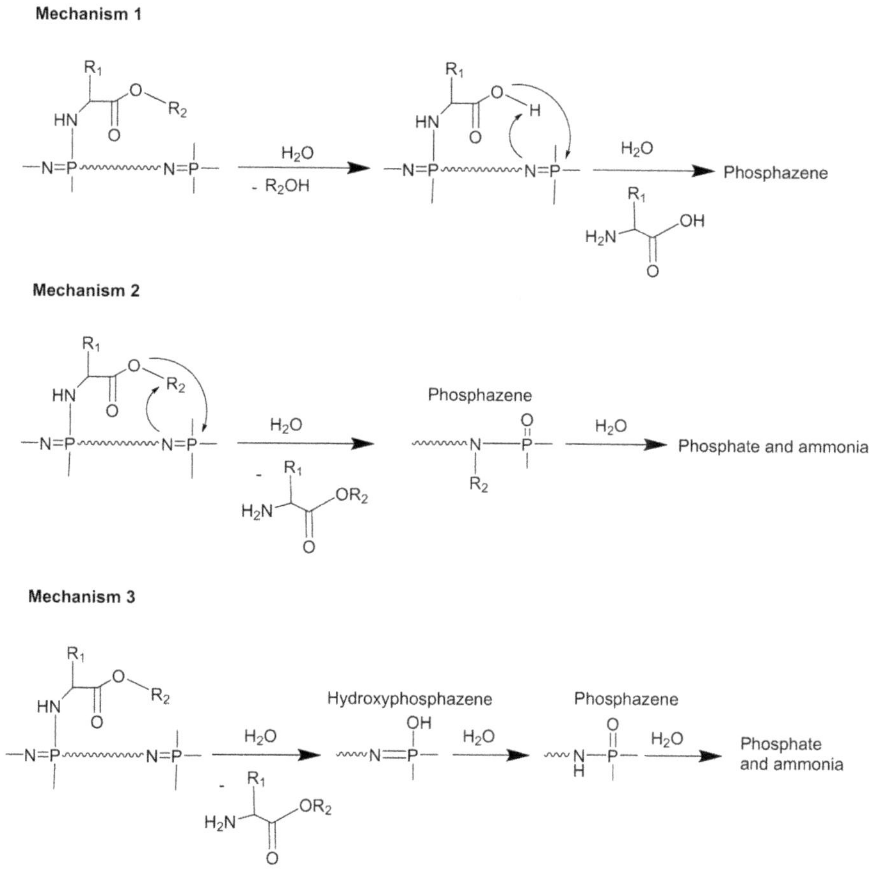

Figure 4. Degradation mechanisms showing three hydrolysis pathways of amino acid ester substituted polyphosphazene.

In addition to the unique neutralizing effect, peptide-based polyphosphazene-PLAGA blends can turn into a 3-D porous structure upon degradation (Figure 6 & 7) (*38*). This erosion mechanism is quite distinct from that exhibited by any biodegradable systems currently available. The mechanism of this erosion was proposed by Deng et al. to occur in three stages: 1. the breakdown of the intermolecular hydrogen bonding between the peptides (of the polyphosphazenes) and PLAGA 2. The formation of intra-molecular hydrogen bonding between the polyphosphazene molecules leading to the reorientation (self-assembly) of the system into spheres 3. Final breakdown of the intra-molecular hydrogen bonding within the polyphosphazene molecules leading to complete degradation of the blend (*38*). As shown in the IR spectra in Figure 6, peak at 1677 cm^{-1} at 0 week correspond to the initial intermolecular hydrogen bonding between the peptides (of the polyphosphazenes) and PLAGA. The disappearance of peak at 1677 cm^{-1} at week 2 and 4 is due to the breakdown of the intermolecular hydrogen bonds in stage 1 of the erosion mechanism. The formation of weak absorption band in 1655 cm^{-1} after 4 weeks (at 7 weeks, 10 weeks, and 12 weeks) corresponds to the intra-molecular hydrogen bonding between the polyphosphazene molecules (*38*).

Our lab is currently working on optimizing the inherent pore-forming properties of the peptide-based polyphosphazene-PLAGA blends. Intrinsic pore-forming properties can be an avenue to preventing the prefabrication of porous structure and promoting cell infiltration, tissue ingrowth, nutrient transport, and oxygen diffusion during implantation. Numerous polyphosphazene blends have shown excellent osteocompatibility and have been investigated as matrices for bone tissue engineering (*10*, *25*).

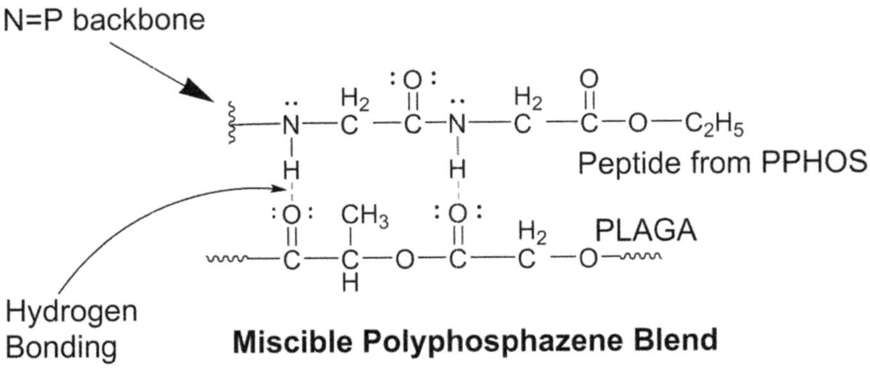

Figure 5. Polyphosphazene-PLAGA miscible blends showing the intermolecular hydrogen interactions between the peptide (glycylglycine ethyl ester) from polyphosphazene and PLAGA.

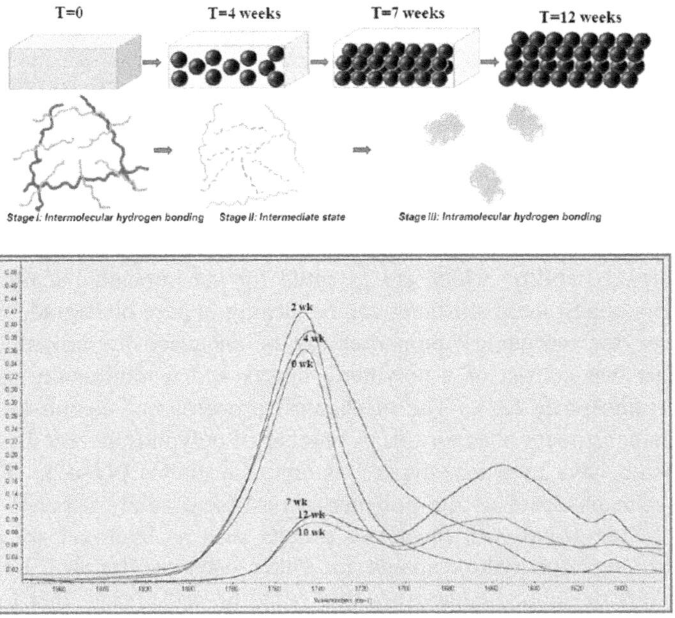

Figure 6. Polyphosphazene-PLAGA unique erosion mechanism that involves three stages. 0 week spectrum with peak at 1677 cm-1 depicts the presence of initial intermolecular hydrogen bonds between peptides of the polyphosphazenes and PLAGA. The disappearances of 1677 cm^{-1} in week 2 & 4 indicate the subsequent breakdown of the intermolecular hydrogen bonds.1655 cm^{-1} peaks at week 7,10, 12 show the formation of intramolecular hydrogen bonds within the polyphosphazene molecules (38). Reproduced with permission from reference (38). Copyright 2010 John Wiley & Sons.

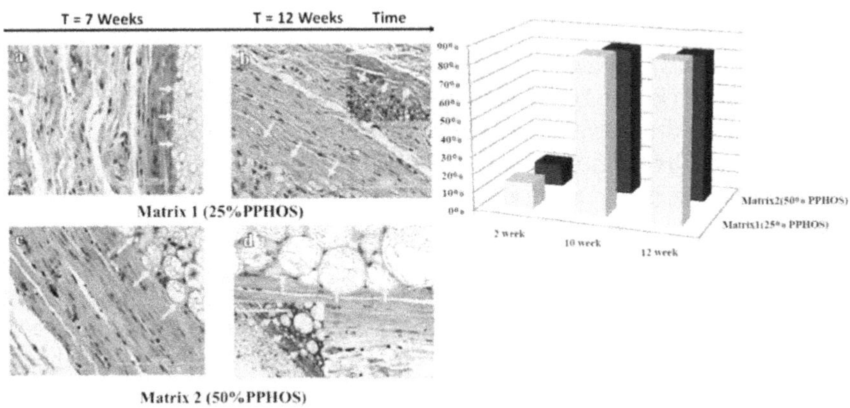

Figure 7. Left: Histology images showing the unique erosion mechanism that result in the formation of polymer spheres with porosity. Right: Graphs showing the In-situ porosity((total volume-polymer volume)/total volume) of polyphosphazene blends during degradation (38). Reproduced with permission from reference (38). Copyright 2010 John Wiley & Sons.

Polyphosphazene Composites

Mechanical competence is one of the criteria a biomaterial must meet to be suitable for bone tissue regeneration (*5*, *11*, *39*). A biodegradable biomaterial should have good initial mechanical properties which are consistent with the anatomical site into which it is to be implanted and must be supportive of the developing tissues during the regeneration process (*10*, *11*, *40*). Being mechanically competent involves having sufficient mechanical integrity and dimensional stability, which are essential for the implanted scaffolds (*11*, *40*, *41*). Sometimes, these attributes can be lacking in pure biodegradable polymers. However, the mechanical properties can be enhanced by designing composite materials that consist of a polymeric matrix and a reinforcing agent such as calcium phosphate (*21*). The mechanical properties of composites are higher than the constituent phases (*42*). A number of polyphosphazene-hydroxyapatite composites have been investigated as bone substitutes (*43–45*). The selection of polyphosphazenes as the polymeric constituent solely depends on the side group. Side group with functional groups such as hydroxyl are preferred as they can nucleate hydroxyapatite for better osteointegration. Laurencin and co-workers developed composites that could be self-setting, that form at room temperature by combining polyphosphazenes with hydroxyapatites (*44*). The polyphosphazene-HAp composites produced were found to mimic essential aspects of bone. They can be constituted in surgery as pastes, placed in a defect and form HAp in vivo as a bone mineral analog under conditions unlikely to be aggressive to surrounding tissues (Figure 8). Ambrosio et al. (*45*) developed polyphosphazene-hydroxyapatite composites to examine its use in bone tissue regeneration. The composite was produced by mixing the polymer solution (dissolved in THF) with the hydroxyapatite particles in a 1:3 ratio and then precipitated out of the solution with hexane. The resulting composite was formed into cylindrical samples for biocompatibility evaluation. Results showed a favorable interaction between the composite material and the MC3T3-E1 cells as cells adhere and proliferate as compared to the polystyrene coated tissue culture plates (TCPS) (Figure 8a).There were improved mechanical properties due to the presence of hydroxyapatite in the polymer matrix (Figure 8b). This improvement in mechanical properties suggests the suitability of polyphosphazene composite for tissue regeneration applications. In a study in 2008, Nukavarapu et al. (*44*) investigated the suitability of polyphosphazene/nano-hydroxyapatite composite microsphere for bone tissue regeneration. In the study, leucine, valine, and phenylalanine amino ester single-substituted polyphosphazenes were produced, and the phenylalanine-based polymer was selected for further fabrication because it exhibited a higher Tg than the physiological temperature. The high Tg will ensure structural integrity in an in vivo environment. The poly[bis(ethyl phenylalanato)phosphazene] (PNEPhA) and 20 wt% nHAp were mixed to form composite microspheres using emulsion/solvent evaporation technique. This microsphere composite was utilized to fabricate 3-D microsphere scaffolds for mechanical, and osteocompatibility evaluation (Figure 9a). Similarly, PLAGA microsphere scaffold with 20 wt% of nHAp were also prepared and used as a control. As shown in Figure 9d & 9e, the three-dimensional polyphosphazene-nHAp composite microsphere scaffolds showed good osteoblast cell adhesion, proliferation, and phenotypic expression as the quantitative cell proliferation and alkaline phosphatase expression for the PNEPhA-20nHAp were similar to those of PLAGA-20nHAp.

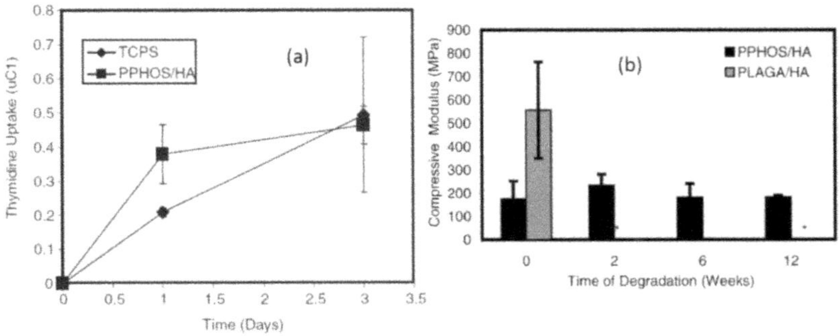

Figure 8. (a) Comparison of MC3T3 Cell adhesion and proliferation for polyphosphazene-hydroxyapatite composites and tissue culture polystyrene (b) Comparison of compressive modulus of PPHOS-HA and PLAGA-HA composites during degradation in PBS at 37 °C (45). Reproduced with permission from reference (45). Copyright 2003 IEEE.

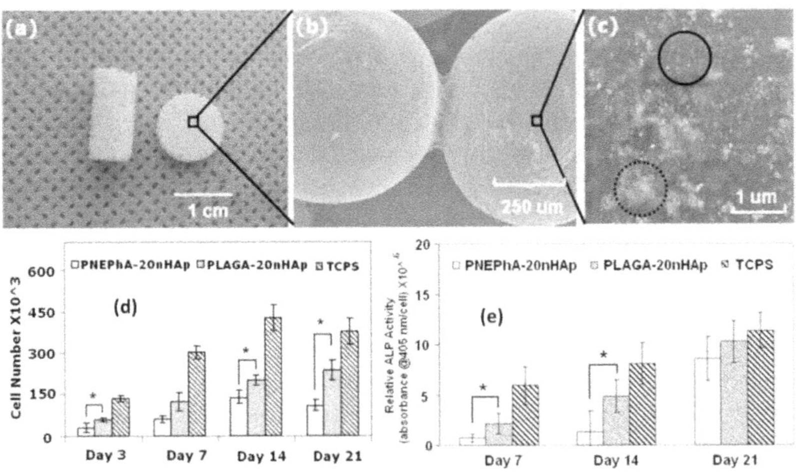

Figure 9. Phenylalanine-substituted polyphosphazene-nano-hydroxyapatite composite in macro, micro, nano structure and its suitability for tissue regeneration. (a) Cylindrical and disk-shaped scaffolds viewed by optical microscopy (b) Fused microspheres of the scaffolds viewed by SEM (c) Dispersion of HAp particle on a microsphere surface (d) Proliferation through MTS assay of Primary rat osteoblast cells (e) phenotypic expression of primary rat osteoblast cells on PNEPhA-20nHAp, PLAGA-20nHAp composite scaffolds and planar TCPS (44). Reproduced with permission from reference (44). Copyright 2008 American Chemical Society.

Suitability of Polyphosphazene for Regenerative Engineering

A material can be suitable for tissue regeneration if it is biocompatible, bioerodible, mechanically competent, osteoconductive, and produces resorbable products upon degradation (*21*, *46*). These criteria are discussed in the following sections, and amino acid esters will mainly be the focus because of their biocompatibility, controllable hydrolytic degradation rates, and non-toxic degradation products.

Biocompatibility

Laurencin et al. carried out the first evaluation of the biocompatibility of amino acid ester substituted polyphosphazene for tissue regeneration (*47*, *48*). The results of one of the studies showed that the rat primary osteoblast adhesion to poly ((ethyl glycinato) phosphazene) (PNEG) were comparable to PLAGA and polyanhydrides. Cell proliferation was not affected by the hydrolysis of PNEG however, the cell growth was enhanced as much as the PLAGA used as a control (*47*). Laurencin et al. demonstrated that a 3-D cell-amino acid containing polyphosphazene matrix systems (poly[(methylphenoxy)(ethyl glycinato) phosphazene]) with an average pore diameter of 165 μm exhibited steady attachment and growth of osteoblast cells throughout 21-day In Vitro study. On the other hand, the cell growth for 2-D made of the same polyphosphazene matrices showed a decline after seven days. This enhanced growth in the 3-D matrices is due to the resemblance (in shape and size) of its interconnected porosity with the natural bone tissue (*48*). In another study by Laurencin's group (*49*), a mixed substituent ethyl glycinato/methyl phenoxy polyphosphazenes were cultured with MC3T3-E1 cells (osteoblast precursor cell line from mice) and with the same exact experimental procedure addressed above. Favorable cell responses to the polyphosphazene polymers were observed, with the response being more pronounced for the polymers with a high composition of ethyl glycinato substituents. The cell adhesion and proliferation were not reduced as compared to the tissue culture plate (TCP) and PLAGA controls. Enhanced cell growth was witnessed for polymers with 50% and higher of the composition of ethyl glycinato substituents as compared to the TCP but was better than the PLAGA which is regarded as the gold standard. Moreover, the polymer with 25% ethyl glycinato substitution was a little bit lower than the TCP in terms of cell adhesion and proliferation characteristics but better than PLAGA. Sethuraman et al. (*50*) demonstrated the in vivo biocompatibility of co-substituted polyphosphazenes with aryloxy and ethyl alanato substituents as side groups. In the study, two polymers such as poly[(ethyl alanato)1(ethyl oxybenzoate)1 phosphazene] (PNEAEOB) and poly[(ethyl alanato)1(propyl oxybenzoate)1phosphazene] (PNEAPOB) were evaluated using a rat model. Results revealed good cell adhesion and proliferation as compared to the controls. PNEAEOB and PNEAPOB can be used for self-setting bone cements. Swiss 3T3 and HepG2 cells were employed to study the cytotoxicity of the degradation solution of poly[bis(ethyl 4-aminobutyro)phosphazene] (*51*). The proliferation of the cell lines was unaffected, signifying cytocompatibility of the novel polymer and its great prospects for in vivo applications (*51*).

The above studies with primary rat osteoblast were promising as the cell adhesion and proliferation were supported as compared to the conventional materials used previously in this kind of study. However, cells from rat and mouse may not be accurate reflections of actual results that may be obtained if it were primary human cells since specie-to-specie variations may play a prominent role in material-cell interactions. It is recommended that the human osteoblasts be employed in this study to get a better understanding of the actual material-cell interactions of the human subjects. In vivo evaluation of alanine amino acid ester-substituted polyphosphazenes was carried out to ascertain the biocompatibility of the polymers. Rat subcutaneous model was employed in this study using three polymers: Poly[(ethyl alanato)polyphosphazene](PNEA), Poly[(ethyl alanato)(p-methylphenoxy)phosphazene](PNEAmPh), and Poly[(ethyl alanato)(p-phenyl phenoxy)phosphazene](PNEAPhPh). The polymers recorded mild to moderate immune response and then it subsided over time. PNEAmPh and PNEAPhPh exhibited a minimal immune response after 12 weeks, and this was supported by the presence of few neutrophils, erythrocytes, and lymphocytes (Figure 10) (*50*). Laurencin et al. (*52*) carried out an in vivo evaluation of Poly[(ethyl glycinato)(p-methylphenoxy)]phosphazene (PNEGmPh) polymers to check their biocompatibility in New Zealand white rabbits. The polymer supported bone growth in the rabbits when compared to PLAGA as the histological examination after one week showed the formation of a thin layer of fibrous tissue on the surface of the implant and the area of the bone in contact with the implant. After 12 weeks of implantation, a lamellar bone was developed, and fat and bone marrow cells were seen on the implant interface. On the other hand, the PLAGA allowed the formation of thin fibrous membrane that is made up of giant cells, lymphocytes, and plasma cells. Furthermore, the PNEGmPh-based material did not show the presence of implant fragmentation, fat necrosis, or granuloma formation during the period of implantation underscoring its excellent biocompatibility. Overall, the biocompatibility studies showed minimal inflammatory responses of polyphosphazene-based biomaterials suggesting that they are suitable for regenerative engineering.

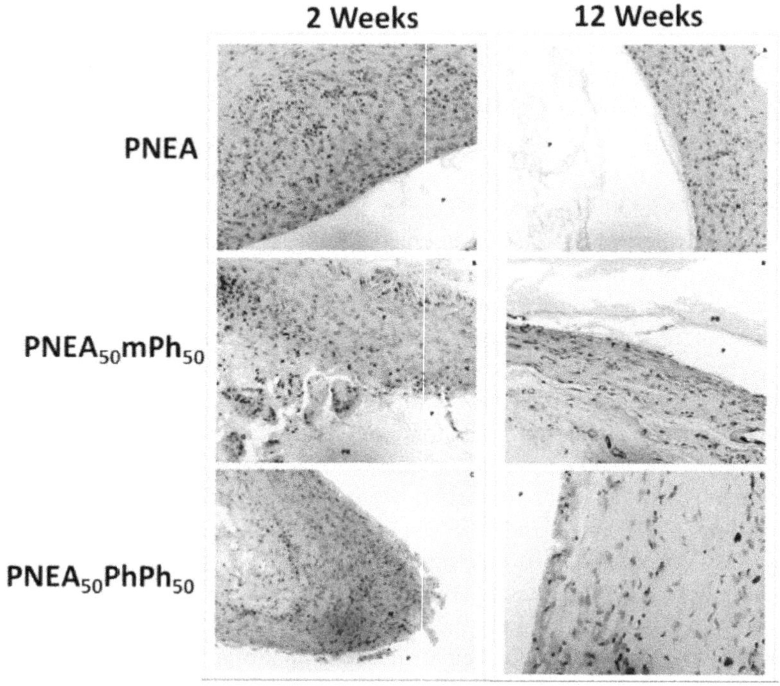

Figure 10. Rat subcutaneous model showing tissue responses to polyphosphazene-based polymers after 2 and 12 weeks of implantation P, polymer; N, neutrophils; F, fibrous tissue; G, giant cells; PR, polymer residue (×20 magnification) (50). Reproduced with permission from reference (50). Copyright 2006 John Wiley & Sons.

Biodegradability

A polymer must be biodegradable and have resorbable degradation products to be suitable for regenerative engineering (*6*, *11*, *12*). For this reason, many investigations have been carried out on the degradation properties of polyphosphazenes. As previously mentioned, polyphosphazene degradation can occur through three pathways. Numerous studies have demonstrated that biodegradable polyphosphazene (amino acid ester-based substituted) degrade into non-toxic products that can be easily excreted or metabolized (*1*). The degradation products which include ammonium phosphate and corresponding side groups could act as a natural buffer that can stabilize the pH of tissue microenvironment. This characteristic is unique and uncommon among the conventional biodegradables polymers (such as polyesters) that are currently available in the market. Allcock et al. (*32*) carried out a qualitative analysis to determine the formation of phosphates during degradation by using silver nitrate or zirconyl chloride to test for the presence of phosphates. In the study, phosphate was confirmed as silver nitrate reacted with phosphate to form a yellow silver phosphate and zirconyl chloride forms a white zirconyl phosphate precipitate on reaction with ammonium phosphate. The rates at which a polymer degrades can play a vital role in determining the overall success of tissue regeneration. In other words, polymeric biomaterial used for scaffold design should have a degradation rate that matches the rate of tissue growth. For instance, if the scaffold material has very fast degradation rates, it becomes mechanically weak to provide sufficient support for nascent developing tissues. If the material has slow or no degradation, it may now require surgical incisions which can cause damage to the tissues and could trigger issues of mechanical mismatch as compared to the native tissues. Poly(amino acid ester)phosphazenes with a wide range of degradation rates can be obtained by changing the type and ratios of side groups. Therefore, the side group chemistry enables the design of a variety of poly(amino acid ester)phosphazene that can match specific tissue types and disease states. The effect of varying the type of esters and amino acids on degradation have been extensively investigated (*32*). glycine-based poly(amino acid ester)phosphazenes such as poly[bis(methyl glycinat-N-yl)phosphazene] (PNMG), poly[bis(ethyl glycinat-N-yl)phosphazene] (PNEG), poly[bis(tert-butyl glycinat-N-yl)phosphazene] (PNtBG), and poly[bis(benzyl glycinat-N-yl)phosphazene] (PNBzG) were employed to study the effects. In the study, the degradation occurs via the breakdown of the molecular weight. The decrease in the molecular weight was in the following order PNBzG < PNtBG < PNEG < PNMG. PNBzG has the least decrease in molecular weight indicating the effect of hydrophobicity/hydrophilicity of side groups on degradation rates. The higher the hydrophobicity, the lower the molecular weight breakdown (from methyl to benzyl). Having a very hydrophobic side group, shield the polymer backbone from hydrolytic attack, and this hinders degradation. In the case of the effect of changing amino acids, poly[bis(methyl glycinat-N-yl)phosphazene] (PNMG), poly[bis(methyl alaninat-N-yl)phosphazene] (PNMA), poly[bis(methyl valinat-N-yl)phosphazene] (PNMV), and poly[bis(methyl phenylalaninat-N-yl)phosphazene] (PNMF) were employed. The molecular weight decreased in the following order: PNMF < PNMV < PNMA < PNMG. Similar to the trends observed for the ester groups, the molecular weight breakdown was highest for PNMG and least for the PNMF as the presence of more hydrophobic group such as phenylalanine at the α-position of amino acid retards hydrolysis. Singh et al. (*30*) also investigated the effect of the types of side groups on the degradation rates of L-alanine cosubstituted polyphosphazenes. PNEA, poly[(ethyl alanato)1 (ethyl glycinato)1phosphazene] (PNEAEG), PNEAmPh, and PNEAPhPh, were used. It was revealed that the molecular weight of ethyl glycinato cosubstituted phosphazene (PNEAEG) experienced the highest breakdown, whereas the molecular weight breakdown of biphenyl cosubstituted phosphazene (PNEAPhPh) was relatively slower. This study was able to demonstrate that with cosubstitution and careful selection of side group substituents, the degradation rates of poly(amino acid ester)phosphazenes can be customized to suit specific requirements. This efficient control over the degradation properties of amino acid ester-based polyphosphazene is an essential criterion in the development of a biomaterial for tissue regeneration.

Crommen et al. (*53*) also investigated the effect of using different side group ratios on the degradation rates of polyphosphazene-based polymers. In this study, mass loss was employed instead of molecular weight loss to determine the rate of degradation. Mass loss does not directly relate to molecular weight loss, but still a good indicator of relative degradation rates (*30*). Ethyl 2-(O-glycyl) lactate and ethyl glycinato cosubstituted polyphosphazenes with different ratios were incubated in PBS (which has similar physiological conditions as the human body). Increasing the composition of ethyl 2-(-O-glycyl) lactate in the macromolecular cosubstitution, facilitated the rate of mass loss and in turn, the degradation of the polymer. This is attributed to the increase in hydrolytic sensitivity as the Ethyl 2-(O-glycyl) is more hydrophilic and less bulky than the ethyl glycinato.

Overall, it has been proven that poly (amino acid ester) phosphazenes are biodegradable and their degradation rates can be tuned by changing a variety of factors, such as type and ratio of side groups making them suitable for regenerative engineering.

Mechanical Properties

As mentioned earlier, mechanical competence constitutes the main attributes that a biomaterial must meet to be clinically viable for use in tissue regeneration applications. It is vital in maintaining the structural stability of biomaterials as the initial mechanical properties of the constructs should match the native tissues. These mechanical properties include compressive strength, tensile strength, dimensional stability, and structural integrity. Problems of mismatch may come into play when the mechanical properties of the tissue-engineered constructs do not correspond to that of native tissues. This often leads to catastrophic failure of the construct (*54*, *55*). Few studies have been conducted to investigate the mechanical properties of amino acid ester-based polyphosphazenes. In one of such studies, Sethuraman et al. (*56*) investigated the effect of the structure of side groups on compressive strength of alanine-based polyphosphazene. In the study, poly(bis(ethyl alanato) phosphazene) (PNEA), poly((50% ethyl alanato) (50% methyl phenoxy) phosphazene) (PNEA(50)mPh(50)), poly((50% ethyl alanato) (50% phenyl phenoxy) phosphazene) (PNEA(50)PhPh(50)) were designed and the mechanical properties compared to PLAGA (85:15). Results of the mechanical testing indicated that the mechanical properties could be influenced by using the side group chemistry which entails the nature and the ratio of the pendant groups attached to the polymer backbone. It was also discovered that the compressive strength of PNEA(50)PhPh(50) (59.24 ± 25.59 MPa) was significantly higher than poly(lactide-co-glycolide) (85:15 PLAGA) (34.9 ± 5.7 MPa) due to the steric hindrance of the phenylphenoxy and restriction on the rotation of the polymer backbone (Figure 11). Also, the compressive strengths of PNEA, and $PNEA_{50}mPh_{50}$ were comparable to 85:15 PLAGA, with compressive strengths of 46.61 ± 17.56 MPa and 24.98 ± 11.26 MPa, respectively. It is important to note that the mechanical properties, just like the degradation rates can be customized via side group optimization to meet specific requirements for tissue regeneration. The observation associated with the compressive strength results holds true for the tensile strength as an increase in the steric bulk for the cosubstituent side groups will amount to an increase in tensile strength and elasticity of the material. This effect is more significant for a change of side chain from a small amino acid like glycine or alanine ethyl esters, such as in PNEAEG and PNEA, to large aromatic substituents, such as in PNEAmPh and PNEAPhPh. This can be attributable to the fact that glass transition temperature and molecular weight can be significantly influenced by the presence of large aryloxy substituents, which consequently affect the mechanical properties of the material. This is further confirmation that cosubstitution of large aromatic groups alongside amino acid esters present a tool for use for the fine-tuning the mechanical properties of polyphosphazenes.

Nevertheless, the mechanical properties and degradation rates of biodegradable polymers can be a tradeoff as an improvement in the mechanical properties of biodegradable polyphosphazenes by cosubstitution may present undesirable problems of increasing or decreasing the degradation rates. Thus, further research effort is required to investigate other means (scaffold fabrication techniques) by which mechanical properties can be enhanced without having to affect the degradation rates.

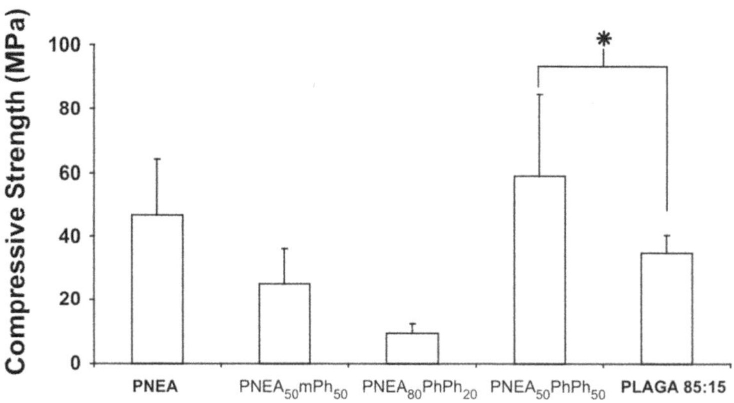

Figure 11. Comparison of the compressive strength of different alanine amino acid ester-based polyphosphazenes with PLAGA (85:15) (56). Reproduced with permission from reference (56). Copyright 2010 Elsevier.

Conclusion

Regenerative engineering is a new and increasingly important approach with high prospects of addressing clinical challenges of musculoskeletal tissue loss or failure. Degradable polymeric biomaterials are an indispensable tool for the success of this approach. Polyphosphazene-based biomaterials constitute a promising class of these biomaterials with unique characteristics. This chapter reviewed the design and suitability of biodegradable polyphosphazene polymers as biomaterials for regenerative engineering. The synthetic flexibility, tunability of degradation rates and mechanical properties, and buffering effect of the degradation products of this unique class of polymers have stimulated tremendous interest in tissue regeneration applications.

Acknowledgments

Support from NIH DP1 AR068147 and the Raymond and Beverly Sackler Center for Biomedical, Biological, Physical and Engineering Sciences, are gratefully acknowledged.

References

1. *Polyphosphazenes for Biomedical Applications (1)*; Wiley: Hoboken, NJ, 2009.
2. Allcock, H. R. The expanding field of polyphosphazene high polymers. *Dalton Trans.* **2016**, *45*, 1856–1862.
3. Andrianov, A. K.; Svirkin, Y. Y.; LeGolvan, M. P. Synthesis and biologically relevant properties of polyphosphazene polyacids. *Biomacromolecules.* **2004**, *5*, 1999–2006.
4. Nair, L. S.; Laurencin, C. T. Biodegradable polymers as biomaterials. *Prog. Polym. Sci.* **2007**, *32*, 762–798.
5. Ulery, B. D.; Nair, L. S.; Laurencin, C. T. Biomedical applications of biodegradable polymers. *J. Polym. Sci., Part B: Polym. Phys.* **2011**, *49*, 832–864.
6. Zhang, Z.; Ortiz, O.; Goyal, R.; Kohn, J. Biodegradable polymers. *Principles Tissue Engineering.* **2014**, 441–473.
7. Rezaie, H. R.; Bakhtiari, L.; Öchsner, A. Application of Biomaterials. In *Biomaterials and Their Applications*; Springer, 2015; pp 19–24.
8. Bhat, S.; Kumar, A. Biomaterials and bioengineering tomorrow's healthcare. *Biomatter* **2013**, *3*, e24717.
9. Oliva, N.; Unterman, S.; Zhang, Y.; Conde, J.; Song, H. S.; Artzi, N. Personalizing biomaterials for precision nanomedicine considering the local tissue microenvironment. *Adv. Healthcare Mater.* **2015**, *4*, 1584–1599.
10. Ogueri, K. S.; Ivirico, J. L. E.; Nair, L. S.; Allcock, H. R.; Laurencin, C. T. Biodegradable polyphosphazene-based blends for regenerative engineering. *Regen. Eng. Transl. Med.* **2017**, *3*, 15–31.
11. Laurencin, C. T.; Khan, Y. Regenerative engineering. *Sci. Transl. Med.* **2012**, *4*, 160ed9.
12. Nair, L. S.; Laurencin, C. T. Polymers as biomaterials for tissue engineering and controlled drug delivery. *Adv Biochem. Eng. Biotechnol.* **2006**, *102*, 47–90.
13. Heinegård, D.; Oldberg, A. Structure and biology of cartilage and bone matrix noncollagenous macromolecules. *FASEB J.* **1989**, *3*, 2042–2051.
14. Gumbiner, B. M. Cell adhesion: the molecular basis of tissue architecture and morphogenesis. *Cell* **1996**, *84*, 345–357.
15. Ikada, Y. Surface modification of polymers for medical applications. *Biomaterials* **1994**, *15*, 725–736.
16. Vroman, I.; Tighzert, L. Biodegradable polymers. *Materials.* **2009**, *2*, 307–344.
17. Sabir, M. I.; Xu, X.; Li, L. A review on biodegradable polymeric materials for bone tissue engineering applications. *J. Mater. Sci.* **2009**, *44*, 5713–5724.

18. Rodrigues, M. T.; Carvalho, P. P.; Gomes, M. E.; Reis, R. L. Biomaterials in Preclinical Approaches for Engineering Skeletal Tissues. In *Translational Regenerative Medicine*; Elsevier, Inc., 2015; pp 127–139.
19. Albertsson, A.-C.; Varma, I. K. Recent developments in ring opening polymerization of lactones for biomedical applications. *Biomacromolecules.* **2003**, *4*, 1466–1486.
20. Schacht, E.; Vandorpe, J.; Dejardin, S.; Lemmouchi, Y.; Seymour, L. Biomedical applications of degradable polyphosphazenes. *Biotechnol. Bioeng.* **1996**, *52*, 102–108.
21. Allcock, H. R.; Morozowich, N. L. Bioerodible polyphosphazenes and their medical potential. *Polym. Chem.* **2012**, *3*, 578–590.
22. Furió-Más, C.; Luisa Calatayud, M.; Guisasola, J.; Furió-Gómez, C. How are the concepts and theories of acid–base reactions presented? Chemistry in textbooks and as presented by teachers. *Int. J. Sci. Educ.* **2005**, *27*, 1337–1358.
23. Deng, M.; Nair, L. S.; Nukavarapu, S. P.; Jiang, T.; Kanner, W. A.; Li, X.; Kumbar, S. G.; Weikel, A. L.; Krogman, N. R.; Allcock, H. R. Dipeptide-based polyphosphazene and polyester blends for bone tissue engineering. *Biomaterials.* **2010**, *31*, 4898–4908.
24. Deng, M.; Nair, L. S.; Nukavarapu, S. P.; Kumbar, S. G.; Jiang, T.; Krogman, N. R.; Singh, A.; Allcock, H. R.; Laurencin, C. T. Miscibility and in vitro osteocompatibility of biodegradable blends of poly [(ethyl alanato)(p-phenyl phenoxy) phosphazene] and poly (lactic acid-glycolic acid). *Biomaterials.* **2008**, *29*, 337–349.
25. Xu, L.-C.; Li, Z.; Tian, Z.; Chen, C.; Allcock, H. R.; Siedlecki, C. A. A new textured polyphosphazene biomaterial with improved blood coagulation and microbial infection responses. *Acta Biomater.* **2017**.
26. Allcock, H. R. Recent developments in polyphosphazene materials science. *Curr. Opin. Solid State Mater. Sci.* **2006**, *10*, 231–240.
27. Dück, K.; Gates, D. P.; Main-Chain, Phosphorus-Based Polymers. In *Main Group Strategies for Functional Hybrid Materials*; Wiley, 2018; pp 329–351.
28. Rothemund, S.; Teasdale, I. Preparation of polyphosphazenes: a tutorial review. *Chem. Soc. Rev.* **2016**, *45*, 5200–5215.
29. Baillargeon, A. L.; Mequanint, K. Biodegradable polyphosphazene biomaterials for tissue engineering and delivery of therapeutics. *BioMed Res. Int.* **2014**, 2014.
30. Singh, A.; Krogman, N. R.; Sethuraman, S.; Nair, L. S.; Sturgeon, J. L.; Brown, P. W.; Laurencin, C. T.; Allcock, H. R. Effect of side group chemistry on the properties of biodegradable L-alanine cosubstituted polyphosphazenes. *Biomacromolecules.* **2006**, *7*, 914–918.
31. Weikel, A. L.; Krogman, N. R.; Nguyen, N. Q.; Nair, L. S.; Laurencin, C. T.; Allcock, H. R. Polyphosphazenes that contain dipeptide side groups: synthesis, characterization, and sensitivity to hydrolysis. *Macromolecules* **2009**, *42*, 636–639.
32. Allcock, H. R.; Pucher, S. R.; Scopelianos, A. G. Poly[(amino acid ester)phosphazenes]: synthesis, crystallinity, and hydrolytic sensitivity in solution and the solid state. *Macromolecules.* **1994**, *27*, 1071–1075.

33. Andrianov, A. K.; Marin, A. Degradation of polyaminophosphazenes: effects of hydrolytic environment and polymer processing. *Biomacromolecules* **2006**, *7*, 1581–1586.
34. Deng, M.; Nair, L. S.; Nukavarapu, S. P.; Kumbar, S. G.; Brown, J. L.; Krogman, N. R.; Weikel, A. L.; Allcock, H. R.; Laurencin, C. T. Biomimetic, bioactive etheric polyphosphazene-poly (lactide-co-glycolide) blends for bone tissue engineering. *J. Biomed. Mater. Res., Part A* **2010**, *92*, 114–125.
35. Kumbar, S.; Laurencin, C.; Deng, M. *Natural and Synthetic Biomedical Polymers*; Newnes, 2014.
36. Wilfert, S.; Iturmendi, A.; Schoefberger, W.; Kryeziu, K.; Heffeter, P.; Berger, W.; Brüggemann, O.; Teasdale, I. Water-soluble, biocompatible polyphosphazenes with controllable and pH-promoted degradation behavior. *J. Polym. Sci., Part A: Polym. Chem.* **2014**, *52*, 287–294.
37. Roach, P.; Farrar, D.; Perry, C. C. Interpretation of protein adsorption: surface-induced conformational changes. *J. Am. Chem. Soc.* **2005**, *127*, 8168–8173.
38. Deng, M.; Nair, L. S.; Nukavarapu, S. P.; Kumbar, S. G.; Jiang, T.; Weikel, A. L.; Krogman, N. R.; Allcock, H. R.; Laurencin, C. T. In situ porous structures: a unique polymer erosion mechanism in biodegradable dipeptide-based polyphosphazene and polyester blends producing matrices for regenerative engineering. *Adv. Funct. Mater.* **2010**, *20*, 2794–2806.
39. Saltzman, W. M. *Tissue Engineering: Engineering Principles for the Design of Replacement Organs and Tissues*; Oxford University Press, 2004.
40. Lanza, R.; Langer, R.; Vacanti, J. P. *Principles of Tissue Engineering*; Academic Press, 2011.
41. Khademhosseini, A.; Vacanti, J. P.; Langer, R. Progress in tissue engineering. *Scientific American* **2009**, *300*, 64–71.
42. Hull, D.; Clyne, T. W. *An Introduction to Composite Materials*; Cambridge University Press, 1996.
43. Bhattacharyya, S.; Kumbar, S. G.; Khan, Y. M.; Nair, L. S.; Singh, A.; Krogman, N. R.; Brown, P. W.; Allcock, H. R.; Laurencin, C. T. Biodegradable polyphosphazene-nanohydroxyapatite composite nanofibers: scaffolds for bone tissue engineering. *J. Biomed. Nanotechnol.* **2009**, *5*, 69–75.
44. Nukavarapu, S. P.; Kumbar, S. G.; Brown, J. L.; Krogman, N. R.; Weikel, A. L.; Hindenlang, M. D.; Nair, L. S.; Allcock, H. R.; Laurencin, C. T. Polyphosphazene/nano-hydroxyapatite composite microsphere scaffolds for bone tissue engineering. *Biomacromolecules* **2008**, *9*, 1818–1825.
45. Laurencin, C.; Ambrosio, A.; Sahota, J.; Runge, C.; Kurtz, S.; Lakshmi, S.; Allcock, H. Novel polyphosphazene-hydroxyapatite composites as biomaterials. *IEEE Engineering in Medicine and Biology Magazine* **2003**, *22*, 18–26.
46. Baillargeon, A. L.; Mequanint, K. Biodegradable polyphosphazene biomaterials for tissue engineering and delivery of therapeutics. *BioMed Res Int.* **2014**, *2014*, 761373.

47. Laurencin, C. T.; Morris, C. D.; Pierre-Jacques, H.; Schwartz, E. R.; Keaton, A. R.; Zou, L. Osteoblast culture on bioerodible polymers: studies of initial cell adhesion and spread. *Polym. Adv. Technol.* **1992**, *3*, 359–364.
48. Laurencin, C. T.; El-Amin, S. F.; Ibim, S. E.; Willoughby, D. A.; Attawia, M.; Allcock, H. R.; Ambrosio, A. A. A highly porous 3-dimensional polyphosphazene polymer matrix for skeletal tissue regeneration. *J. Biomed. Mater. Res., Part A* **1996**, *30*, 133–138.
49. Laurencin, C. T.; Norman, M. E.; Elgendy, H. M.; El-Amin, S. F.; Allcock, H. R.; Pucher, S. R.; Ambrosio, A. A. Use of polyphosphazenes for skeletal tissue regeneration. *J. Biomed. Mater. Res., Part A* **1993**, *27*, 963–973.
50. Sethuraman, S.; Nair, L. S.; El-Amin, S.; Farrar, R.; Nguyen, M. T. N.; Singh, A.; Allcock, H. R.; Greish, Y. E.; Brown, P. W.; Laurencin, C. T. In vivo biodegradability and biocompatibility evaluation of novel alanine ester based polyphosphazenes in a rat model. *J. Biomed. Mater. Res., Part A* **2006**, *77*, 679–687.
51. Gümüşderelioglu, M.; Gür, A. Synthesis, characterization, in vitro degradation and cytotoxicity of poly [bis (ethyl 4-aminobutyro) phosphazene]. *React. Funct. Polym.* **2002**, *52*, 71–80.
52. Laurencin, C.; Ambrosio, A.; Bauer, T.; Allcock, H.; Attawia, M.; Borden, M.; Gorum, W.; Frank, D. In *Tilte1998*; Society for Biomaterials.
53. Crommen, J. H.; Schacht, E. H. Synthesis and evaluation of the hydrolytical stability of ethyl 2-(α-amino acid) glycolates and ethyl 2-(α-amino acid) lactates. *Bulletin Des Societes Chimiques Belges* **1991**, *100*, 747–758.
54. Sarkar, S.; Salacinski, H.; Hamilton, G.; Seifalian, A. The mechanical properties of infrainguinal vascular bypass grafts: their role in influencing patency. *European Journal of Vascular and Endovascular Surgery* **2006**, *31*, 627–636.
55. Konig, G.; McAllister, T. N.; Dusserre, N.; Garrido, S. A.; Iyican, C.; Marini, A.; Fiorillo, A.; Avila, H.; Wystrychowski, W.; Zagalski, K. Mechanical properties of completely autologous human tissue engineered blood vessels compared to human saphenous vein and mammary artery. *Biomaterials* **2009**, *30*, 1542–1550.
56. Sethuraman, S.; Nair, L. S.; El-Amin, S.; Nguyen, M.-T.; Singh, A.; Krogman, N.; Greish, Y. E.; Allcock, H. R.; Brown, P. W.; Laurencin, C. T. Mechanical properties and osteocompatibility of novel biodegradable alanine based polyphosphazenes: side group effects. *Acta Biomater.* **2010**, *6*, 1931–1937.

Chapter 4

Polyphosphazene-Based Nanoparticles as Contrast Agents

Maryam Hajfathalian,[1] Mathilde Bouché,[1] and David P. Cormode*,[1,2,3]

[1]Department of Radiology,
[2]Department of Bioengineering,
[3]Medicine, Division of Cardiovascular Medicine,
University of Pennsylvania, 3400 Spruce St., 1 Silverstein,
Philadelphia, Pennsylvania 19104, United States
***E-mail: david.cormode@uphs.upenn.edu. Tel.: 215-746-1382.**
Fax: 240-368-8096.

Advancements in nanotechnology have led to significant changes to what is possible with medical imaging techniques in the past decades; however, there are still enormous incentives to develop novel imaging contrast agents that could facilitate detection of cancer at early stages, have improved renal clearance, have longer circulation half-lives, or otherwise provide complementary information. A vast variety of nanomaterials have been developed to enhance the contrast of medical images. Among them, polymers such as polyphosphazenes (PPPs) have recently gained considerable interest due to their excellent properties such as biocompatibility, biodegradability, synthetic flexibility, high versatility, hydrophilicity, and nontoxicity. Here, we present the synthetic methods, properties, and applications of PPPs. We describe imaging modalities such as computed tomography (CT), magnetic resonance imaging (MRI), photoacoustics (PA) and fluorescence, along with the advantages of using polymeric nanoparticles in these imaging techniques. Moreover, multimodal PPPs-based agents that enhanced the imaging accuracy, renal clearance and cytotoxicity of conventional contrast agents are discussed.

Introduction

Nanotechnology is a field of science that involves the manufacturing and engineering of functional structures or devices that are typically less than 100 nanometers (nm) in one or more dimensions. The small size of nanoparticles introduces the possibility of outstanding diagnostics and/or therapeutic properties due to their penetration into disease sites with greater specificity than most non-nanosized entities (*1*). Therefore, nanostructures have potential for use in advanced applications such as sensing, imaging, therapeutics, drug delivery and tissue engineering (*2*). Contrast agents have played a crucial role in the development of novel imaging modalities and have been extensively investigated in the past few decades. Inorganic nanoparticle contrast agents have attracted substantial attention because their characteristics are different to those of small molecule contrast agents. For example, they can be easily synthesized, functionalized, and coated such that they have longer circulation times, accumulate in disease sites, or target specific cell types. The high payloads of nanoparticles improve sensitivity for these applications. In comparison conventional small molecule agents are rapidly excreted, diffuse into healthy tissues increasing background signal and have low payloads. Moreover, inorganic nanoparticles can have unique contrast generation properties that are not available for small molecules (*3*). Nevertheless, there can be drawbacks to these nanoparticles such as their biocompatibility or long-term organ retention, which have been major hurdles for their FDA-approval (*4–6*).

Many different materials have now been used as coatings or carriers for contrast agents such as small molecules, lipids, emulsions, proteins and polymers (*5*, *7*, *8*). The invention of polymeric biomaterials has changed the field tremendously, since they can address some the issues of inorganic nanoparticles. Polymers can improve the biocompatibility of inorganic nanoparticles and help evade the immune system. Moreover, biocompatible polymers have been widely used in therapeutic applications due to their potential for drug loading and tunable release of those drugs, leading to spatial and temporal control over drug release (*9–12*).

Poly (methyl methacrylate) was the first polymer used in biomedical applications when an ophthalmologist, Harold Ridley, used it for intraocular lenses in cataract patients (*13*). Subsequent work has led to the development of many additional polymers for biomedical applications such as dextran, alginate, polyacetals, polyketals, polyglutamic acid, polyphosphoesters, poly-(glycolic acid), poly (lactic acid), poly (lactic-glycolic acid), poly-(caprolactone), polyurethanes, polyphosphazenes (PPP) and many others (*14*, *15*). Among these polymers, PPPs have attracted much attention due to the straightforward and easy functionalization of their phosphorous backbone, giving access to a large range of properties. Furthermore, the phosphorous bonds are reactive toward hydrolysis, therefore allowing degradation of these polymers into harmless products over controllable periods of time. This characteristic is an appealing feature for a carrier platform for contrast agents. PPPs contain nitrogen and phosphorous in their backbone along with two side groups attached to each phosphorus atom (Figure 1). They have properties such as chain flexibility, high temperature stability, biodegradability, hydrophilicity and potential for renal clearance that explains their use in tissue engineering for bone regeneration, as well as drug, gene and contrast generating material delivery (*16*, *17*).

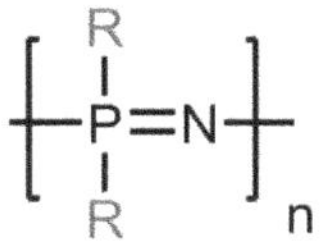

Figure 1. General chemical structure of PPPs.

More than 700 different PPP structures have been synthesized by modifying their side chains, molecular weight and organic substituents. These structural differences can greatly alter their properties. For instance, the presence of multiple amine side groups rendered poly [(2-dimethylamino ethylamino) phosphazene suitable for gene delivery. Galactose functionalized PPPs have been also used in drug delivery and gene delivery to tumors due to the ability of galactose to target hepatic tumors (*18*). Poly[bis(carboxyphenoxy)phosphazene] (PCPP) and poly[bis(carboxyethylphenoxy)phosphazene] are examples of PPPs that have been used for vaccine delivery and immunomodulation. Although tissue engineering and drug delivery have been the most investigated bioapplications of PPPs, their bioerodibility, water-solubility and ionic conductivity offer excellent potential to be used as coatings for contrast agents (*19*, *20*).

In this chapter, we provide background information on some of the diagnostic modalities such as CT, MRI, PA, and fluorescence imaging, as well as contrast agent candidates for them. In addition, we will describe the synthesis methods of PPPs, their various designs and their biomedical applications. Last, we will focus on PPPs-based contrast agents that have been developed by assembly of a PPPs and contrast generating nanocrystals (i.e. gold, iron oxide, quantum dots, and nanophosphors) for various medical imaging techniques.

Imaging Modalities

Since the discovery of X-rays in 1895, medical imaging has found broad clinical use and now encompasses a plethora of different techniques. To give the reader a better understanding of topics covered later this chapter and of the requirements needed for nanoparticle contrast agents, we briefly review several imaging techniques in this section i.e. CT, PA, MRI and fluorescence imaging.

Computed Tomography

X-ray computed tomography (CT) is a whole-body imaging method. It is one of the most commonly used medical imaging techniques and is widely used in trauma, to monitor tumors and for cardiovascular disease detection amongst many other applications in other conditions. An X-ray source and a detector are the base constituents of a CT scanner. The X-ray generator emits X-rays into the patient's body, where some of them are absorbed by the tissues, bones or air, while the remainder pass through the patient's body and can be absorbed by X-ray detectors. In some configurations of CT scanners, the X-ray generator rotates around the patient and the detector arrays are positioned on the opposite side to the source. Therefore, in these scanners the absorption of X-rays by the patient from all angles is achievable (Figure 2A). Some of the major benefits of using CT include fast scan times, high spatial resolution, linearity of contrast, low cost, and wide clinical availability. The drawbacks of CT include low soft-tissue contrast, low sensitivity to contrast agents and exposure to ionizing radiation (*21–23*).

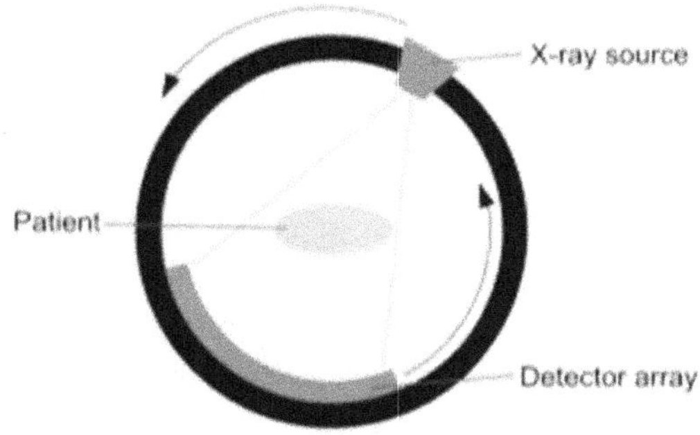

Figure 2. Schematic depiction of a CT scanner. (Adapted with permission from ref. (23). Copyright 2014 John Wiley and Sons.)

The X-ray attenuation of different materials varies widely. Heavy elements such as iodine, barium, and tungsten have very high X-ray attenuation and can act as CT contrast agents. These contrast media are valuable for disease diagnosis. For example, iodine-based contrast agents are used for cardiovascular angiography and barium sulfate agents are used for imaging gastrointestinal tract. Although these contrast agents are well-established for clinical use, they have several disadvantages. For example, iodinated contrast agent suffer from very short circulation half-lives, a lack of specificity and can cause contrast-induced nephropathy in patients with renal insufficiency. Researchers have consequently started to develop nanoparticle CT contrast agents that are hypothesized to be able to address some of these issues (*24*, *25*).

Over the past decade, many different types of nanoparticle CT contrast agents have been studied, such as polymeric nanoparticles, solid metal nanoparticles, micelles, lipoproteins and so forth (*26–28*). The appeal of nanoparticles as CT contrast agents focuses on their long-circulation times, potential for targeted imaging and high payloads. In addition to iodine, nanoparticle CT contrast agents have been reported that use elements such as gold, bismuth, tantalum, silver, ytterbium and others (*6*, *23*). For instance, Perera et al. reported a bismuth-iron inorganic coordination polymer that was coated with polyvinylpyrrolidone as a potential CT contrast agent (*29*). Despite the usefulness of these agents, their biocompatibility, degradability, toxicity and excretion need more investigation prior to clinical approval.

Photoacoustic Imaging

PA is a hybrid optical and acoustic imaging technique where substances that absorb light in the NIR region can be detected. It is a noninvasive technique that is based on the thermoelastic expansion effect of a contrast agent or a tissue using pulsed laser excitation. The thermoelastic expansion leads to the generation of ultrasonic waves that can be detected and converted into images (*30*). PA can provide high spatial resolution and functional information such as hemoglobin and blood oxygenation properties, reasonable tissue penetration, real-time imaging, and does not involve ionizing radiation. However, it is not a whole body imaging technique, has limited endogenous contrast and a lack of reliable phantom assessment for quality control in both small animals and human studies have limited its progress towards FDA approval.

PA contrast agents have been studies such as dyes and metal nanoparticles. For example, indocyanine-green (ICG) is a dye that is FDA-approved for other applications, but absorbs light in the range of 650–950 nm and therefore is suitable for PA applications (*31*). Plasmonic noble metals such as AuNP are also attractive for PA because they can be synthesized such that they have strong scattering and absorption in the NIR, where biological tissues absorb the light the least. Altering the shape, size or composition of AuNP can drastically change their optical behavior. For instance, small spherical AuNP with diameter around 5 nm have adsorption around 500 nm, whereas the absorbance peak of 100 nm AuNP is around 600 nm. Anisotropic shapes such as rods also present extinction spectra with multiple peaks from ultraviolet to NIR. Murphy et al. reported that increasing the aspect ratio of gold nanorods from 1.3 to 4.4 led to a shift in the plasmon peak from 600 to 900 nm (*32*). The disadvantages of using these nanoparticles as a PA contrast agent include lack of excretion and possible changes in their morphology (and hence absorbance) upon irradiation. For example, the morphologies of some plasmonic AuNP with complex shapes such as nanorods can be changed to spheres when exposed to high laser power (*33*). However, Chen and et al. have found that photothermal stability of AuNP can be enhanced by applying coatings such as silica and PEG on gold nanorods (*34*).

Fluorescence Imaging

Fluorescence imaging is extensively used in pre-clinical settings for molecular imaging of a variety of diseases. Fluorescence occurs when specific types of molecules (such as polyaromatic hydrocarbons or heterocycles) or nanoparticles (such as quantum dots) absorb light at a certain wavelength, putting the agent into an excited state from which they can emit the excitation energy as light (*35*). Fluorescence imaging is yet to be FDA-approved in any form, however inter-operative fluorescence imaging systems used to aid tumor resection are being explored in clinical trials (*36*).

The first medical application of fluorescence imaging was in 1924 when autofluorescence of endogenous porphyrins in tumors was observed. Then in 1948, the first attempt of using fluorescein to improve the detection of brain tumors was reported (*37*). Today, with fluorescence imaging systems being widespread, it is commonly used in cancer research and other diseases. This technology offers many advantages compared to other methods, such as availability, sensitivity, ease of use, low cost, and possibility of imaging multiple components at the same time. In addition to visualizing the uptake of fluorophores specific cell types can be studied if they express fluorescent reporter proteins (*38*).

Fluorophores with characteristics such as resistance to photobleaching, absorbance and emission wavelengths in NIR, water solubility and biocompatibility, are desirable for biological applications. These agents can be small molecules and can be attached to antibodies, peptides or other ligands for targeting imaging applications. Small molecule fluorophores can also be attached to or loaded into nanoparticles. For example, Mieszawska and et al. modified poly(lactic co glycolic) acid with AuNP and formed this polymer into nanoparticles by coating it with phospholipids. This nanoparticle was loaded with doxorubicin, a cytotoxic drug, in the polymer core and an anti-angiogenic drug sorafenib in the lipidic corona. A NIR Cy7 dye was attached to the distal end of PEG groups on the lipid coating (*39*). This platform therefore has the potential for being detected with both CT and NIR fluorescence imaging. This has the advantage that fluorescence imaging systems are more frequently available, are higher throughput and are more sensitive than CT. Furthermore, the fluorophore allows microscopic analysis of the cellular localization of the nanoparticle.

Quantum dots (QDs) are a class of nanoparticles with unique optical properties, such as excitation spectra ranging from ultraviolet to NIR, high quantum yields, and symmetric emission spectra. It has been demonstrated that the QDs can be used with fluorescent imaging for cancer diagnosis due to their sensitivity and multicolor fluorescence imaging capability (*40*). Furthermore, the use of QDs allowed high spatial resolution fluorescence imaging of tissue vasculature (*41*).

Magnetic Resonance Imaging

MRI is a non-invasive technique that generates anatomical images. MRI does not use ionizing radiation, and thus it is suitable for frequently repeated imaging. MRI uses very high magnetic fields that contain gradients. Under such fields, protons can be excited by radiowaves. As the protons return to their ground state radiowaves are released, which are recorded and used to form images. The relaxation rates of protons, known as T_1 and T_2, are highly dependent on their environment. Minor changes in the environment of protons in different organs result in differences in signal and therefore MRI produces good soft tissue contrast. MRI's drawbacks include its high cost, long image acquisition times, lower availability and incompatibility with some types of patients (e.g. those with metal implants or claustrophobia) (*3*, *42*).

MRI contrast agents are classified as T_1 (resulting in increases in image contrast, known as positive contrast) or T_2 (resulting in reductions of intensity in images, known as negative contrast) contrast agents, and are mostly gadolinium (Gd) or iron-based agents, respectively. Research has been ongoing for Gd-based contrast agents for many years, and therefore several gadolinium chelates are FDA-approved for use in patients. Gd is a toxic element, however chelated

Gd is broadly considered as safe. In addition, several iron oxide nanoparticles have been approved for human use, typically with some kind of dextran-derived coating. A variety of applications have been reported for nanoparticle-based MRI contrast agents. For example, various types of Gd nanoparticles have been used in vascular imaging and also to image organs such as liver (*43*). The FDA-approved iron oxide nanoparticles are mainly used for distinguishing liver tumors from normal tissues. A variety of biological disease markers, such as macrophages, VCAM-1, and VEGFr have been imaged using targeted iron oxide nanoparticles. They have also been used in many multifunctional platforms, such as drug or gene delivery agents (*44*, *45*).

Polyphosphazenes

PPPs are a class of compounds that has been widely investigated for multiple bio-related purposes ranging from tissue engineering to drug delivery (*46*, *47*). PPPs can be referred to as a hybrid polymer since they have an inorganic, alternating phosphorous-nitrogen backbone to which are appended organic side groups (*48*). The phosphorous atom is pentavalent with an oxidation state of +V.

While the first synthetic procedures to form PPP derivatives were reported by Allcock and co-workers in the early sixties, control over the degradability of the backbone has been explored more recently, and bioerodible polymers have mostly been developed over the past two decades (*49*, *50*). The degradability of polymers is a highly desirable feature for biomedical use, although specific degradation profiles are required for individual applications. These polymers are usually degraded by hydrolysis and ultimately release harmless compounds, namely metabolizable phosphates and ammonia, which can be readily excreted. A highly diverse range of chemical structures and associated properties have been developed so that PPPs find application in numerous fields. Among the most common properties, PPPs have been designed to have water solubility, strong mechanical and thermal stability, water swelling properties and high ionic conductivity. Numerous PPPs have been designed to integrate acidic units such as carboxylic acid or sulfonic acid moieties, thus displaying great proton conductivity, which prompted their use as membranes in fuel cells (*51*). A large number of PPP derivatives have been designed for biomedical applications to display features such as high water solubility and controlled degradation profiles (*52*). Such properties, when combined with pH dependent degradability and ionic sensitivity, make PPPs that are interesting as immunoadjuvants for vaccine (*53*). In addition, the mechanical strength, biocompatibility and hydrolytic degradability of PPPs has proved of great interest for the development of implants and for regenerative engineering. Thus, PPPs allowed the development of matrices that could slowly be resorbed by the body and excreted safely (*54*, *55*). Finally, several contrast agents based on PPP carriers have been developed and will be discussed in detail in the next section (*56–58*). However, first we will provide more background on PPPs. There are two major strategies to synthesize PPPs, direct synthesis of PPPs by polymerization and post-polymerization substitution on the poly(dichloro)phosphazene core. These two methods are briefly discussed below.

Direct Synthesis of PPPs by Polymerization

The most common strategy to synthesize PPPs remains the ring-opening polymerization of hexachlorocyclotriphosphazene (*49*). This process is carried out at high temperatures of up to 250 °C in order to promote the cleavage of chlorine moieties, formation of a cationic phosphazenium intermediate and to initiate the polymerization. Interestingly, the higher the temperature, the more likely cross-linking is to occur. It can be limited by lowering the reaction temperature, although this consequently reduces the polymerization kinetics, unless using an additional Lewis acid as catalyst. Furthermore, the ring opening polymerization strategy is hard to control once the ring opening is initiated and often gives rise to polydisperse and high molecular weight PPPs. Consequently, alternative synthetic procedures have been developed, particularly living cationic polymerization to access PPP derivatives with lower molecular weights and more uniform size dispersion (*59, 60*). This strategy requires an initiation step through reaction of one monomer of trichlorophosphoranimine with two units of PCl_5 to promote the formation of an intermediate $[Cl_3PNCl_3]PCl_6$ species. Finally, this cationic intermediate further reacts with other $Cl_3PNSiMe_3$ units until complete reagent consumption, which marks the end of chain growth. The necessity to use high purity trichlorophosphoranimine reagent is an issue and therefore its replacement by chlorinated tertiary phosphines of a general structure R_3PCl_2, which are more widely available commercially, allows easier access to more diverse PPP structures. The synthesis of PPPs by direct polymerization of either chloro-alkyl or chloro-aryl- phosphoranimines of the general formula $ClR_2PNSiMe_3$ by initiation with trimethyl phosphite has recently attracted significant attention (*48*), since it can be done under mild reaction conditions in water, avoiding the use of organic solvents, which are highly desirable for biomedical use due to possible toxicity.

Post-Polymerization Substitution on the Poly(dichloro)phosphazene Core

As already mentioned, a significant advantage in the use of PPPs is the varied and straightforward functionalization possible of the phosphorous atom, which allows the functional groups along the polymer backbone to be fine-tuned to access desired properties for specific applications (*46*). Post-functionalization is typically achieved using the poly(dichloro)phosphazene compound where the highly reactive P-Cl bonds can either be hydrolysed when exposed to moisture, or the chlorine atom can easily be substituted by either oxygen-based nucleophiles as aryloxides and alkyloxides or primary amines (Figure 3) (*61*). The improved thermodynamic stability of the newly formed P-N or P-O bonds and thus of the corresponding functionalized PPP is the driving force for this reaction (*62*).

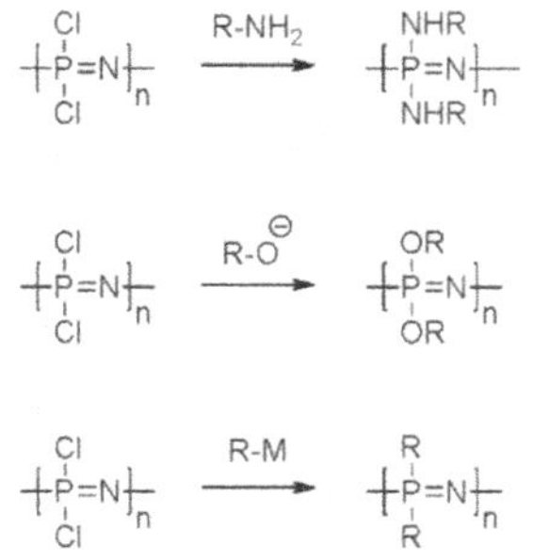

Figure 3. Post-functionalization of the poly(dichloro)phosphazene backbone by nucleophilic substitution. (Adapted with permission from ref. (61). Copyright 2013 John Wiley and Sons.)

The post-functionalization strategy has become very common since a large diversity of structures are achievable by modifying both the nature of the pendant groups and the substitution ratio (*63*, *64*). Indeed, depending on the reaction conditions used, PPPs functionalized with either single- or mixed-substituents can be produced in a controlled manner. This type of synthetic approach has allowed careful study of the influence of the pendant groups on the polymer properties (*48*). A report from Allcock and co-workers illustrated this concept by demonstrating the structure-activity relationship existing among PPPs bearing different cycloalkanoxy groups (*65*). In this study, they showed that the glass transition phase temperature depended on the functionalization of the polymer, with the bulkiest side groups at the phosphorous atom promoting a glass transition at higher temperatures compared to polymers bearing smaller side groups.

Structural Diversity for Selected Applications

Modification of the PPPs Architecture

The polymerization techniques previously detailed have allowed the synthesis of a large range of functionalized PPP derivatives, which display a broad spectrum of properties. PPP properties can be finely tuned by controlling several key parameters, therefore granting access to unusual architectures (*66*). Both the shape and size have a critical influence toward the fate of polymers in the body, being either its transport or excretion, and have thus been the topic of intensive research (*67*, *68*). In this respect, a wide range of PPP architectures have been reported to date such as the common linear backbone, branched polymers, dendritic scaffolds and helical structures (*69*).

One of the most investigated PPP derivatives, PCPP (Figure 4), has a high water solubility and a high degradation rate, which prompted its use as vaccine adjuvant (*70*, *71*). This polyelectrolyte is very sensitive to pH, with slow hydrolytic degradation occurring at neutral pH, while acidic media increases its degradation rate, with the protonation of the backbone being the rate limiting step. Moreover, PCPP is highly sensitive to ionic media, similarly to other polyelectrolytes and therefore it self-assembles in the presence of inorganic salts, and can be used to encapsulate cargoes for delivery applications. PCPP particles formation has also been found when it is exposed to proteins, thanks to ionic interactions and hydrogen bonding. Similarly, to most PPP derivatives, PCPP degrades into harmless byproducts and was suggested to be biocompatible both *in vitro* and *in vivo* (*72*). Several formulations successfully associated antigens with PCPP as a vaccine adjuvant based on the strong interaction of the polymer with the biomaterial and displayed promising immunostimulant properties (*73*).

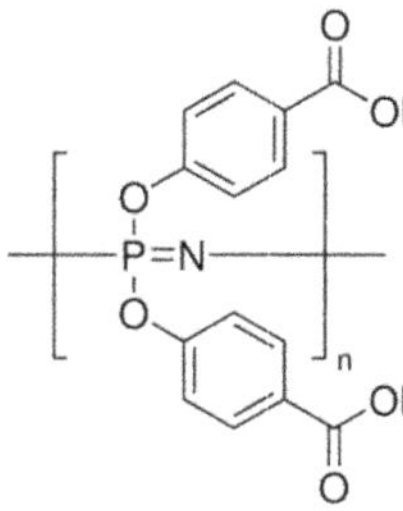

Figure 4. Chemical structure of poly[di(carboxylatophenoxy)phosphazene].

Several recent reports focus on the synthesis of star-shaped PPPs where a central core serves as a backbone for grafting of multiple arms (*74*). The ring opening polymerization of chlorophosphoranimine being difficult to control, this synthetic pathway is not used to form PPPs of complex architecture (*75*). However, the use of living cationic polymerization has been highlighted as an effective synthetic pathway to access star-shape brushes in a controlled manner (*76*, *77*). Remarkably, Allcock and co-workers reported star-like PPPs obtained by atom transfer radical polymerization, which avoided unwanted cross-linking and resulted in narrow polymer size distribution (*76*). Similarly, star-shaped PPP derivatives could be obtained by a three step procedure starting from the 1,1,1-tris(diphenylphosphino)methane acting as the structure's core, subsequent chlorination by hexachloroethane and finally living cationic polymerization of PPP units (*77*). Further macromolecular substitution with Jeffamine of the P-Cl bond at the phosphorous backbone of the polymer finally afforded the desired second-generation star-like brushes with a narrow polydispersity and increased hydrodynamic diameters.

Chirality in self-assembled polymers is a desirable feature for biomedical applications considering the widely spread chiral recognition of biomolecules. Therefore, the combination of the highly flexible and versatile PPP backbone with a helical shape that could help in recognizing biomolecules of interest has led to several novel PPP derivatives displaying twisted morphologies (*78*). Interestingly, the preferred formation of one helical sense has been shown to be possible by synthesizing a block copolymer where an optically active binophtaxyphosphazene unit is alternated with a diphenylphosphazene unit (*78*). This alternation induces a twisting of the non-chiral block that appears to favour the formation of a left-handed helical structure as supported by ^{31}P NMR and TEM. Remarkably, the helical orientation was transferred from single molecules to nanoparticle assemblies of multiple molecules of this polymer and paved the way for novel PPP based devices intended for the molecular recognition of biomolecules. Similarly, the grafting of optically active amines on an achiral PPP core proved to selectively promote the formation of a left handed-helical structure over its isomer as supported by circular dichroism measurements (*78*, *79*).

The versatility of PPP derivatives enabled the formation of various types of supramolecular structures such as hydrogels and polymersomes obtained by self-assembly, which are promising platforms for the development of drug delivery biomaterials (*48*). An amphiphilic PPP derivative functionalized with both ethyl-p-aminobenzoate and polyethyleneglycol moieties could be obtained in a straightforward manner by ring opening polymerization of hexachlorocyclotriphosphazene and further sequential substitution of the chloride groups at the phosphorous atom (Figure 5) (*80*). This amphiphilic polymer is then capable of self-aggregation in water into polymersomes and could be used as vehicle for the delivery of water soluble anticancer drugs.

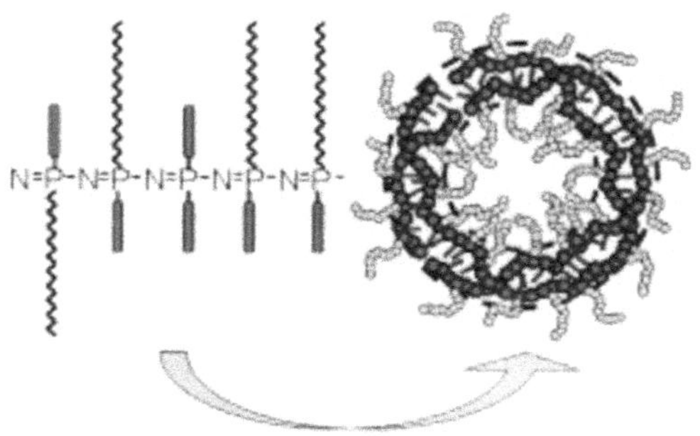

Figure 5. Self-aggregation of an amphiphilic PPPs for drug delivery. (Adapted with permission from ref. (80). Copyright 2009 Elsevier.)

One-dimensional nanomaterials exclusively composed of polymer are of interest due to their high surface area available for functionalization, which offers benefits for applications such as drug delivery or chemical sensing. Examples of one-dimensional PPP structures are rare, however, nanorods made of poly(cyclotriphosphazene-co-4,4′-sulfonyldiphenol) (phosphazene) could be obtained by assembly along a sacrificial silver nanowire (*81*). To do so, the functionalized PPP compound was obtained by acidic polycondensation of co-monomers hexachlorocyclotriphosphazene and 4,4′-sulfonyldiphenol followed by in situ reduction of silver nitrate to afford nanorods coated with PPPs. The PPP stabilization of silver nanorods was confirmed by transition electron microscopy (TEM) and scanning electron microscopy. A purely polymeric nanotube was formed by careful acidic treatment that resulted in the controlled degradation of the inner silver rod. The same group developed a similar technique for the synthesis of hollow spherical PPP nanoparticles using polystyrene nanoparticles as a sacrificial template, creating structures that could be relevant for drug delivery (*82*).

Mixed PPPs-Inorganic Materials

The large diversity of PPP structures and properties is of interest for the stabilization and encapsulation of metal nanoparticles. For example, the self-assembly of PPPs has been used by Xu and co-workers to form tubular carriers for silver nanoparticles of 5 to 20 nm diameter (*83*). Multiple characterization techniques confirmed the successful grafting of silver particles at the surface of the PPP nanotubes, namely TEM, energy dispersive X-ray spectroscopy (EDS), SEM and atomic absorption spectrometry. These mixed nanorods were tested for the catalytic reduction of 4-nitrophenol and proved to be an efficient catalyst with turnover frequency up to 101.4 h^{-1}, even after being reused 5 times. Similarly, the growth of either small AuNP of 5-8 nm diameter or mixed Au/AgNPs at the surface of a PPP derivative or the loading of 15-25 nm magnetic Fe_3O_4 nanoparticles has been developed and tested as catalyst for the 4-nitrophenol reduction or as support for catalysts (*84–86*). Although these composites were not investigated for biomedical applications, given the contrast generating properties of AuNPs and iron oxides, these studies pointed towards the potential of PPPs for delivery of nanocrystals in vitro or in vivo.

Polyphosphazenes as Carriers for Contrast Agents

PPPs have been extensively investigated in the course of polymeric based biomaterials and biomedical devices development. The diversity of synthetically available PPPs and their broad spectrum of properties such as water solubility, biocompatibility and biodegradability make them very versatile and promising polymers. Accordingly, their use in the development of biocompatible platforms in combination with inorganic contrast agents has been studied for medical imaging (Table 1). Here, we will describe the results to date where PPPs have been used as carriers of contrast generating materials for imaging modalities such as CT, MRI, PA and fluorescence imaging.

As previously discussed in detail, the high tunability of the physico-chemical properties and bioerodability of PPPs make them appealing candidates for the development of implants. The degradation and lifetime monitoring for these systems usually involves X-ray imaging techniques. Most PPPs are radiotransparent, similar to other polymers, and the most common method to achieve radio-opaque implants is to incorporate inorganic salts as a composite together with the polymers. However, it is uncertain if the release of those inorganic materials truly correlates with the degradation of the polymer. To address this issue, Allcock and co-workers developed novel PPPs functionalized with iodine atoms to achieve radio-opaque bioerodible polymers (*87*). A range of iodoaryloxy-PPPs was synthesized by nucleophilic substitution of chloride atoms along the poly(dichlorophosphazene) precursor with iodo-amino acid derivatives. Single- and mixed substituents functionalized with both diiodotyrosine group and non-iodinated amino acid esters were obtained by substitution of the P-Cl bonds. The same procedure was used for the synthesis of non-iodinated amino acid PPPs for a control and to allow evaluation of the influence of the iodine atoms. The iodinated and iodine-free PPPs were formulated as films of identical thickness and tested for hydrolytic degradation at 37°C. Good degradation profiles were observed by ^{1}H and ^{31}P NMR and quantification of the phosphates and ammonia release was achieved by silver nitrate and ninhydrin tests. Their radio-opacity was confirmed under X-ray irradiation even for PPPs functionalized with one iodine atom per repeating unit, although, as expected, greater incorporation of iodine atoms afforded greater opacity of the material under X-ray irradiation. The use of X-rays of higher energy confirmed these results and no additional degradation of the polymers was observed under medical X-ray conditions, thus supporting their potential for use as a biomaterial in implants.

Hu et al recently reported the use of PPPs as a capping ligand for the stabilization of superparamagnetic iron oxide nanoparticles (SPIONs) for use as a negative contrast agent in MRI (Figure 6) (*56*). Despite the large body of work on SPIONs as MRI contrast agents, new coatings that can improve their stability and safety are appealing. Consequently, poly(cyclotriphosphazene-co-4,4′-sulfonyldiphenol) phosphazene was explored as a coating for the stabilization of SPIONs. The numerous alcohol functions of this polymer, as well as the presence of phosphorous, nitrogen and sulphur atoms, were expected to promote H-bonding with neighbouring water molecules and positively impact the time for water residence close to these superparamagnetic nanoparticles.

Table 1. Imaging modalities applied to PPP based carriers for contrast agents

Imaging modality	*PPP derivatives*	*Payload*	*References*
X-rays	PPPs of the general formula $[PR_1R_2{=}N]_n$ with side groups R_1 and/or R_2 being: OPh-I, I_2-Tyr, I-Phe, OPh, trifluoroethoxy, ethoxy, Gly, Ala, Phe	Iodine	(*86*)
CT	PCPP	AuNPs	(*57, 58*)
MRI	PCPP	Fe_3O_4 NPs	(*58*)
	Poly(cyclotriphosphazene-co-4,4′-sulfonyldiphenol)		(*87*)
Fluorescence	PCPP	CdS	(*58*)
PA	PCPP	AuNPs	(*58*)

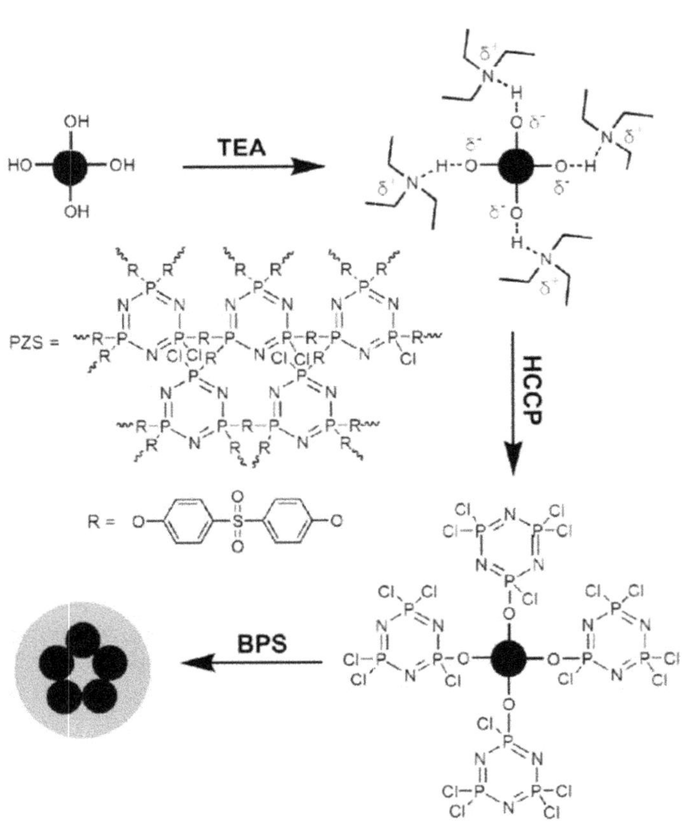

Figure 6. One-pot polycondensation of HCCP and BPS on Fe_3O_4 nanoparticles. (Adapted with permission from ref. (87). Copyright 2010 Royal Society of Chemistry.)

Triethyleneglycol coated SPIONs were synthesized by thermal decomposition of a $Fe(acac)_3$ precursor and further one-pot polycondensation of hexachloro-cyclotriphosphazene (HCCP) with 4,4′-sulfonyldiphenol (BPS) activated by ultrasound to afford stable and homogenous PPP particles loaded with 8.2 nm Fe_3O_4 nanoparticles. Successful iron oxide incorporation was determined by TEM (Figure 7), UV-vis spectroscopy and X-ray diffraction. The formation of a PPP was confirmed by Fourier-transform infrared spectroscopy thanks to the characteristic P=N, P-N and O=S=O vibration bands.

An appealing feature of the use of PPPs as coating for metal nanoparticles lie in its degradability in water into harmless byproducts, in this case phosphates, ammonia and 4,4′-sulfonyldiphenol. The PPP-SPIONs were found to very slowly degrade at both physiological and endosomal pH with a loss of weight of 37% and 44%, respectively, after 100 days. *In vitro* testing of the PPP-SPIONs on Hela cells confirmed their good biocompatibility since no adverse effect on cell viability was observed. In addition, TEM images of the treated HeLa cells indicated intracellular uptake of the iron oxide nanoparticles. The magnetic properties were evaluated for three different loadings of SPIONs in the PPP particles and the negligible hysteresis observed indicates that the superparamagnetic nature of the SPIONs was unaffected by the PPP coating. Substantial T_2 relaxivity was observed for the three different PPPs-Fe_3O_4 formulations tested, with relaxivity found to correlate with increases in iron oxide loading. Greater SPION aggregation was proposed to be the reason for the observed T_2 relaxivity increase. Negative contrast produced in MR images of solutions of these nanoparticles agreed with the results from relaxivity measurements, confirming the potential of these PPPs-SPIONs particles as efficient negative contrast agents for MRI (*56*).

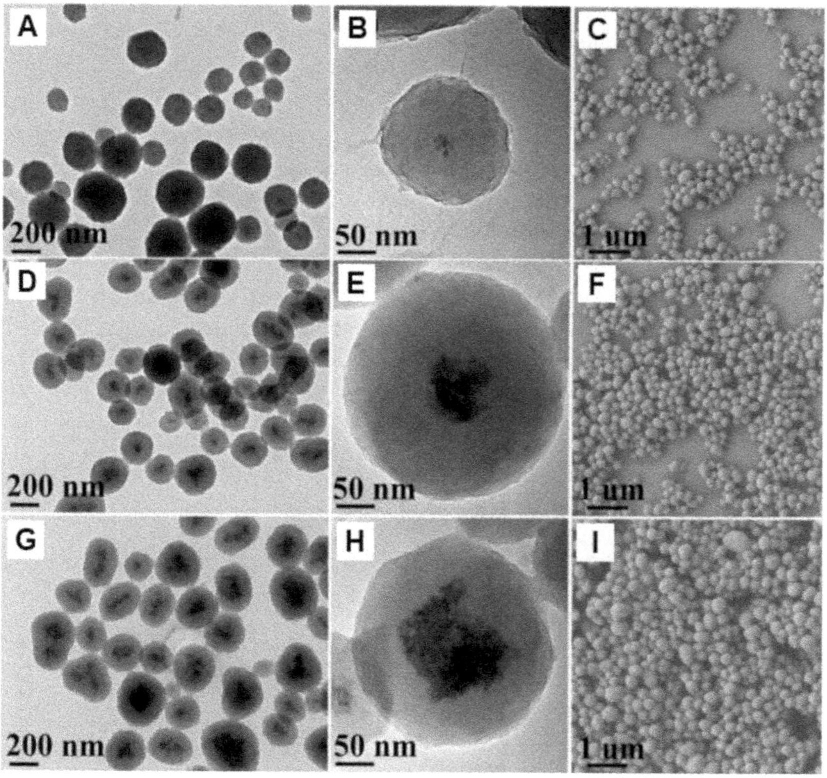

Figure 7. TEM images of PPPs particules loaded with (A,B) 5 mg Fe_3O_4, (D,E) 10 mg Fe_3O_4, and (G,H) 20 mg Fe_3O_4. FE-SEM images of PPPs particules loaded with (E) 5 mg Fe_3O_4, (F) 10 mg Fe_3O_4, and (I) 20 mg Fe_3O_4. (Adapted with permission from ref. (56). Copyright 2013 American Chemical Society.)

As we discussed in the last section, one of the most important PPPs is PCPP, which has been used for encapsulation, as a vaccines carrier, protein delivery and as an immunological adjuvant (*47*). Here we summarize the results on inorganic nanoparticle encapsulation in PCPP polymers and their application as contrast agents.

Imaging techniques such as PA need metal nanoparticles with surface plasmon resonance in the NIR. AuNP that are of relatively large size, such as rods or shells have absorbances in the NIR, but are too large for renal clearance. However, AuNP that are small enough for renal clearance (i.e. below 6 nm in diameter) do not absorb in the NIR and are not effective for imaging applications that require long circulation times or accumulation in disease sites. To address these problems, a biodegradable AuNP formulation was reported by Cheheltani and et al. (*57*) This platform utilizes small (<6 nm) AuNP that are incorporated into the biodegradable PCPP polymer. Over time, the biodegradable polymer will break down into harmless byproducts, while the small AuNP will be released for renal clearance. Other key features of this platform are the ability to control the size of these nanoparticles and to tune their absorbance into the NIR by altering the PCPP nanoparticle size and loading.

In this research, the AuNP were synthesized via reduction of gold chloride with sodium borohydride and capped with biocompatible ligands such as glutathione and 11-MUA to create AuNP that were stable in biological fluids. For synthesis of PCPP nanoparticles that were loaded with the afore-mentioned AuNP (termed Au-PCPP), a microfluidic device was used to mix PCPP and AuNP with spermine and PEG-PLL. Figure 8 shows the TEM of Au-PCPP that were synthesized using the microfluidic device. 0.5-5 mg Au payloads of glutathione coated AuNP could be incorporated per mg of polymer (Figure 8 A-C). Furthermore, the size of Au-PCPP nanoparticles can be tuned from 40-500 nm by controlling the relative amount of PEG-PLL to PCPP (Figure 8 D-F).

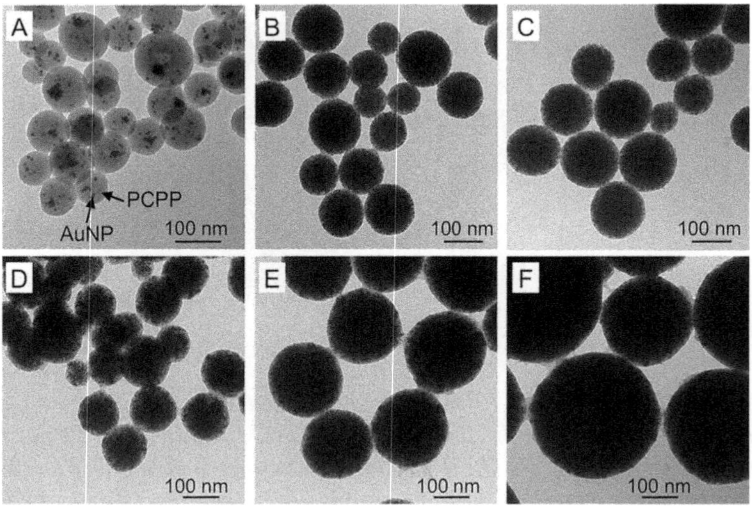

Figure 8. TEM of Au-PCPP that have either different AuNP loadings or different nanoparticle sizes. A-C) Au-PCPP loaded with glutathione coated AuNP at (A) 0.5, (B) 2.5 or (C) 5 mg Au per mg polymer. D-F) Au-PCPP loaded with 11-MUA coated AuNP and their size varied by inclusion of (D) 25, (E) 12.5 or (F) 6.25 μg PEG-PLL per mg PCPP. (Adapted with permission from ref. (57). Copyright 2016 Elsevier.)

The authors found that encapsulating AuNP in PCPP (PCPP-NP) resulted in a red shift of absorbance into the NIR, and that the absorbance could be tuned by varying the payload and nanoparticle size. For example, Figure 9A shows the UV/vis absorption spectra of free AuNP, Au-PCPP and PCPP-NP where the red shift of the AuNP plasmon peak into the near infrared by incorporation into PCPP nanoparticles can be seen. Since light in the NIR region (e.g., 650-900 nm) can penetrate soft tissues relatively well, Au-PCPP could be a good candidate for bioapplications such as PA or photothermal ablation. The effect on cell viability of Au-PCPP nanoparticles incubated with hepatocytes (HepG2), endothelial cells (SVEC4-10EHR1) and renal cells (Renca) cell lines is shown in Figure 9B. Au-PCPP did not reduce cell viability under any condition tested, indicating the excellent biocompatibility of this formulation.

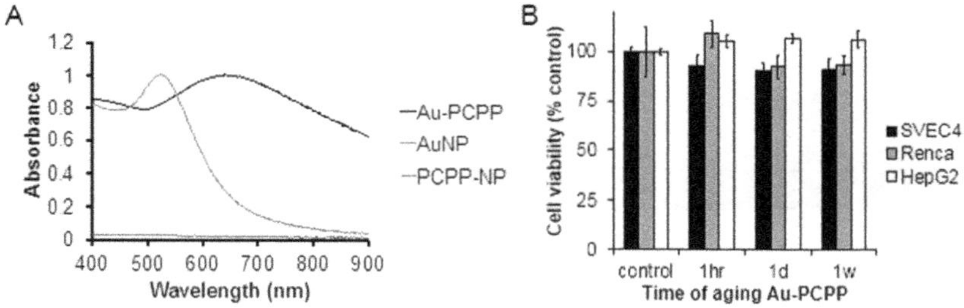

Figure 9. A) The optical extinction spectra of AuNP, Au-PCPP, and PCPP-NP. B) Effect of Au-PCPP on the viability of several cell lines. (Adapted with permission from ref. (57). Copyright 2016 Elsevier.)

Due to the promising properties of Au-PCPP, in vitro and in vivo experiments were carried out to assess the CT and PA contrast of these nanoparticles (Figure 10). CT imaging of phantoms containing these nanoparticles showed their attenuation rate to be 52.3 HU/mg (Figure 10A). Pre-injection and post-injection CT scans (0.75 mg Au in the thigh muscle of mice) are shown in Figure 10B. Strong CT contrast (4350 ± 1698 HU) was observed after injection of this small dose of Au-PCPP, which confirmed that the biocompatible Au-PCPP can be used as a CT contrast agent. These novel nanoparticles were also assessed for PA contrast properties. Samples of Au-PCPP at different concentrations were imaged in vitro using a laser frequency of 700 nm. Au-PCPP produced strong contrast due to their absorption in the NIR, however low signal was detected for encapsulated AuNP since their absorption at the laser wavelength was low (Figure 10C).

A low dose of Au-PCPP was injected in the thigh muscles of mice to assess in vivo PA contrast. PA images of the injected leg vs a non-injected leg are shown in Figure 10D and shows the PA signal produced by Au-PCPP. Overall, the data indicates that incorporating AuNP in PCPP nanoparticles created an excellent contrast agent. The synthesis of these nanoparticles is straightforward and tunable, they are potentially excretable, allow control over NIR absorption and can be used for both CT and PA imaging. These nanoparticles may have potential additional uses for applications such as radiosensitization or photothermal ablation.

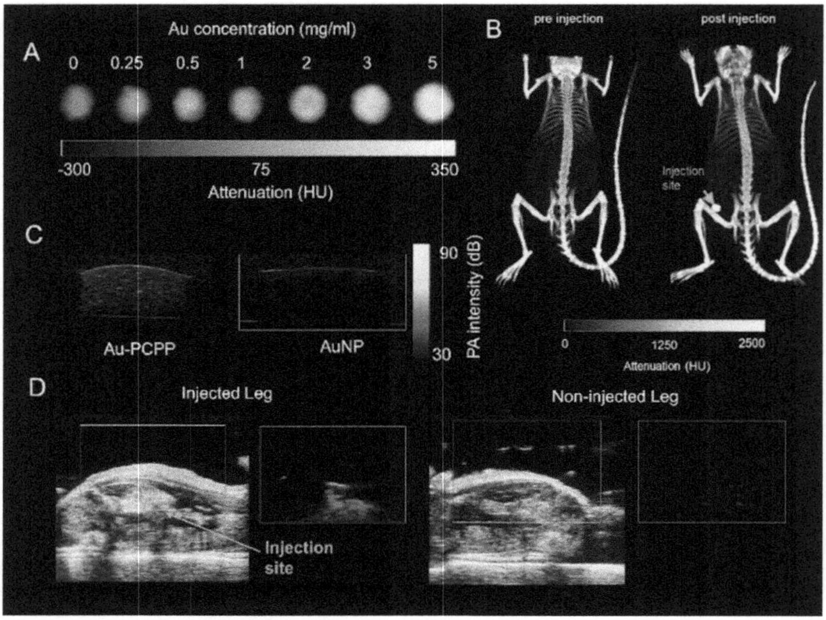

Figure 10. CT and PA imaging with Au-PCPP. A) CT images of Au-PCPP samples at different concentrations. B) The CT contrast of Au-PCPP before and after injection into the thigh muscle of a mouse. C) PA images of samples of Au-PCPP and AuNP. D) PA images of the leg of a mouse injected with Au-PCPP or uninjected. (Adapted with permission from ref. (57). Copyright 2016 Elsevier.)

We have also shown that PCPP nanoparticles (PCPP-NP) can be modified to include other types of inorganic nanoparticle payloads such as iron oxide nanoparticles (*58*). These different payloads allow this platform to be used as a contrast agent for other imaging modalities such as MRI or fluorescence imaging, as well as therapeutic applications. We reported the self-assembly of PCPP and nanocrystals that can be located on the surface of the polymer nanoparticle or in its core (Figure 11). The surface loading was carried out on nanocrystals that ranged in size, shape and composition, i.e., iron oxide, quantum dots (CdS), and nanophosphors (LF_3). Furthermore, surface loading could be done with PCPP polymers whose molecular weights ranged from 30 to 3000 kDa. The key to this surface loading was found to be coating the inorganic nanoparticles with phospholipids.

The PCPP-NP were synthesized in essentially the same method as described above. Using dimercaptosuccinic acid coated iron oxides (IO-DMSA) as a payload resulted in core loading of PCPP-NP, whereas MHPC micelle encapsulated iron oxide nanoparticles (IO-MHPC) preferentially adhered to the surface of PCPP-NP (Figure 11B and C). Moreover, when carboxylic acid coated AuNP and IO-MHPC were used in the synthesis at the same time, the IO-MHPC were localized to the surface and the AuNP were localized to the core of the particle (Figure 11D). This combination makes this platform a great candidate for dual imaging techniques such as MRI and CT since AuNP can result in strong CT attenuation, while the iron oxide can produce MRI contrast.

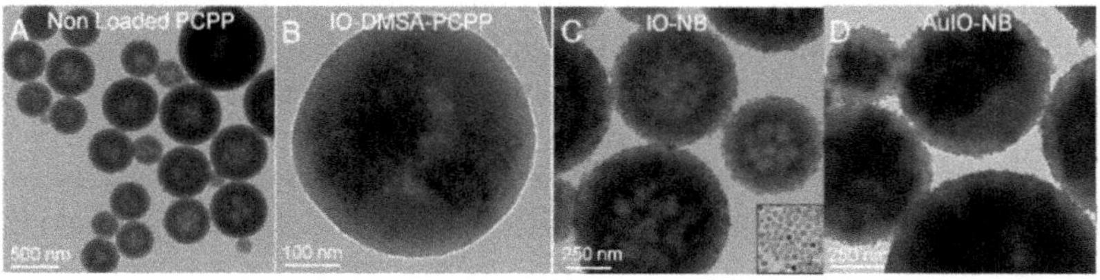

Figure 11. TEM of different PCPP nanospheres. A) Non-loaded PCPP nanospheres, B) IO-DMSA loaded PCPP nanospheres, C) IO-MHPC surface loaded PCPP, D) Loading of both IO-MHPC and AuNP into PCPP nanospheres. (Adapted with permission from ref. (58). Copyright 2014 American Chemical Society.)

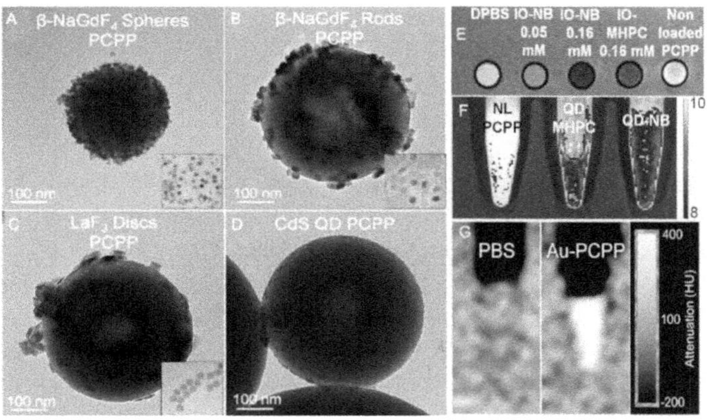

Figure 12. TEM images and applications of surface loaded PCPP nanospheres. A) β-$NaGdF_4$ spheres, B) β-$NaGdF_4$ rods, C) LaF_3 discs, and D) CdS quantum dots surface loaded onto PCPP nanospheres (QD-NB). E) MR image of a phantom that contains samples of IO-MHPC loaded PCPP-NP and controls, F) Fluorescence image of PCPP-NP samples, some of which were loaded with quantum dots, G) CT phantom image of AuNP loaded PCPP-NP. (Adapted with permission from ref. (58). Copyright 2014 American Chemical Society.)

Other diagnostically active nanocrystals were also loaded into PCPP-NP. TEM of oleic acid coated quantum dots (CdS spheres) and nanophosphors (β-$NaGdF_4$ spheres and rods and LaF_3 discs) incorporated into MHPC micelles and loaded into PCPP-NP are shown in Figure 12. This data shows that nanocrystals of a variety of sizes, shapes and compositions can be loaded into PCPP-NP. To evaluate the diagnostic properties of surface loaded PCPP nanoparticles, MRI and fluorescence phantom images were acquired. MRI phantom imaging showed that IO-MHPC loaded PCPP-NP resulted in signal reduction, or darkening in T_2

weighted MR images (Figure 12A) indicating the potential of these nanoparticles as a contrast agent for MRI. Furthermore, fluorescent imaging of samples of quantum dots coated with MHPC (QD-MHPC) and or QD-MHPC loaded onto PCPP-NP revealed similar signal production (Figure 12F). CT images displayed in Figure 12G indicate that AuNP loaded PCPP-NP generated significant contrast compared with PBS. These experiments confirmed the potential of PPPs in a range of medical imaging techniques.

In summary, PPPs are an attractive class of polymers for biomedical applications due to their diverse properties, namely versatility, degradability, biocompatibility and water solubility. Their diversity has prompted their exploration in numerous biomedical applications such as tissue engineering, biocompatible matrices for prosthetics, and drug delivery. Here, we described some of the novel applications of PPP carriers loaded with contrast generating materials for CT, PA, MRI and fluorescence imaging techniques. These polymers provided for valuable characteristics such as control over nanocrystal spatial distribution, degradation and tuning NIR absorption. We expect to see further novel developments of PPP based carriers for contrast agents in the coming years, as well as additional uses as therapeutics.

References

1. Cuenca, A. G.; Jiang, H.; Hochwald, S. N.; Delano, M.; Cance, W. G. *Cancer* **2006**, *107*, 459–466.
2. Daniel, M. C.; Astruk, D. *Chem. Rev.* **2004**, *104*, 293–346.
3. Cormode, D. P.; Sanchez-Gaytan, B. L.; Mieszawska, A. J.; Fayad, Z. A.; Mulder, W. J. *NMR Biomed.* **2013**, *7*, 766–780.
4. Cormode, D. P.; Naha, P. C.; Fayad, Z. A. *Contrast Media Mol. Imaging* **2014**, *9*, 37–52.
5. Yoon, S. J.; Mallidi, S.; Tam, J. M.; Tam, J. O.; Murthy, A.; Johnston, K. P.; Sokolov, K. V.; Emelianov, S. Y. *Opt. Lett.* **2010**, *35*, 3751–3753.
6. Cormode, D. P.; Roessl, E.; Thran, A.; Skajaa, T.; Gordon, R. E.; Schlomka, J.-P.; Fuster, V.; Fisher, E. A.; Mulder, W. J. M.; Proksa, R.; Fayad, Z. A. *Radiology.* **2010**, *256*, 774–782.
7. Thaxton, C. S.; Rink, J. S.; Naha, P. C.; Cormode, D. P. *Adv. Drug Delivery Rev.* **2016**, *106*, 116–131.
8. Jarzyna, P. A.; Skajaa, T.; Gianella, A.; Cormode, D. P.; Sabmer, D. D.; Dickson, S. D.; Chen, W.; Griffioen, A. W.; Fayad, Z. A.; Mulder, W. J. *Biomaterials* **2009**, *36*, 6947–6954.
9. Wilfert, S.; Iturmendi, A.; Schoefberger, W.; Kryeziu, K.; Heffeter, P.; Berger, W.; Brüggemann, O.; Teasdale, I. *J. Polym. Sci., Part A: Polym. Chem.* **2014**, *52*, 287–294.
10. Markovsky, E.; Baabur-Cohen, H.; Eldar-Boock, A.; Omer, L.; Tiram, G.; Ferber, S.; Ofek, P.; Polyak, D.; Scomparin, A.; Satchi-Fainaro, R. *J. Controlled Release* **2012**, *161*, 446–460.
11. Duncan, R.; Vicent, M. J. *Adv. Drug Delivery Rev.* **2010**, *62*, 272–282.
12. Kadajji, V. G.; Betageri, G. V. *Polymers* **2011**, *3*, 1972–2009.

13. Apple, D. J. *Sir Harold Ridley and His Fight for Sight*; SLACK Incorporated: Thorofare, NJ, 2006.
14. Naha, P. C.; Zaki, A. A.; Hecht, E.; Chorny, M.; Chhour, P.; Blankemeyer, E.; Yates, D. M.; Witschey, W. R.; Litt, H. I.; Tsourkas, A.; Cormode, D. P. *J. Mater. Chem. B* **2014**, *46*, 8239–8248.
15. Conde, J.; Doria, G.; Baptista, P. *J. Drug Delivery* **2012**, *2012*, 751075.
16. Watson, B. M.; Kasper, F. K.; Mikos, A. G. *Biomed. Mater.* **2014**, *2*, 025014–025024.
17. Luo, Q.; Hickey, R. J.; Park, S. J. *ACS Macro Lett.*. **2013**, *2*, 107–111.
18. Yang, Y.; Zhang, Z.; Chen, L.; Gu, W.; Li, Y. *Biomacromolecules* **2010**, *4*, 927–933.
19. Ullah, R. S.; Wang, L.; Yu, H.; Abbasi, N. M.; Akram, M.; Saleem, M.; Haroon, M.; Khan, R. U. *RSC Adv.* **2017**, *7*, 23363–23391.
20. Singh, A.; Krogman, N. R.; Sethuraman, S.; Nair, L. S.; Sturgeon, J. L.; Brown, P. W.; Laurencin, C. T.; Allcock, H. R. *Biomacromolecules* **2006**, *7*, 914–918.
21. Lee, N.; Choi, S. H.; Hyeon, T. *Adv. Mater.* **2013**, *25*, 2641–2660.
22. Baptista, P. V. *Expert Rev. Mol. Diagn.* **2012**, *12*, 541–543.
23. Cormode, D. P.; Naha, P. C.; Fayad, Z. A. *Contrast Media Mol. Imaging* **2014**, *9*, 37–52.
24. Solomon, R.; Dumouchel, W. *Invest. Radiol.* **2006**, *8*, 651–660.
25. Stacul, F.; van der Molen, A. J.; Reimer, P.; Webb, J. A.; Thomsen, H. S.; Morcos, S. K.; Almén, T.; Aspelin, P.; Bellin, M. F.; Clement, O.; Heinz-Peer, G. Contrast media safety committee of european society of urogenital radiology (ESUR). *Eur. Radiol.* **2011**, *12*, 2527–2541.
26. Cormode, D. P.; Naha, P. C.; Fayad, Z. A. *Contrast Media Mol. Imaging* **2014**, *1*, 37–52.
27. Cormode, D. P.; Skajaa, T.; van Schooneveld, M. M.; Koole, R.; Jarzyna, P.; Lobatto, M. E.; Calcagno, C.; Barazza, A.; Gordon, R. E.; Zanzonico, P.; Fisher, E. A.; Fayad, Z. A.; Mulder, W. J. *Nano Lett.* **2008**, *11*, 3715–3723.
28. Kim, J.; Chhour, P.; Hsu, J.; Litt, H. I.; Ferrari, V. A.; Popovtzer, R.; Cormode, D. P. *Bioconjugate Chem.* **2017**, *6*, 1581–1597.
29. Perera, V. S.; Hao, J. H.; Gao, M.; Gough, M.; Zayalij, P. Y.; Flask, C.; Basilion, J. P.; Huang, S. D. *Inorg. Chem.* **2011**, *50*, 7910–7912.
30. Favazza, C. P.; Wang, L. V.; Jassim, O.; Cornelius, L. A. *J. Biomed. Opt.* **2011**, *16*, 161–166.
31. Wu, D.; Huang, L.; Jiang, M. S.; Jiang, H. *Int. J. Mol. Sci.* **2014**, *15*, 23616–23639.
32. Murphy, C. J.; Sau, T. K.; Gole, A. M.; Orendorff, C. J.; Gao, J.; Gou, L.; Hunyadi, S. E.; Li, T. *J. Phys. Chem. B* **2005**, *109*, 13857–13870.
33. Chang, S. S.; Shih, C. W.; Chen, C. D.; Lai, W. C.; Wang, C. C. *Langmuir* **1999**, *15*, 701–709.
34. Chen, Y. S.; Frey, W.; Kim, S. *Opt. Exp.* **2010**, *18*, 8867–8878.
35. Haugland, R. P. *Handbook of Fluorescent Probes and Research Chemicals*; Molecular Probes, Inc.: Eugene, OR, 1996; pp 1–4.
36. Predina, J. D.; Newton, A.; Despande, C.; Low, P.; Singhal, S. *J. Biomed. Opt.* **2018**, *1*, 1–4.

37. Jobsis, F. F. *Science* **1977**, *198*, 1264–1267.
38. Tatsuta, M.; Iishi, H.; Ichii, M.; Baba, M.; Yamamoto, R.; Okuda, S.; Kikuchi, K. *Lasers Surg. Med.* **1989**, *9*, 422–426.
39. Mieszawska, A. J.; Kim, Y.; Gianella, A.; van Rooy, I.; Priem, B.; Labarre, M. P.; Ozcan, C.; Cormode, D. P.; Petrov, A.; Langer, R.; Farokhzad, O. C. *Bioconjugate Chem.* **2013**, *24*, 1429–1434.
40. Gao, X.; Cui, Y.; Levenson, R. M.; Chung, L. W.; Nie, S. *Nat. Biotechnol.* **2004**, *8*, 969–976.
41. Larson, D. R.; Zipfel, W. R.; Williams, R. M.; Clark, S. W.; Bruchez, M. P.; Wise, F.; Webb, W. W. *Science* **2003**, *5624*, 1434–1436.
42. Shahbazi-Gahrouei, D.; Williams, M.; Rizvi, S.; Allen, B. J. *J. Magn. Reson. Imaging* **2001**, *14*, 169–174.
43. Alric, C.; Taleb, J.; Le Duc, G.; Mandon, C.; Billotey, C.; Le Meur-Herland, A.; Brochard, T.; Vocanson, F.; Janier, M.; Perriat, P.; Roux, S. *J. Am. Chem. Soc.* **2008**, *130*, 5908–5915.
44. Schellenberger, E. A.; Bogdanov, A. Jr.; Högemann, D.; Tait, J.; Weissleder, R.; Josephson, L. *Mol. Imaging* **2002**, *2*, 102–107.
45. Reichardt, W.; Hu-Lowe, D.; Torres, D.; Weissleder, R.; Bogdanov, A. Jr. *Neoplasia* **2005**, *9*, 847–853.
46. Payne, L. G.; Andrianov, A. K. *Adv. Drug Delivery Rev.* **1998**, *31*, 185–196.
47. Andrianov, A. K.; Cohen, S.; Visscher, K. B.; Payne, L. G.; Allcock, H. R.; Langer, R. *J. Controlled Release* **1993**, *27*, 69–77.
48. Rothemund, S.; Teasdale, I. *Chem. Soc. Rev.* **2016**, *45*, 5200–5215.
49. Allcock, H. R. In *Chemistry and Applications of Polyphosphazenes*; Wiley: Hoboken, NJ, 2002; pp 1−725.
50. Allcock, H. R.; Morozowich, N. L. *Polym. Chem.* **2012**, *3*, 578–590.
51. Allcock, H. R. *Curr. Opin. Solid State Mater. Sci.* **2006**, *10*, 231–240.
52. Ullah, R. S.; Wang, L.; Yu, H. J.; Abbasi, N. M.; Akram, M.; Zain-ul-Abdin; Saleem, M.; Haroon, M.; Khan, R. U. *RSC Adv.* **2017**, *7*, 23363–23391.
53. Eng, N. F.; Garlapati, S.; Gerdts, V.; Potter, A.; Babiuk, L. A.; Mutwiri, G. K. *Curr. Drug Delivery* **2010**, *7*, 13–20.
54. Li, Y. L.; Rodrigues, J.; Tomas, H. *Chem. Soc. Rev.* **2012**, *41*, 2193–2221.
55. Ogueri, K. S.; Ivirico, J. L. E.; Nair, L. S.; Allcock, H. R.; Laurencin, C. T. *Regenerative Eng. Transl. Med.* **2017**, *3*, 15–31.
56. Hu, Y.; Meng, L.; Niu, L.; Lu, Q. *Langmuir* **2013**, *29*, 9156–9163.
57. Cheheltani, R.; Ezzibdeh, R. M.; Chhour, P.; Pulaparthi, K.; Kim, J.; Jurcova, M. M.; Hsu, J. C.; Blundell, C.; Litt, H. I.; Ferrari, V. A.; Allcock, H. R. *Biomaterials* **2016**, *102*, 87–97.
58. Chhour, P.; Gallo, N.; Cheheltani, R.; Williams, D.; Al-Zaki, A.; Paik, T.; Nichol, J. L.; Tian, Z.; Naha, P. C.; Witschey, W. R.; Allcock, H. R. *ACS Nano.* **2014**, *8*, 9143–9153.
59. Allcock, H. R.; Chester, A. C.; Morrissey, C. T.; Nelson, J. M.; Reeves, S. D. *Macromolecules* **1996**, *29*, 7740–7747.
60. Teasdale, I.; Wilfert, S.; Nischang, I.; Brüggermann, O. *Polym. Chem.* **2011**, *2*, 828–834.
61. Henke, H.; Wilfert, S.; Iturmendi, A.; Brüggemann, O.; Teasdale, I. *J. Polym. Sci., Part A: Polym. Chem.* **2013**, *51*, 4467–4473.

62. Allcock, H. R. Current Status of Polyphosphazene Chemistry. In *Inorganic and Organometallic Polymers: Macromolecules Containing Silicon, Phosphorus, and Other Inorganic Elements*; Zaldin, M., Wynne, K. J., Allcock, H. R., Eds.; ACS Symposium Series 360; American Chemical Society: Washington, DC, 1988; pp 250−267.
63. Mark, J. E.; Allcock, H. R.; West, R. In *Inorganic Polymers*; Prentice Hall: Englewood Cliffs, NJ, 1992; Vol.1; pp 1−272.
64. Teasdale, I.; Waser, M.; Wilfert, S.; Falk, H.; Brüggemann, O. *Monatsh. Chem. Chem. Mon.* **2012**, *143*, 355–360.
65. Zhicheng, T.; Hess, A.; Fellin, C. R.; Nulwala, H.; Allcock, H. R. *Macromolecules* **2015**, *48*, 4301–4311.
66. Mueller, A. H. E., Matyjaszewski, K. In *Controlled and Living Polymerizations: From Mechanisms to Applications*; Wiley-VCH: Weinheim, 2009; Vol. 1; pp 1−612.
67. Fox, M. E.; Szoka, F. C.; Frechet, J. M. J. *Acc. Chem. Res.* **2009**, *42*, 1141–1151.
68. Reichert, S.; Welker, P.; Calderon, M.; Khandare, J.; Mangoldt, D.; Licha, K.; Kainthan, R. K.; Brooks, D. E.; Haag, R. *Small* **2011**, *7*, 820–829.
69. Frechet, J. M. J.; Hawker, C. J.; Wooley, K. Patent WO 1993021259, 1991.
70. Andrianov, A. K.; Decollibus, D. P.; Marin, A.; Webb, A.; Griffin, Y.; Webby, R. J. *J. Pharm. Sci.* **2011**, *4*, 1436–1443.
71. Cayatte, C.; Marin, A.; Rajani, G. M.; Schneider-Ohrum, K.; Bennett, A. S.; Marshall, J. D.; Andrianov, A. K. *Mol. Pharm.* **2017**, *7*, 2285–2293.
72. Kumbar, S. G.; Bhattacharyya, S.; Nukavarapu, S. P.; Khan, Y. M.; Nair, L. S.; Laurencin, C. T. *J. Inorg. Organomet. Polym. Mater.* **2006**, *16*, 365–385.
73. Cayatte, C.; Marin, A.; Rajani, G. M.; Schneider-Ohrum, K.; Bennett, A. S.; Marshall, J. D.; Andrianov, A. K. *Mol. Pharm.* **2017**, *14*, 2285–2293.
74. Deng, Y.; Zhang, S.; Lu, G.; Huang, X. *Polym. Chem.* **2013**, *4*, 1289–1299.
75. Allcock, H. R.; Powell, E. S.; Maher, A. E.; Prange, R. L.; de Denus, C. R. *Macromolecules* **2004**, *37*, 3635–3641.
76. Liu, X.; Tian, Z.; Chen, C.; Allcock, H. R. *Macromolecules* **2012**, *45*, 1417–1426.
77. Henke, H.; Posch, S.; Brüggemann, O.; Teasdale, I. *Macromol. Rapid Commun.* **2016**, *37*, 769–774.
78. Suárez-Suárez, S.; Carriedo, G. A.; Presa Soto, A. *Chem. Eur. J.* **2015**, *21*, 14129–14139.
79. Yashima, E.; Maeda, K.; Yamanaka, T. *J. Am. Chem. Soc.* **2000**, *122*, 7813–7814.
80. Zheng, C.; Qui, L.; Zhu, K. *Polymer* **2009**, *50*, 1173–1177.
81. Fu, J.; Chen, J.; Chen, Z.; Xu, Q.; Huang, X.; Tang, X. *New J. Chem.* **2012**, *34*, 599–602.
82. Wang, X.; Fu, J.; Chen, Z.; Li, Q.; Wu, X.; Xu, Q. *RSC Adv.* **2015**, *5*, 33720–33728.
83. Wang, M.; Fu, J.; Huang, D.; Zhang, C.; Xu, Q. *Nanoscale* **2013**, *5*, 7913–7919.
84. Wang, M.; Fu, J.; Chen, Z.; Wang, X.; Xu, Q. *Mater. Lett.* **2015**, *143*, 201–204.

85. Wang, X.; Wang, M.; Fu, J.; Zhang, C.; Xu, Q. *J. Nanopart. Res.* **2013**, *15*, 1853–1863.
86. Yan, Y.; Fu, J.; Wang, M.; Liu, S.; Xin, Q.; Chen, Z.; Xu, Q. *RSC Adv.* **2016**, *6*, 24921–24928.
87. Hindlang, M. D.; Soudakov, A. A.; Imler, G. H.; Laurencin, C. T.; Nair, L. S.; Allcock, H. R. *Polym. Chem.* **2010**, *1*, 1467–1474.

Chapter 5

Fluorinated Polyphosphazene Coatings Using Aqueous Nano-Assembly of Polyphosphazene Polyelectrolytes

Victoria Albright,[1] Victor Selin,[1] Hanna Hlushko,[1] Anbazhagan Palanisamy,[1] Alexander Marin,[2] Alexander K. Andrianov,[*,2] and Svetlana A. Sukhishvili[*,1]

[1]Department of Materials Science & Engineering, Texas A&M University, College Station, Texas 77843, United States
[2]Institute for Bioscience and Biotechnology Research, University of Maryland, Rockville, Maryland 20850, United States
[*]E-mails: aandrianov@ibbr.umd.edu (A.K.A.); svetlana@tamu.edu (S.A.S.).

Fluorinated polyphosphazenes, such as poly[bis(trifluoro ethoxy)phosphazene], PTFEP, attract significant interest as hydrophobic and biocompatible materials for industrial and life science applications. A novel class of fluorinated polyphosphazenes, which include ionic and fluorine containing moieties, is gaining recognition as a macromolecular system capable of nano-assembly into highly hydrophobic materials. The present chapter reviews synthetic approaches to ionic fluoropolymers with polyphosphazene backbones and summarizes their solution behavior. It then focuses on nano-assembly of fluorinated polyions using the layer-by-layer (LbL) deposition technique - a route to creating high-precision fluorinated surfaces for biomaterials applications. This approach enables highly controlled linear film growth with superior persistence of layering and affords a high degree of hydrophobicity to the resultant coating. Moreover, these nano-films display superb biocompatibility, as demonstrated by the hemolysis of whole blood and protein adsorption studies in vitro, and display the ability to self-heal at near physiological conditions. Finally, nano-assembled fluorinated coatings enable controlled encapsulation and release of bioactives. The multi-functionality of fluorinated polyphosphazene nano-films, along with their precision supramolecular architecture, suggests their broad potential applications as coatings for cardiovascular stents, catheters, and various implants.

Introduction

Polyphosphazenes with fluorinated organic side groups, which comprise some of the most hydrophobic polymers known (*1*, *2*), attract significant interest as materials for a variety of industrial and biomedical applications (*3–5*). In particular, poly[bis(trifluoroethoxy)phosphazene], PTFEP and its copolymers have been extensively studied as high-performance elastomers, membranes, and surface coatings (*2–8*). As with conventional fluorinated polymers, the high hydrophobicity of PTFEP is owed to fluorine-containing groups, such as the trifluoromethyl group, which tend to minimize the interfacial free energy (*9*). However, the inherent flexibility of the polyphosphazene backbone (*3*) may also play an important role in surface hydrophobicity by allowing the fluoroalkoxy side groups to orient toward the surface and dominate the interfacial properties of the polymer (*10*). This assumption is supported by very low glass transition temperatures of PTFEP (-66 °C) (*2*) and its mixed substituent copolymers (as low as -90 °C) (*4*, *5*). Hydrophobic properties of typical solution cast PTFEP surfaces are characterized with contact angles exceeding 100° (*10*). Superhydrophobic surface properties of PTFEP can be attained with rough surfaces, for example through forming electrospun nanofiber mats of PTFEP, which display contact angles as high as 159° (*10*). In the life sciences, PTFEP based coatings are recognized for their outstanding biocompatibility (*11*) and are important constituents of clinically-validated injectable microspheres (*12*, *13*) and medical devices (*14*, *15*).

PTFEP coating deposition is typically achieved via heat molding or solution casting (*2*, *11*, *16*, *17*), however, these methods allow only limited control over uniformity, thickness, and functionality of the deposited film (*18*). In contrast, the layer-by-layer (LbL) deposition technique enables coating substrates of a variety of shapes and surface chemistries with nanoscopically structured films (*19*, *20*). The capability of the LbL technique to generate conformal coatings with programmed thicknesses and diverse compositions has been widely used for functionalization of biomedical devices (*11*, *21*, *22*). Moreover, the use of water as a solvent (*23*) makes LbL assembly environmentally benign and aids in preserving the native structure and activity of biomolecules incorporated into films (*24*). While water solubility is a common feature of a wide range of ionic macromolecules, endowing this capability to fluorinated polyphosphazenes has been challenging.

The present chapter describes the synthesis and solution characteristics of water-soluble ionic polyphosphazenes containing hydrophobic fluorinated groups, investigates the feasibility of LbL deposition of fluorinated nanofilms, and explores some of the key features of these coatings.

Synthetic Approaches to Fluorinated Polyphosphazene Polyacids

Polyacids appear to be the only class of fluorinated polyelectrolytes with a phosphazene backbone synthesized to date (*25*). In general, synthetic routes to polyphosphazene polyacids involve macromolecular substitution (*3*) of reactive precursor - poly(dichlorophosphazene), PDCP, with appropriate nucleophiles (Figure 1). As acidic groups can potentially interfere with the process, which can lead to undesired cross-linking and degradation of the polymer, there is a need for their protection during the substitution reaction (*26–28*). For polymers containing carboxylic acid groups, such as poly[di(carboxylatophenoxy)phosphazene], PCPP (Figure 1, route "A"), this is usually achieved through covalent protection with subsequent hydrolysis of ester group under basic conditions (*26–29*). This pathway provides reliable control over molecular weight characteristics and was employed for the current Good Manufacturing Practice (cGMP) production of clinical grade PCPP (*27*, *30*).

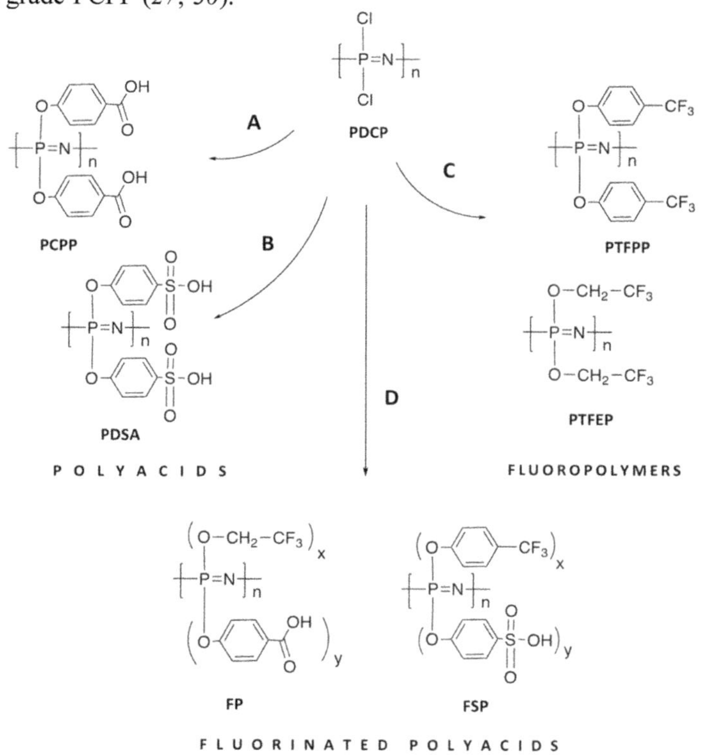

Figure 1. Synthetic pathways to polyphosphazene polyacids, fluorinated polyphosphazenes, and fluorinated polyphosphazene polyacids.

Sulfonic acid derivatives, such as poly[diphenoxyphosphazenedisulfonic acid] (PDSA) (Figure 1, route "B"), can be prepared by direct "post-substitution" sulfonation of polyphosphazenes containing phenoxy side groups (*31*, *32*), however, this may not be practical when precise control over the composition is required or for mixed substituent copolymers. Alternatively, alkane sulfonation can be carried out through the multi-step process of synthesizing polyaminophosphazenes and then reacting them with propanesultone, yet this results in zwitterionic polymers (*33*). A direct pathway to PDSA was developed by reacting PDCP with the corresponding nucleophile converted to a hydrophobic alkylammonium salt (*28*). The technique was successfully applied to the synthesis of both homopolymers and mixed substituent polyphosphazenes (*28*). This "non-covalent protection" approach resulted in a high-level control over the polymer composition (*28*).

Non-ionic fluorinated polyphosphazenes (Figure 1, route "C") can be generally synthesized through a single-step macromolecular substitution reaction with corresponding alkoxides (PTFEP) or phenoxides (PTFPP) (*6*). Their synthesis and properties are the subject of several comprehensive reviews (*3*, *5*). Synthesis of ionic fluoropolymers with phosphazene backbone was first reported for copolymers of PCPP and PTFEP (Figure 1, copolymers FPs, route "D") (*25*). Polyphosphazenes containing 60 (FP60), 86 (FP86), and 97 (FP97) % (mol.) of fluorinated groups were prepared through sequential addition of propyl paraben and sodium trifluoroethoxide, followed by alkaline hydrolysis of ester moieties. The composition of copolymers was controlled through the amount of ester-containing nucleophile, which reacted with PDCP almost quantitatively (*25*). Fluorinated polyphosphazenes containing sulfonic acid groups (FSPs) were also synthesized using hydrophobic ammonium salts of hydroxybenzenesulfonic acid and sodium 4-(trifluoromethyl)phenoxide (*34*).

Positively charged polyphosphazene containing amines in their side groups have been synthesized by macromolecular substitution reaction (*35–45*). However, synthesis of their fluorinated derivatives is yet to be accomplished.

Solution Behavior of Fluorinated Polyphosphazene Polyacids

Some important solution properties of fluorinated polyacids are summarized in Table 1. Remarkably, the sodium salt of polyphosphazene polyacid containing 60% (mol.) of hydrophobic trifluoroethoxy groups (FP60) was fully soluble in aqueous buffers at pH 7.4 (physiological pH). Moreover, even polyphosphazene with 97% (mol.) of fluorinated groups was soluble in water-ethanol mixtures. It is tempting to speculate that the hydrophilic backbone, mainly due to the presence of the nitrogen lone pair electrons, contributed to this rather unexpected water-solubility. These results suggest that fluorinated polyacids may be compatible with aqueous based coating techniques, such as LbL polyelectrolyte assembly.

Environmental sensitivity of fluorinated polyacids, which was evaluated through their ability to undergo phase separation in the presence of various salts, generally followed those of ionic homopolymer, PCPP (Table 1). PCPP displays calcium-induced gelation also known as "egg-box" cross-linking, which is commonly observed for ionic polysaccharides like alginate and pectin (*26, 46–48*). Importantly, adding fluorinated moieties (as in FP60) does not prevent this gelation. Moreover, PCPP and all fluorinated derivatives can be self-assembled into nanoparticulates in the presence of multivalent cations, such as spermine (*49*). An interesting difference is the lack of "salting out" effect in the presence of sodium chloride for polymers with a higher content of hydrophobe (FP86 and FP97). Moreover, even FP60, despite the presence of hydrophobic groups, which can be expected to facilitate phase separation, demonstrated a higher threshold of precipitation with sodium chloride compared to PCPP (*25*). As was discussed previously, steric factors may play an important role in condensation of sodium ions with PCPP and introduction of substituents in this polyacid, may result in interference with this process (*27*).

Table 1. Solubility and Ionic Sensitivity of Fluorinated Polyphosphazene Polyacids

Polymers	*PCPP*	*FP20*	*FP60*	*FP86*	*FP97*
COOH, % (mol.)	*100*	*80*	*40*	*14*	*3*
Solubility					
Aqueous Buffer, pH 7.4	+	+	+	-	-
Water-Ethanol Mixtures	+	+	+	+	+
Salt Induced Phase Separation					
Sodium Chloride	+	+	+	-	-
Calcium Chloride	+	+	+	+	+
Spermine Chloride	+	+	+	+	+

To assess the potential of ionic, fluorinated polyphosphazenes as water-soluble polymers, the initial evaluation of hydrophobic-hydrophilic properties of amphiphilic FPs was conducted through the assessment of their water-sorption characteristics – a commonly used technique for polymeric materials (*50, 51*). To that end, water sorption within solution-cast films of acidic FP60 exposed to air saturated with water vapors was similar to PTFEP films (Figure 2a), which emphasizes the hydrophobic nature of FP60 (*25*). Figure 2a also shows that sodium salt form of the same polymer (FP60) displays strikingly larger water uptake, resulting in 144 % (w/w) film swelling versus 10 % swelling of PTFEP. This result suggests that water uptake in fluorinated polyphosphazenes is driven by ionization of carboxylic acid groups. Figure 2b shows that copolymers containing 86% and 97% of trifluoroethoxy groups displayed little uptake of water even in their sodium salt form.

Therefore, fluorinated ionic polyphosphazenes display the desired solubility in aqueous solutions and maintain some of the key functionalities of polyelectrolytes, such as ion complexation and inotropic gelation. Nevertheless, when converted into a non-ionized form, these polymers display highly hydrophobic behavior characteristic of this class of fluorinated polymers. For those reasons, they may constitute an exceptional macromolecular system for controlled LbL nano-assembly of hydrophobic fluorinated surfaces from aqueous solutions.

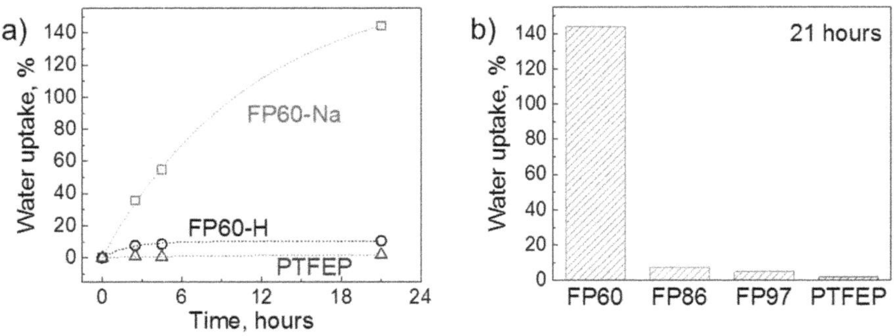

Figure 2. Water-sorption of polyphosphazenes spun cast on glass substrates: (a) of PTFEP, FP60 in sodium salt (FP60-Na) and acid form (FP60-H) versus time and (b) of PTFEP, and fluorinated polyions: FP60, FP86, and FP97 in their sodium salt form after 21 hours (saturated water vapor, 22°C).

Fluorinated Nano-Coatings

LbL Polymer/Polymer Assemblies

As discussed above, thus far, fluorinated polyphosphazenes containing carboxylic and sulfonic groups have been successfully synthesized, both of which can provide negative charges to utilize for electrostatic LbL deposition with polycations. Electrostatic deposition of multilayers is controlled by many factors including the concentration of salt in deposition solution (*52*), charge density of polymers (*53*, *54*), and other secondary interactions such as hydrophobicity and hydrogen bonding (*55*). In general, it is accepted that there are two modes of growth of LbL films, linear, in which polymers are highly layered due to tight binding, and supralinear or exponential, in which polymer "layers" are highly intermixed due to weak interactions. Mode of growth can be regulated by a large variety of factors including deposition pH (*56*), layer number (*57*) or deposition method (spin- vs. dip-assisted) (*58*), among others. Below, the assembly and properties of LbL assemblies of carboxylic acid containing, fluorinated polyphosphazenes are discussed. LbL assemblies with sulfonic acid derivatives of fluorinated polyphosphazenes remain unexplored. However, the differences in LbL assemblies due to changing interaction strength (carboxylic vs. sulfonic acid) will be interesting to study in future work.

Effects of Fluorination Degree on Assembly Thickness and Water Uptake

PCPP is a weak polyelectrolyte that is almost fully ionized in physiological conditions (*59*) and thus can easily be used for LbL deposition. Interestingly, fluorinated polyphosphazenes with 86 % (mol) fluorine-containing side groups and 14% carboxylic acid moieties have sufficient charge density at physiological pH (pH 7.4) to enable layer-by-layer binding (*60*). Ionic fluorinated and non-fluorinated polyphosphazenes are able to form layers with a variety of polycations including synthetic polymers, such as branched polyethyleneimine (BPEI), and quaternized poly(2-(dimethylamino)ethyl methacrylate) and biodegradable polymers including poly-(L)-lysine (*60*). Regardless of polycation partner, the degree of fluorination of ionic polyphosphazenes had a dramatic effect on film growth. Non-fluorinated PCPP formed thicker assemblies with BPEI (M_w = 25 kDa) compared to layers of highly fluorinated counterparts (FP60 and FP86) with BPEI, as determined by ellipsometry of dry LbL films (Figure 3). For the highly fluorinated polyelectrolytes, the increments of film thickness increase were similar (3.9 to 3.8 nm per bilayer for FP60 and FP86, respectively), but twice thicker films were formed with non-fluorinated PCPP. Such dramatic differences in film thickness per bilayer suggest the largely different strength of polymer-polymer interactions between assemblies of non-fluorinated and fluorinated polyphosphazenes, which is attributed to the likely enhancement of electrostatic pairing by hydrophobic groups of fluorinated polyions – FP60 and FP86. When deposition of BPEI with a polyphosphazene that has a low degree of fluorination (FP20) is explored, exponential growth similar to that of PCPP/BPEI is observed. This suggests that a certain threshold exists for fluorination degree to induce film growth to transition from linear to exponential. Importantly, this trend of decreasing film thickness with increasing fluorination degree suggests the ability to control film growth and strength of interactions within the film through film component selection.

The strength of polymer-polymer interactions often dictates film properties/behavior. For example, increased strength of ionic pairing has been shown to enhance stability of electrostatic films in salt solutions (*58*). The strength of polymer-polymer interactions exhibits itself in the degree of film swelling. PCPP/BPEI films swelled significantly in water (~45% film thickness increase due to water uptake), while water content in FP86/BPEI multilayers was extremely low, and did not exceed ~6% of their mass when exposed to phosphate buffer at pH 7.4 (Figure 3b). Weaker bound PCPP/BPEI multilayers formed 'fluffier' layers, which were easily swollen in water due to the intrinsic hydrophilicity of the unbound polymer units, the osmotic pressure created by the counterions, and the electrostatic repulsion between excessive charges in the loops. In contrast, hydrophobic FP60 and F86 strongly contributed to expulsion of water from the film by creating a low-dielectric-constant environment (*61*).

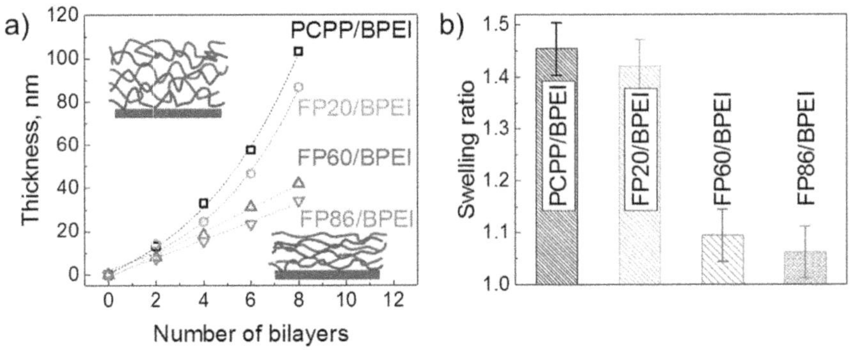

Figure 3. Deposition of LbL films with polyphosphazenes of varied hydrophilicity/hydrophobicity on silicon substrates (a), and comparison of LbL film swelling in 0.01 M phosphate buffer at pH 7.4 (b).

Film Hydrophobicity and Deposition on Various Substrates

For biomedical applications, coatings need to be deposited onto substrates of various chemistries and arbitrary shapes with control over surface hydrophobicity. However, controlling surface hydrophobicity is challenging due to the tendency of hydrophobic groups to reorient into films upon exposure to water (*62*). Surprisingly, fluorinated multilayers showed large oscillations of water contact angle depending on capping layer (*60*) and fluorinated content. When a polyphosphazene layer was on top of a polyphosphazene/BPEI film, contact angles of ~95, 80, and 60° were observed for FP86, FP60, and FP20 assemblies, respectively. All of these values are much higher than the value (40°) observed for films capped with PCPP on silicon wafers. On the other hand, when BPEI was on top, contact angles dropped significantly to ~55° for multilayers containing fluorinated polyphosphazenes. By increasing the content of fluorinated side groups, the amplitude of contact angle oscillations can be controlled.

Importantly, fluorinated polyphosphazenes can be coated onto materials such as titanium and polyester felt, which are used commonly for orthopedic implants (*63*) and medical device applications (*64*), respectively. Similarly to those deposited on silicon substrates, FP-capped films coated on other substrates demonstrated high hydrophobicity. While the contact angle for an FP60/BPEI film (capped with FP60) deposited on a titanium foil substrate was similar to that for a film on a silicon wafer (~75°), the film on the porous surgical mesh had an enhanced value of contact angle (130°) as a result of the unique porous structure of the polyester surgical felt, which enhanced substrate roughness (Figure 4) (*65*). Thus, the developed coatings can be used to efficiently tune the wettability of surfaces.

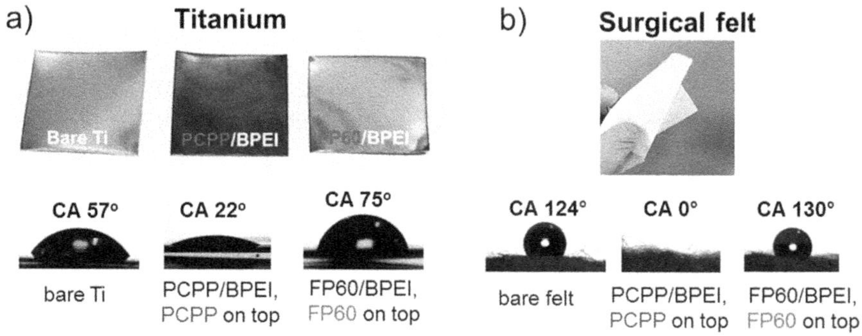

Figure 4. Images and water contact angles of LbL films of PCPP/BPEI and FP60/BPEI deposited on titanium foil (a) and polyester surgical felt (b).

Functional Properties of FP LbL Nanofilms: Biocompatibility and Self-Healing

Good biocompatibility is the first major prerequisite for materials intended for life science applications (*66–68*). To explore the potential of LbL nanocoatings of ionic fluorinated polyphosphazenes for applications in cardiovascular stents or catheters, the hemocompatibility of 200 nm films was evaluated using the common hemolysis test with dilute whole rabbit blood (*66*, *69*). Hemolysis %, which represents an increase in the content of free hemoglobin in blood after incubation with the tested material, was dependent on fluorination degree of polyphosphazene (Figure 5a). Importantly, LbL films containing FP86 demonstrated hemolysis extent comparable to that of a PTFEP coating. Hemocompatibility of clinically validated PTFEP (*13–15*) has been correlated with its ability to selectively and irreversibly adsorb human serum albumin (HSA) from blood plasma (*11*). Moreover, irreversible adsorption of HSA has been previously related to passivation of surfaces against platelet adhesion (*11*). Selective adsorption of HSA from human plasma onto the polyphosphazene-containing multilayers was dramatically increased on FP/BPEI multilayers. Importantly, these coatings outperform not only non-fluorinated PCPP coatings but also nonionic PTFEP coatings. Therefore, ultrathin LbL films of fluorinated polyphosphazenes constructed in aqueous solutions can potentially achieve similar to or even higher hemocompatibility than that observed for clinically validated solution-cast nonionic PTFEP coatings.

Self-healing is a property often displayed by living tissues but hard to achieve for biomaterials. Self-healing of bulk materials, such as hydrogels, has been commonly reported (*70–72*), however, healing of thin films remains challenging. Self-healing thin films of LbL of polyelectrolyte complexes (*73*) and LbL containing enzymes (*74*) have been reported, yet those approaches did not include biocompatible polyelectrolytes. The macroscopic self-healing of thin films arises from the mobility of polymer chains, which is influenced by a variety of factors including glass transition temperature, T_g. While non-ionic PTFEP is a well-known elastomer with a T_g of~-66 °C (*2*), the presence of ionic groups and their binding with polymeric counterparts increases glass transition temperature of the assemblies (*75*). Nevertheless, a ~195 nm thick FP20/BPEI film completely healed a penetrating scratch of 0.3-1 μm width and 25-80 nm depth made with an AFM tip after only two hours of exposure to PBS (pH 7.5, 0.15 M NaCl) (Figure 5b). This suggests that the polymer chain flexibility along with the presence of fluorinated moieties supports self-healing behavior of LbL assemblies of these polymers.

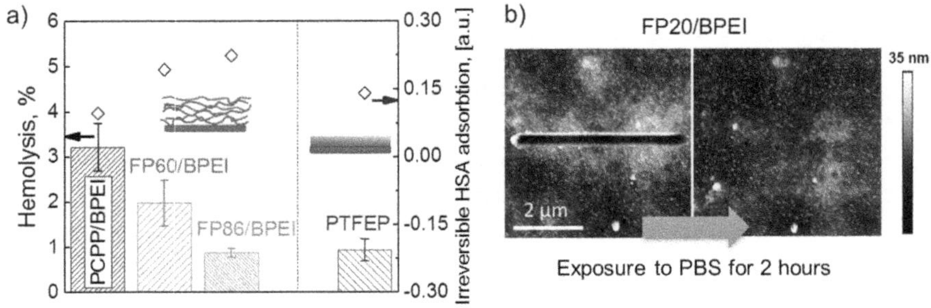

Figure 5. Functional properties of LbL films of fluorinated polyphosphazenes: (a) Hemolysis percentage of whole rabbit blood (columns, left axis) and irreversible adsorption of HSA from human plasma (dots, right axis) on FP/BPEI multilayers and control solution-cast PTFEP coatings; (b) AFM measurements of self-healing of a scratch (made by AFM tip) in a FP20/BPEI assembly in the presence of PBS for two hours at room temperature.

Therefore, fluorinated polyphosphazenes with ionic carboxylic groups can be assembled into thin films from aqueous solutions using the LbL technique. Film properties can be modulated by the degree of fluorination of polyphosphazene as well as by the capping layer of the film. These films display a high degree of hydrophobicity, are biocompatible, and are able to self-heal. The next section will explore LbL assembly of these same polymers with small molecules.

LbL Coatings with Small Molecules

While depositing polymers using the LbL technique is commonly successful because of strong multisite interactions between polymer partners, trapping small molecules within LbL films has been challenging due to the smaller number of binding sites between the small molecules and polymers. To combat this, unique strategies such as complexing small molecules directly with other small molecules (*76*) or depositing small molecules in tetralayers with three other polymers were explored (*77, 78*). The tetralayer strategy was suggested to work by a combination of electrostatic and non-electrostatic interactions (*77, 78*). In our work, fluorinated ionic polyphosphazenes exhibited high efficiency in binding and trapping small molecule cationic antibiotics within LbL nanocoatings, a unique capability that is likely due to the high degree of flexibility of polyphosphazene backbones. The use of ionic hydrophobic polyphosphazenes for trapping small molecules within LbL coatings was successful even in the case of gentamicin – a hydrophilic small antibiotic with only ~3.5 positive charges at pH 7.5 (*79*). One notable difference between polymer-polymer assemblies and coatings with small molecules is the thickness deposited per bilayer. While FP60/BPEI coatings displayed a thickness of ~ 4.8 nm per bilayer, FP60/gentamicin coatings displayed only a thickness of ~1.3 nm per bilayer. To confirm that gentamicin was in the coatings, x-ray photoelectron spectroscopy (XPS), a highly sensitive method that enables resolution of the chemical composition of a coating within a penetration depth of 3-10 nm, was used with the reference peak of carbon set to 284.9 eV. Comparing the nitrogen (N 1s) binding energies of a spun-cast monolayer of FP60 (Figure 6a) and a 20-bilayer FP60/gentamicin film (Figure 6b) clearly shows the emergence of two new peaks at ~400 and ~402 eV corresponding to $-NH_2$ and $-NH_3^+$, respectively, which confirms the presence of gentamicin in the coatings. Activity of gentamicin in the LbL coatings was tested via bacterial cultures of *Staphylococcus aureus,* a gram-positive strain known to form biofilms on biomaterial implants and devices (*80*). LbL-assemblies containing gentamicin were able to inhibit growth of ~3.5 x 10^3 CFU/cm^2 (Figure 6c).

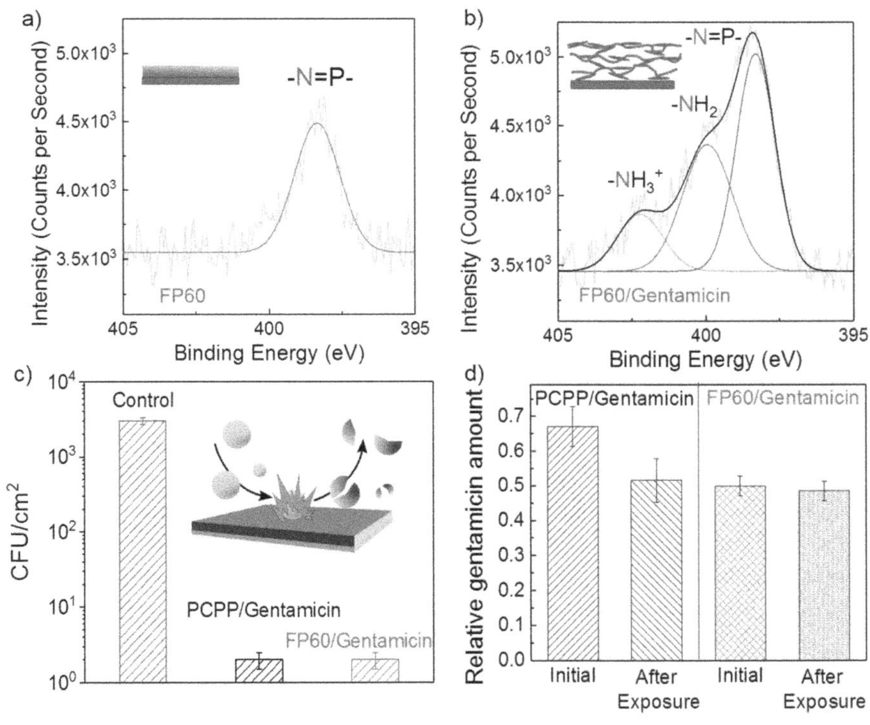

Figure 6. XPS analysis showing nitrogen (N 1s) binding energy in an FP60 monolayer (a) and in 20-bilayers of FP60/Gentamicin (b). (c) 35 nm PCPP/Gentamicin and FP60/Gentamicin films are able to prevent the growth of 10^3 colony forming units (CFU) of Staphylococcus aureus per cm^2. (d) Comparison of peak areas of nitrogen from XPS analysis reveals ~20% loss of gentamicin from PCPP/Gentamicin films after 1 hour exposure to PBS (pH 7.5, 0.15 M NaCl) while FP60/Gentamicin retains gentamicin.

One common problem with electrostatic trapping of small molecules in LbL coatings has been the rapid release of antibiotics on exposure to salt containing solutions (*81*, *82*). However, assemblies of gentamicin with fluorinated ionic polyphosphazenes were able to retain antibiotics after one-hour exposure to PBS (pH 7.5, 0.15 M NaCl), as demonstrated for 20-bilayer FP60/gentamicin coatings using XPS (Figure 6d). On the other hand, assemblies of gentamicin and non-fluorinated ionic polyphosphazene (PCPP) lost ~20% of gentamicin after exposure to the same conditions. Such a difference is likely due to the additional hydrophobic interactions brought by fluorinated groups, which enabled retention of antibiotics even in salt solutions. These data suggest a facile way to incorporate a large variety of cationic small molecules into LbL assemblies at physiological pH without loss of efficacy. This strategy is highly attractive as fluorinated polyphosphazenes have been shown to be biocompatible (*11*, *14*) and polyphosphazene polyelectrolytes can be engineered to degrade within a certain time frame (*83*, *84*). For example, to increase degradation rate, pendant pyrrolidone groups can be added to enhance hydrophilicity and biodegradation (*84*).

Conclusions

Polyphosphazenes with fluorine-containing side groups are fiber-, film- and elastomer forming materials, which display high hydrophobicity and superb biocompatibility. A novel class of fluorinated polyphosphazenes has been introduced, which incorporates ionic functionalities in their macromolecular structure. These mixed-substituent copolymers are water-soluble in their ionized salt form, but become highly hydrophobic when converted to a non-ionic state. Importantly, fluorinated copolymers maintain some of the most important features of polyelectrolytes, such as the ability to form complexes with ionic polymers of the opposite charge or produce ionotropic hydrogels through cross-linking with multivalent ions. The exceptional combination of properties of these macromolecules enables application of advanced aqueous-based methods for the deposition of highly hydrophobic and biocompatible fluorinated coatings. Here, the LbL technique was employed to engineer nanostructured films of fluorinated polyphosphazene polyelectrolytes. The resulting hydrophobic coatings displayed in vitro biocompatibility, which was comparable or better than biocompatibility of a clinically validated fluorinated polyphosphazene - PTFEP. Yet, LbL films of polyphosphazenes display significant advantages over nonionic PTFEP coatings in terms of both environmentally friendly processing conditions and multifunctionality of the resultant films. Specifically, LbL films have easily controlled film thickness and hydrophobicity, can easily be deposited on a wide range of substrates, and can be fabricated to include bioactive molecules, including bactericidal agents. Moreover, nano-layered fluorinated coatings demonstrated the ability to self-heal at near physiological conditions. The versatility of fluorinated polyelectrolytes suggests a wide spectrum of potential applications in life sciences ranging from biocompatible nanoparticle delivery systems to coatings for coronary stents and various implants.

References

1. Allcock, H. R.; Steely, L. B.; Singh, A. Hydrophobic and superhydrophobic surfaces from polyphosphazenes. *Polym. Int.* **2006**, *55*, 621–625.
2. Allcock, H. R.; Steely, L.; Singh, A.; Hindenlang, M. Hydrophobic and superhydrophobic polyphosphazenes. *J. Adhes. Sci. Technol.* **2009**, *23*, 435–445.
3. Allcock, H. R. *Chemistry and Applications of Polyphosphazenes*; Wiley: Hoboken, NJ, 2002; pp 725.
4. Gleria, M.; Bertani, R.; De Jaeger, R. Fluorinated polyphosphazenes: A survey. *J. Inorg. Organomet. Polym.* **2004**, *14*, 1–28.
5. Gleria, M.; Bertani, R.; Jaeger, R. D.; Lora, S. Fluorine containing phosphazene polymers. *J. Fluorine Chem.* **2004**, *125*, 329–337.
6. Allcock, H. R.; Kugel, R. L. Synthesis of high polymeric alkoxy- and aryloxyphosphonitriles. *J. Am. Chem. Soc.* **1965**, *87*, 4216–4217.
7. Tur, D. R.; Koršak, V. V.; Vinogradova, S. V.; Dobrova, N. B.; Novikova, S. P.; Il'ina, M. B.; Sidorenko, E. S. Investigation of the thermostability of poly[bis(trifluoroethoxy)phosphazene]. *Acta Polym.* **1985**, *36*, 627–631.
8. Tur, D. R.; Koršak, V. V.; Vinogradova, S. V.; Slonimskij, G. L.; Il'Ina, M. N.; Dubovik, I. I.; Provotorova, N. P.; Dobrova, N. B.; Smurova, E. V. Investigation of the effects of biological environment on the properties of poly-bis-(trifluoroethoxy)phosphazene (r). *Acta Polym.* **1986**, *37*, 203–208.
9. Nishino, T.; Meguro, M.; Nakamae, K.; Matsushita, M.; Ueda, Y. The lowest surface free energy based on −CF3 alignment. *Langmuir* **1999**, *15*, 4321–4323.
10. Singh, A.; Steely, L.; Allcock, H. R. Poly[bis(2,2,2-trifluoroethoxy)phosphazene] superhydrophobic nanofibers. *Langmuir* **2005**, *21*, 11604–11607.
11. Welle, A.; Grunze, M.; Tur, D. Plasma protein adsorption and platelet adhesion on poly[bis(trifluoroethoxy)phosphazene] and reference material surfaces. *J. Colloid Interface Sci.* **1998**, *197*, 263–274.
12. Smeets, A. J.; Nijenhuis, R. J.; van Rooij, W. J.; Lampmann, L. E. H.; Boekkooi, P. F.; Vervest, H. A. M.; De Vries, J.; Lohle, P. N. M. Embolization of uterine leiomyomas with polyzene F–coated hydrogel microspheres: Initial experience. *J. Vasc. Interv. Radiol.* **2010**, *21*, 1830–1834.
13. Stampfl, U.; Radeleff, B.; Sommer, C.; Stampfl, S.; Dahlke, A.; Bellemann, N.; Kauczor, H.-U.; Richter, G. M. Midterm results of uterine artery embolization using narrow-size calibrated embozene microspheres. *Cardiovasc. Intervent. Radiol.* **2011**, *34*, 295–305.
14. Styllou, P.; Silber, S. A case report of the new Polyzene™-F COBRA PzF™ Nanocoated Coronary Stent System (NCS): Addressing an unmet clinical need. *Cardiovasc. Revasc. Med.* **2016**, *17*, 209–211.
15. Cutlip, D. E.; Garratt, K. N.; Novack, V.; Barakat, M.; Meraj, P.; Maillard, L.; Erglis, A.; Jauhar, R.; Popma, J. J.; Stoler, R.; Silber, S. 9-Month clinical and angiographic outcomes of the COBRA polyzene-F nanocoated coronary stent system. *Cardiovasc. Interventions* **2017**, *10*, 160–167.

16. Lanzer, P.; Sternberg, K.; Schmitz, K.-P.; Kolodgie, F.; Nakazawa, G.; Virmani, R. Drug-eluting coronary stent very late thrombosis revisited. *Herz Kardiovaskuläre Erkrankungen* **2008**, *33*, 334–342.
17. Lei, L.; Guo, S.-R.; Chen, W.-L.; Rong, H.-J.; Lu, F. Stents as a platform for drug delivery. *Expert Opin. Drug Delivery* **2011**, *8*, 813–831.
18. Saylor, D. M.; Guyer, J. E.; Wheeler, D.; Warren, J. A. Predicting microstructure development during casting of drug-eluting coatings. *Acta Biomater.* **2011**, *7*, 604–613.
19. Decher, G.: Layer-by-Layer Assembly (Putting Molecules to Work). In *Multilayer Thin Films*; Wiley-VCH Verlag GmbH & Co. KGaA, 2012; pp 1–21.
20. Decher, G.; Hong, J.-D. Buildup of ultrathin multilayer films by a self-assembly process, 1 consecutive adsorption of anionic and cationic bipolar amphiphiles on charged surfaces. *Makromol. Chem., Macromol. Symp.* **1991**, *46*, 321–327.
21. Pavlukhina, S.; Sukhishvili, S. Polymer assemblies for controlled delivery of bioactive molecules from surfaces. *Adv. Drug Delivery Rev.* **2011**, *63*, 822–836.
22. Liu, X.; Han, F.; Zhao, P.; Lin, C.; Wen, X.; Ye, X. Layer-by-layer self-assembled multilayers on PEEK implants improve osseointegration in an osteoporosis rabbit model. *Nanomed. Nanotechnol. Biol. Med.* **2017**, *13*, 1423–1433.
23. Schönhoff, M. Layered polyelectrolyte complexes: physics of formation and molecular properties. *J. Phys.: Condens. Matter* **2003**, *15*, R1781.
24. Pace, C. N.; Trevino, S.; Prabhakaran, E.; Scholtz, J. M. Protein structure, stability and solubility in water and other solvents. *Philos. Trans. R. Soc., B* **2004**, *359*, 1225–1235.
25. Andrianov, A. K.; Marin, A.; Peterson, P.; Chen, J. Fluorinated polyphosphazene polyelectrolytes. *J. Appl. Polym. Sci.* **2007**, *103*, 53–58.
26. Allcock, H. R.; Kwon, S. An ionically cross-linkable polyphosphazene: Poly[bis(carboxylatophenoxy)phosphazene] and its hydrogels and membranes. *Macromolecules* **1989**, *22*, 75–79.
27. Andrianov, A. K.; Svirkin, Y. Y.; LeGolvan, M. P. Synthesis and biologically relevant properties of polyphosphazene polyacids. *Biomacromolecules* **2004**, *5*, 1999–2006.
28. Andrianov, A. K.; Marin, A.; Chen, J.; Sargent, J.; Corbett, N. Novel route to sulfonated polyphosphazenes: Single-step synthesis using "noncovalent protection" of sulfonic acid functionality. *Macromolecules* **2004**, *37*, 4075–4080.
29. Andrianov, A. K.; Marin, A.; Chen, J. Synthesis, properties, and biological activity of poly[di(sodium carboxylatoethylphenoxy)phosphazene]. *Biomacromolecules* **2006**, *7*, 394–399.
30. Andrianov, A. K.; Chen, J.; LeGolvan, M. P. Poly(dichlorophosphazene) as a precursor for biologically active polyphosphazenes: Synthesis, characterization, and stabilization. *Macromolecules* **2004**, *37*, 414–420.

31. Allcock, H. R.; Fitzpatrick, R. J.; Salvati, L. Sulfonation of (aryloxy)-and (arylamino) phosphazenes: Small-molecule compounds, polymers, and surfaces. *Chem. Mater.* **1991**, *3*, 1120–1132.
32. Guo, Q.; N. Pintauro, P.; Tang, H.; O'Connor, S. Sulfonated and crosslinked polyphosphazene-based proton-exchange membranes. *J. Membr. Sci.* **1999**, *154*, 175–181.
33. Allcock, H. R.; Klingenberg, E. H.; Welker, M. F. Alkanesulfonation of cyclic and high polymeric phosphazenes. *Macromolecules* **1993**, *26*, 5512–5519.
34. Andrianov, A. K.; Marin, A.; University of Maryland: Unpublished.
35. Kwon, S.-K. Synthesis of water-soluble aminoaryloxy-methylamino cosubstituted polyphosphazenes as carrier species for biologically active agents. *Bull. Korean Chem. Soc.* **2001**, *22*, 1243–1247.
36. de Wolf, H.; de Raad, M.; Snel, C.; van Steenbergen, M.; Fens, M.; Storm, G.; Hennink, W. Biodegradable poly(2-dimethylamino ethylamino)phosphazene for in vivo gene delivery to tumor cells. Effect of polymer molecular weight. *Pharm. Res.* **2007**, *24*, 1572–1580.
37. de Wolf, H. K.; Luten, J.; Snel, C. J.; Oussoren, C.; Hennink, W. E.; Storm, G. In vivo tumor transfection mediated by polyplexes based on biodegradable poly(DMAEA)-phosphazene. *J. Controlled Release* **2005**, *109*, 275–287.
38. Yang, Y.; Xu, Z.; Chen, S.; Gao, Y.; Gu, W.; Chen, L.; Pei, Y.; Li, Y. Histidylated cationic polyorganophosphazene/DNA self-assembled nanoparticles for gene delivery. *Int. J. Pharm.* **2008**, *353*, 277–282.
39. Yang, Y.; Xu, Z.; Jiang, J.; Gao, Y.; Gu, W.; Chen, L.; Tang, X.; Li, Y. Poly(imidazole/DMAEA)phosphazene/DNA self-assembled nanoparticles for gene delivery: Synthesis and in vitro transfection. *J. Controlled Release* **2008**, *127*, 273–279.
40. Yang, Y.; Zhang, Z.; Chen, L.; Gu, W.; Li, Y. Galactosylated poly(2-(2-aminoethyoxy)ethoxy)phosphazene/DNA complex nanoparticles: In vitro and in vivo evaluation for gene delivery. *Biomacromolecules* **2010**, *11*, 927–933.
41. Yang, Y.; Zhang, Z.; Chen, L.; Gu, W.; Li, Y. Urocanic acid improves transfection efficiency of polyphosphazene with primary amino groups for gene delivery. *Bioconjugate Chem.* **2010**, *21*, 419–426.
42. Luten, J.; van Nostrum, C. F.; De Smedt, S. C.; Hennink, W. E. Biodegradable polymers as non-viral carriers for plasmid DNA delivery. *J. Controlled Release* **2008**, *126*, 97–110.
43. Luten, J.; van Steenbergen, M. J.; Lok, M. C.; de Graaff, A. M.; van Nostrum, C. F.; Talsma, H.; Hennink, W. E. Degradable PEG-folate coated poly(DMAEA-co-BA)phosphazene-based polyplexes exhibit receptor-specific gene expression. *Eur. J. Pharm. Sci.* **2008**, *33*, 241–251.
44. Luten, J.; Van Steenis, J. H.; Van Someren, R.; Kemmink, J.; Schuurmans-Nieuwenbroek, N. M. E.; Koning, G. A.; Crommelin, D. J. A.; Van Nostrum, C. F.; Hennink, W. E. Water-soluble biodegradable cationic polyphosphazenes for gene delivery. *J. Controlled Release* **2003**, *89*, 483–497.

45. Allcock, H. R.; McIntosh, M. B.; Klingenberg, E. H.; Napierala, M. E. Functionalized polyphosphazenes: Polymers with pendent tertiary trialkylamino groups. *Macromolecules* **1998**, *31*, 5255–5263.
46. Andrianov, A. K.; Chen, J.; Payne, L. G. Preparation of hydrogel microspheres by coacervation of aqueous polyphosphazene solutions. *Biomaterials* **1998**, *19*, 109–115.
47. Cohen, S.; Bano, M. C.; Visscher, K. B.; Chow, M.; Allcock, H. R.; Langer, R. Ionically crosslinkable polyphosphazene: A novel polymer for microencapsulation. *J. Am. Chem. Soc.* **1990**, *112*, 7832–7833.
48. Lee, K. Y.; Mooney, D. J. Alginate: Properties and biomedical applications. *Prog. Polym. Sci.* **2012**, *37*, 106–126.
49. Andrianov, A. K.; Chen, J. Polyphosphazene microspheres: Preparation by ionic complexation of phosphazene polyacids with spermine. *J. Appl. Polym. Sci.* **2006**, *101*, 414–419.
50. Firestone, B. A.; Siegel, R. A. Kinetics and mechanisms of water sorption in hydrophobic, ionizable copolymer gels. *J. Appl. Polym. Sci.* **1991**, *43*, 901–914.
51. Shafee, E. E.; Naguib, H. F. Water sorption in cross-linked poly(vinyl alcohol) networks. *Polymer* **2003**, *44*, 1647–1653.
52. Ladam, G.; Schaad, P.; Voegel, J. C.; Schaaf, P.; Decher, G.; Cuisinier, F. In situ determination of the structural properties of initially deposited polyelectrolyte multilayers. *Langmuir* **2000**, *16*, 1249–1255.
53. Shiratori, S. S.; Rubner, M. F. pH-Dependent thickness behavior of sequentially adsorbed layers of weak polyelectrolytes. *Macromolecules* **2000**, *33*, 4213–4219.
54. Yoo, D.; Shiratori, S. S.; Rubner, M. F. Controlling bilayer composition and surface wettability of sequentially adsorbed multilayers of weak polyelectrolytes. *Macromolecules* **1998**, *31*, 4309–4318.
55. Hammond, P. T. Recent explorations in electrostatic multilayer thin film assembly. *Curr. Opin. Colloid Interface Sci.* **1999**, *4*, 430–442.
56. Xu, L.; Pristinski, D.; Zhuk, A.; Stoddart, C.; Ankner, J. F.; Sukhishvili, S. A. Linear versus exponential growth of weak polyelectrolyte multilayers: correlation with polyelectrolyte complexes. *Macromolecules* **2012**, *45*, 3892–3901.
57. Porcel, C.; Lavalle, P.; Decher, G.; Senger, B.; Voegel, J. C.; Schaaf, P. Influence of the polyelectrolyte molecular weight on exponentially growing multilayer films in the linear regime. *Langmuir* **2007**, *23*, 1898–1904.
58. Zhuk, A.; Selin, V.; Zhuk, I.; Belov, B.; Ankner, J. F.; Sukhishvili, S. A. Chain conformation and dynamics in spin-assisted weak polyelectrolyte multilayers. *Langmuir* **2015**, *31*, 3889–3896.
59. DeCollibus, D. P.; Marin, A.; Andrianov, A. K. Effect of environmental factors on hydrolytic degradation of water-soluble polyphosphazene polyelectrolyte in aqueous solutions. *Biomacromolecules* **2010**, *11*, 2033–2038.
60. Selin, V.; Albright, V.; Ankner, J. F.; Marin, A.; Andrianov, A. K.; Sukhishvili, S. A. Biocompatible nanocoatings of fluorinated

polyphosphazenes through aqueous assembly. *ACS Applied Materials & Interfaces* **2018**.
61. Hougham, G.; Tesoro, G.; Viehbeck, A. Influence of free volume change on the relative permittivity and refractive index in fluoropolyimides. *Macromolecules* **1996**, *29*, 3453–3456.
62. Yasuda, H.; Sharma, A. K.; Yasuda, T. Effect of orientation and mobility of polymer molecules at surfaces on contact angle and its hysteresis. *J. Polym. Sci., Polym. Phys. Ed.* **1981**, *19*, 1285–1291.
63. Brunette, D. M. *Titanium in Medicine: Material Science, Surface Science, Engineering, Biological Responses, and Medical Applications*; Springer, 2001.
64. Soler, M.; Verhaeghe, P. J.; Stoppa, R. Polyester (Dacron®) Mesh. In *Abdominal Wall Hernias: Principles and Management*; Bendavid, R., Abrahamson, J., Arregui, M. E., Flament, J. B., Phillips, E. H., Eds.; Springer: New York, 2001; pp 266−271.
65. Jiang, S. X.; Qin, W. F.; Guo, R. H.; Zhang, L. Surface functionalization of nanostructured silver-coated polyester fabric by magnetron sputtering. *Surf. Coat. Technol.* **2010**, *204*, 3662–3667.
66. Braune, S.; Lendlein, A.; Jung, F. 3 −- Developing Standards and Test Protocols for Testing the Hemocompatibility of Biomaterial. In *Hemocompatibility of Biomaterials for Clinical Applications*; Siedlecki, C. A., Ed.; Woodhead Publishing, 2018; pp 51−76.
67. Xu, L.-C.; Bauer, J. W.; Siedlecki, C. A. Proteins, platelets, and blood coagulation at biomaterial interfaces. *Colloids Surf., B* **2014**, *124*, 49–68.
68. Stanisławska, A. Biomaterials and implants in cardiac and vascular surgery - Review. *Adv. Mater. Sci.* **2014**, *14*, 5–17.
69. Henkelman, S.; Rakhorst, G.; Blanton, J.; van Oeveren, W. Standardization of incubation conditions for hemolysis testing of biomaterials. *Mater. Sci. Eng.: C Mater. Biol. Appl.* **2009**, *29*, 1650–1654.
70. Hsieh, F.-Y.; Tao, L.; Wei, Y.; Hsu, S.-h. A novel biodegradable self-healing hydrogel to induce blood capillary formation. *NPG Asia Mater.* **2017**, *9*, e363.
71. Li, Q.; Liu, C.; Wen, J.; Wu, Y.; Shan, Y.; Liao, J. The design, mechanism and biomedical application of self-healing hydrogels. *Chin. Chem. Lett.* **2017**, *28*, 1857–1874.
72. Phadke, A.; Zhang, C.; Arman, B.; Hsu, C.-C.; Mashelkar, R. A.; Lele, A. K.; Tauber, M. J.; Arya, G.; Varghese, S. Rapid self-healing hydrogels. *Proc. Natl. Acad. Sci. USA* **2012**, *109*, 4383–4388.
73. Wu, M.; An, N.; Li, Y.; Sun, J. Layer-by-layer assembly of fluorine-free polyelectrolyte–surfactant complexes for the fabrication of self-healing superhydrophobic films. *Langmuir* **2016**, *32*, 12361–12369.
74. Gaddes, D.; Jung, H.; Pena-Francesch, A.; Dion, G.; Tadigadapa, S.; Dressick, W. J.; Demirel, M. C. Self-healing textile: Enzyme encapsulated layer-by-layer structural proteins. *ACS Appl. Mater. Interfaces* **2016**, *8*, 20371–20378.

75. Deng, M.; Kumbar, S. G.; Wan, Y.; Toti, U. S.; Allcock, H. R.; Laurencin, C. T. Polyphosphazene polymers for tissue engineering: An analysis of material synthesis, characterization and applications. *Soft Matter* **2010**, *6*, 3119–3132.
76. Zhuk, I.; Jariwala, F.; Attygalle, A. B.; Wu, Y.; Libera, M. R.; Sukhishvili, S. A. Self-defensive layer-by-layer films with bacteria-triggered antibiotic release. *ACS Nano* **2014**, *8*, 7733–7745.
77. Moskowitz, J. S.; Blaisse, M. R.; Samuel, R. E.; Hsu, H.-P.; Harris, M. B.; Martin, S. D.; Lee, J. C.; Spector, M.; Hammond, P. T. The effectiveness of the controlled release of gentamicin from polyelectrolyte multilayers in the treatment of Staphylococcus aureus infection in a rabbit bone model. *Biomaterials* **2010**, *31*, 6019–6030.
78. Shukla, A.; Avadhany, S. N.; Fang, J. C.; Hammond, P. T. Tunable vancomycin releasing surfaces for biomedical applications. *Small* **2010**, *6*, 2392–2404.
79. Josepovitz, C.; Pastoriza-Munoz, E.; Timmerman, D.; Scott, M.; Feldman, S.; Kaloyanides, G. J. Inhibition of gentamicin uptake in rat renal cortex in vivo by aminoglycosides and organic polycations. *J. Pharmacol. Exp. Ther.* **1982**, *223*, 314–321.
80. Barth, E.; Myrvik, Q. M.; Wagner, W.; Gristina, A. G. In vitro and in vivo comparative colonization of Staphylococcus aureus and Staphylococcus epidermidis on orthopaedic implant materials. *Biomaterials* **1989**, *10*, 325–328.
81. Albright, V.; Zhuk, I.; Wang, Y.; Selin, V.; van de Belt-Gritter, B.; Busscher, H. J.; van der Mei, H. C.; Sukhishvili, S. A. Self-defensive antibiotic-loaded layer-by-layer coatings: Imaging of localized bacterial acidification and pH-triggering of antibiotic release. *Acta Biomater.* **2017**, *61*, 66–74.
82. Pavlukhina, S.; Lu, Y.; Patimetha, A.; Libera, M.; Sukhishvili, S. Polymer multilayers with pH-triggered release of antibacterial agents. *Biomacromolecules* **2010**, *11*, 3448–3456.
83. DeCollibus, D. P.; Marin, A.; Andrianov, A. K. Effect of environmental factors on hydrolytic degradation of water-soluble polyphosphazene polyelectrolyte in aqueous solutions. *Biomacromolecules* **2010**, 487–492.
84. Martinez, A. P.; Qamar, B.; Fuerst, T. R.; Muro, S.; Andrianov, A. K. Biodegradable "smart" polyphosphazenes with intrinsic multifunctionality as intracellular protein delivery vehicles. *Biomacromolecules* **2017**, *18*, 2000–2011.

Drug Delivery

Chapter 6

Biodegradable "Scaffold" Polyphosphazenes for Non-Covalent PEGylation of Proteins

Andre P. Martinez,[1,2] Bareera Qamar,[3] Alexander Marin,[1] Thomas R. Fuerst,[1,3] Silvia Muro,[1,4] and Alexander K. Andrianov*,[1]

[1]Institute for Bioscience and Biotechnology Research, University of Maryland, 9600 Gudelsky Dr., Rockville, Maryland 20850, United States
[2]Present address: DSM Biomedical 735 Pennsylvania Drive, Exton Pennsylvanis 19341, United States
[3]Department of Cell Biology and Molecular Genetics, 1109 Microbiology Building, University of Maryland, College Park, Maryland 20742, United States
[4]Fischell Department of Bioengineering, 2330 Jeong Kim Building, University of Maryland, College Park, Maryland 20742, United States
***E-mail: aandrianov@umd.edu.**

Biodegradable "scaffold" polyphosphazenes were investigated as carriers for a single-step non-covalent PEGylation of proteins. These water-soluble polymers include poly(ethylene glycol), PEG pendant groups as a hydrophilic neutral component for steric stabilization of protein payload and ionic moieties to facilitate self-assembly with proteins. PEGylated polyphosphazenes (PPEGs) were capable of forming non-covalent complexes with model cargo protein in aqueous solutions demonstrating their potential as drug delivery carriers. Increase in the content of PEG side groups in polyphosphazenes reduced or completely eliminated environmentally triggered polymer aggregation and provided for greater stability of avidin-PPEG complexes at high protein loadings. Non-covalent PEGylation of avidin using PPEGs did not cause any disruption of protein conformation or functionality and resulted in up to 10-fold increase in hydrodynamic diameter and reduced protein antigenicity, as determined by ELISA. Despite high content of PEG side groups, polyphosphazenes facilitated cell-surface interaction followed by time-dependent internalization of a model protein cargo into cancer cells. These results suggest the potential of PPEG approach for both prolonging protein half-life and intracellular delivery of proteins. PEGylated polyphosphazenes are hydrolytically degradable in aqueous solutions at neutral pH, but degradation rates are reduced at lower temperatures, indicating potentially acceptable solution shelf life.

Introduction

Protein and peptide therapeutics possess high biological activity and specificity, and their clinical use is growing (*1–5*). However, these macromolecular drugs are generally characterized by suboptimal pharmacokinetic profiles, in particular short in vivo half-life, and undesirable inherent immunogenicity (*3*, *4*, *6*). PEGylation – a technology, which is based on a covalent attachment of polyethylene glycol (PEG) to the protein surface, is currently one of the most successful techniques addressing these shortcomings (*7–11*). Protein bound PEG imparts a steric barrier, which masks immunogenic sites on the protein thereby shielding it from the body's immune system (*12*, *13*). In addition, PEGylated proteins are characterized by increased hydrodynamic volume, which results in reduced renal clearance (*12*, *14*, *15*). Although PEGylation has been proven efficient in prolonging in vivo half-life and achieved a significant commercial success, the technology still suffers from a number of limitations. In particular, covalent conjugation can lead to modification of protein active sites resulting in a reduction of therapeutic efficacy (*16*). Also, PEGylation processes produce heterogeneous mixtures containing unreacted materials and by-products of activating agents and require time consuming and expensive multistep purifications (*16–18*).

A non-covalent PEGylation approach has been recently suggested as an alternative, which can potentially address these limitations. PEGs containing dansyl-, L-tryptophan–, phenylbutylamino-, benzyl- and cholesteryl end groups were synthesized and investigated for complexation with proteins through ionic or hydrophobic interactions (*17*, *19*, *20*). Some of those non-covalently bound PEGs reduced aggregation of salmon calcitonin and lysozyme in vitro (*17*, *19*, *20*). Supramolecular PEGylation of insulin was also achieved using PEGylated cucurbit[7]uril, which was synthesized by "click" chemistry between a cucurbit[7]uril supramolecular host molecule bearing a single azide moiety and a dibenzocyclooctyne-functional PEG polymer (*21*). Strong non-covalent binding between macrocyclic host and amino acid side chains on the protein endowed proteins with PEG functionality (*21*). Non-covalent PEGylation using metal coordination bonds between PEG functionalized with a chelating agent, nitrilotriacetic acid, and histidine-rich proteins was described (*16*). However, the association, which was based on a single point attachment of PEG to the protein, was determined to be not stable in vivo (*16*).

PEG anchoring functionalities, which are suitable for more stable, multivalent attachment through non-covalent bonding, were also investigated. PEGylated semi-synthetic polyanions - pentosan polysulfate and dextran sulfate were complexed with Keratinocyte Growth Factor-2 in order to investigate their effect on thermal stability of the protein (*22*). The suggested branched polymer architecture, which is still uncommon in the PEGylation technology, can be more efficient in preventing the approach of anti-protein antibodies and immunocompetent cells (*12*). Although, polymers were effective in improving thermal stability of the protein, no data is available related to the effect on protein antigenicity or half-life. A non-covalent attachment of PEGs through multiple electrostatic interactions was also realized using a synthetic block copolymer of poly(ethylene glycol) and poly(N,N-dimethylaminoethylmethacrylate) and L-asparaginase (*23*, *24*). The formation of a water-soluble protein–polyelectrolyte complex did not lead to the loss of secondary structure or enzyme activity and successfully inhibited protease digestion of the enzyme. However, the use of non-biodegradable polyacrylates may limit further development of this system for clinical applications.

Polyphosphazenes - synthetic polymers, which are based on a biodegradable phosphorus-nitrogen backbone and organic side groups, may provide an appealing approach for developing PEGylated "scaffold" polymers for further non-covalent assembly with proteins. Polyphosphazene macromolecules are characterized by tunable hydrolytic degradation, which can be controlled by careful selection of pendant groups. Dual side-group structure of the repeating unit provides support for creating high charge density, which enables multivalent ionic interactions with protein, and still allows for extensive PEGylation. Finally, the macromolecular substitution pathway, which relies on the use of reactive macromolecular precursor, polydichlorophosphazene (PDCP) and organic nucleophiles, empowers versatile derivatization chemistry.

A number of polyphosphazenes containing PEG side groups of various chain length have been synthesized. They include either amphiphilic or hydrophilic neutral macromolecules designed to serve as carriers for the delivery of small molecule drugs. In particular, polyphosphazene copolymers bearing hydrophilic methoxy PEG and hydrophobic dipeptide ethyl esters side groups were prepared as conjugate prodrugs for the delivery of platinum based anticancer agents (*25*, *26*). Furthermore, polyphosphazene containing PEG and hydrophobic ethyl tryptophan groups was synthesized with the aim of constructing doxorubicin loaded micelles for cancer therapy (*27*). Polyphosphazene homopolymers grafted with poly(ethylene oxide-co-propylene oxide) (PEO-co-PO), either through various amino acid spacers or directly through the terminal aminogroup of PEO-co-PO, were synthesized in order to modulate biocompatibility and biodegradability of these macromolecular carriers (*28*). The latter were then conjugated with platinum (IV) (*29*) or organoruthenium and organorhodium complexes (*30*) to yield anticancer prodrugs, as well as with imidazoquinoline derivatives for use as immunotherapies (*31*). PEGylated polyphosphazenes were also employed for assembling hydrogels based on the inclusion complex between the PEG side groups and α-cyclodextrin (*32*).

Recently, we have introduced a new class of PEGylated polyphosphazenes - polyelectrolytes, which contain carboxylic acid moieties along with pendant PEG groups (PPEGs) (*33*). Present paper investigates these polyions for their ability to bind protein cargo – an important feature that can potentially provide a convenient pathway to a single-step non-covalent PEGylation of macromolecular drugs. It also explores environmental sensitivity of these polyphosphazenes, confirms their hydrolytic biodegradability, and evaluates the ability of these polymers to reduce antigenicity of protein cargo. The potential of PEGylated polyphosphazenes to facilitate cellular uptake of proteins is also established in cell culture using cancer cells.

Materials and Methods

Materials

Sodium phosphate monobasic dihydrate, methoxypolyethylene glycol amine (5 kDa), PEG-NH_2, bovine serum albumin, BSA, egg white avidin and avidin conjugated to fluorescein isothiocyanate, FITC (Sigma-Aldrich, Saint Louis, MO), bis(2-methoxyethyl) ether, diglyme (Acros Organics, Morris Plains, NJ), ethanol (Warner-Graham, Cockeysville, MD), hydrochloric acid, potassium hydroxide (Alfa Aesar, Haverhill, MA), HyPure™WFI Quality Water (GE Life Sciences, Pittsburgh, PA), sodium carbonate (Amresco, Solon, OH), methyl 3(4-hydroxyphenyl)propionate, MHP (TCI, Portland, OR), polysorbate 20, Tween-20 (Spectrum Chemical, Gardena, CA), acetonitrile (EM Science, Darmstadt, Germany), phosphate buffered saline pH 7.4, PBS (Life Technologies, Carlsbad, CA), poly(acrylic acid) standards (American Polymer Standards, Mentor, OH), avidin monoclonal antibody, biotinylated horseradish peroxidase (Invitrogen, Rockford, IL), TMB peroxidase EIA substrate kit (Bio-Rad Laboratories, Hercules, CA), porcine red blood cells (Innovative Research, Novi, MI), sodium chloride (Fisher Scientific, Waltham, MA), sodium phosphate dibasic heptahydrate, sodium bicarbonate (VWR, Radnor, PA), biotinylated mouse IgG (BD Biosciences PharminGen, San Jose, CA), Texas Red goat anti-mouse IgG (Life Technologies, Carlsbad, CA), and Dulbecco's Modified Eagle's Medium, DMEM with 4.5g/L glucose, L-glutamine and sodium pyruvate (Corning Life Sciences, Tewksbury, MA) were used as received. Phosphonitrilic chloride trimer, hexachlorocyclotriphosphazene was generously donated by Fushimi Pharmaceutical Co. Ltd. (Kagawa, Japan). Polydichlorophosphazene (PDCP) was synthesized by a ring-opening polymerization reaction in a pressure reactor (*34*). PPEGs were synthesized via subsequent addition of nucleophiles PEG-NH_2 and MHP to PDCP followed by hydrolysis of an ester bearing copolymer to yield polyphosphazene polyacid.

Characterization

Gel permeation chromatography, GPC was performed using a Hitachi HPLC system with L-2450 diode array detector, L-2130 pump, and L-2200 autosampler (Hitachi LaChrom Elite system, Hitachi, San Jose, CA) and Ultrahydrogel Linear size exclusion column (Waters Corporation, Milford, MA). PBS, pH 7.4 with 10% of acetonitrile was employed as a mobile phase with a flow rate of 0.5 mL/min. Samples were prepared at a concentration of 0.5 mg/mL in PBS, pH 7.4 and were filtered using Millex 0.22 µm filters (EMD Millipore, Billerica, MA) prior to the analysis. Molecular weights were calculated using EZ-Chrome Elite software (Agilent Technologies, Santa Clara, CA). Calibration curve was obtained using narrow poly(acrylic acid) standards (American Polymer Standards Corporation, Mentor, OH).

Dynamic light scattering, DLS was carried out using a Malvern Zetasizer Nano series, ZEN3600 and analyzed using Malvern Zetasizer 7.10 software (Malvern Instruments Ltd., Worcestershire, UK). Samples were prepared in PBS, pH 7.4 and filtered using Millex 0.22 µm filters prior to the analysis.

UV-Vis readings for hemolysis assays were performed using a Thermo Scientific Multiscan Spectrum spectrophotometer (Thermo Fisher Scientific, Waltham, MA). Data was analyzed using SkanIt 2.4.4 software (Thermo Fisher Scientific, Waltham, MA).

Circular dichroism (CD) measurements were carried out using a Chiroscan qCD spectrometer (Applied Photophysics, Leatherhead Surrey, United Kingdom) in a 1 mm cell at ambient temperature.

Asymmetric Flow Field Flow Fractionation, AF4 was performed using a Postnova AF2000 MT series (Postnova Analytics GmbH, Landsberg, Germany). The system was equipped with two PN1130 isocratic pumps, PN7520 solvent degasser, PN5120 injection bracket and UV-Vis detector (SPD-20A/20AV, Shimadzu Scientific Instruments, Columbia, MD). A regenerated cellulose membrane with molecular weight cutoff of 10 kDa (Postnova Analytics GmbH, Landsberg, Germany) and a 350 µm spacer were used in a separation micro-channel employing both laminar and cross flows of an eluent - PBS (pH 7.4). The collected data was processed using AF2000 software (Postnova Analytics GmbH).

Multi-angle static light scattering, MALS analysis was performed using a Wyatt miniDAWN TREOS MALS detector (Wyatt Technology Corporation, Santa Barbara, CA) assembled in line with Optilab T- rEX refractometer (Wyatt Technology Corporation, Santa Barbara, CA), 1200 series isocratic pump, 1200 series variable wavelength detector (Agilent Technologies, Santa Clara, CA) and Ultrahydrogel column (Waters Corporation, Milford, MA). PBS, pH 7.4 was used as a mobile phase. Samples were prepared at a concentration of 0.5 mg/mL in PBS, pH 7.4 and filtered using Millex 0.22 µm filters prior to the analysis. Molecular weights were determined using ASTRA V 5.3.4.14 software (Wyatt Technology Corporation, Santa Barbara, CA).

For studies on hydrolytic degradation of PPEGs, polymers were dissolved to a concentration of 0.50 mg/mL in 1xPBS then filtered through a 0.22 µm membrane.

Solutions were stored at 4°C, ambient temperature, 37°C, and 65°C. Samples were taken for GPC analysis at various time intervals.

Non-Covalent PPEG–Protein Binding and Functional Activity of Complexes

AF4 was used to evaluate binding between PPEG and avidin. PPEG and lyophilized FITC-avidin were dissolved in PBS, which was also used as an eluent. PPEG and FITC-avidin samples were run individually and in complex at a molar ratio of 1:10 (PPEG to FITC-avidin) based on molecular weight determined by MALS. 50 μl of sample were injected and AF4 profiles were recorded at 280 nm and 495 nm. Functional activity of avidin in a noncovalent complex with copolymer was evaluated by examining its affinity to the fluorescently labeled substrate FITC-biotin. PPEG-16 and avidin solutions in PBS were mixed to form a complex with a molar ratio of 1:10 (polymer to avidin). Next, FITC-biotin was added to the solution to achieve a molar concentration 4-fold that of avidin. A control sample was prepared by premixing solutions of avidin and FITC-biotin followed by addition of PPEG-16, generating the same final concentrations. Concentrations were chosen to achieve 1:1 ratio between FITC-biotin and avidin binding sites. Samples were then analyzed by AF4 as described above.

The stability of polymer–protein complexes in the presence of serum was evaluated using AF4. Samples of 0.5 mg/mL avidin in PBS, 0.5 mg/mL avidin in 50% rabbit serum (v/v) in PBS, and 2.55 mg/mL PPEG-1 with 0.50 mg/mL avidin in 50% rabbit serum (v/v) in PBS were made and incubated at 37°C. Samples were taken at various time intervals and diluted 10 fold with FITC-biotin in PBS to a final concentration of 2 μg/mL FITC-biotin. 50 μl of these solutions were injected and PBS was used as an eluent. AF4 profiles were recorded at 495 nm.

Evaluation of Hemolytic Activity

The membrane disruptive activity of multifunctional carriers was tested as described previously (*35–37*). First, 300 μL of fresh porcine red blood cells (RBCs) as a 10% suspension in PBS was resuspended in 1.2 mL of PBS. Then, 50 μL of resuspended RBCs was added to 925 μL of 50 mM PBS at the appropriate pH and vortexed, and to this mixture was added 25 μL of 2 mg/mL polymer in PBS followed by vortexing. In the case of polymer–protein complexes, 25 μL of a solution of 4 mg/mL PPEG with 1.48 mg/mL, 3.64 mg/mL, or 0.40 mg/mL avidin for PPEG-1, PPEG-5, or PPEG-16 samples respectively was used. Samples were incubated at 37 °C for 1 h on a shaker table. Cells were then centrifuged at 14,000 rpm for 5 min, and the absorbance of the supernatant was then measured at 541 nm using a Multiskan Spectrum microplate spectrophotometer (ThermoFisher Scientific, Waltham, MA). To determine 100% hemolysis, RBCs were suspended in distilled water and lysed by ultrasound (Branson Sonifier, Model 450). All hemolysis experiments were conducted in triplicate.

Antigenicity of Protein in PPEG Complexes

The amount of avidin available for interaction with antibody was measured using an enzyme-linked immunosorbent assay (ELISA). 5 µL of avidin monoclonal antibody was mixed with 10.5 mL 0.05 M carbonate-bicarbonate buffer (pH 9.6). 100 µL aliquots of this solution were added to a 96-well plate and incubated overnight at 4°C. Next, the solution was removed and the plate was washed with PBS. To prevent non-specific interaction, 300 mL of blocking buffer (1% BSA in PBS) was added to each well and incubated for 1 h at room temperature. The plate then was rinsed with washing buffer (0.05% Tween-20 in PBS). Formulations containing 0.01 mg/mL avidin with various concentrations of PPEG in PBS were diluted to a final concentration of 0.05 µg/mL avidin. 100 µL of these solutions were added to each well and incubated for 1 h at room temperature. The plate was then washed with washing buffer, 100 µL of biotinylated horseradish peroxidase (0.5 µg/mL in PBS containing 0.5% BSA and 0.05% Tween) was added to each well and incubated for 30 min at room temperature, then rinsed with washing buffer. 100 µL of TMB peroxidase EIA substrate kit solution was added into each well and incubated for 20 min. The reaction was stopped by adding of 100 µL 1 M sulfuric acid. Optical density at 450 nm was measured by Multiscan Spectrum microplate spectrophotometer (ThermoFisher Scientific, Waltham, MA). The data were presented as a percent of inhibition (I), which was calculated using the following equation: $I = (OD_0 - OD_{Poly}) / OD_0 \times 100$, where OD_0 and OD_{Poly} are the optical densities of the solution without polymer and in the presence of polymer.

Binding and Uptake in Cell Culture

Oral adenosquamous carcinoma Cal27 cells (American Type Culture Collection) were cultured on 12-mm^2 glass coverslips at 37°C, 5% CO_2 and 95% relative humidity in DMEM supplemented with 10% fetal bovine serum and 1% Penicillin/Streptomycin. Cells were incubated in serum-supplemented medium containing PPEG16/FITC-avidin complex (0.2 mg/ml polymer carrier and 0.1 mg/ml avidin cargo) or control FITC-avidin (0.1 mg/ml) for 30 minutes, 2 hours or 5 hours. Cell were then washed to remove unbound materials and fixed with 2% paraformaldehyde to stop any further uptake. To distinguish cell-surface versus internalized materials, cells were then incubated with biotin-conjugated mouse IgG, which can only access surface-bound FITC-avidin but not internalized FITC-avidin. Surface-bound counterparts were finally stained using a Texas Red-labeled secondary antibody and nuclei were stained with 4′,6-diamidino-2-phenylindole (DAPI). Using this protocol, total FITC-avidin associated with cells fluoresces in green, while avidin at the cell-surface fluoresces in both red and green. Samples were imaged using an Olympus IX81 microscope (Olympus, Inc., Center Valley, PA), 60× oil immersion objective (UPlanApo, Olympus, Inc., Center Valley, PA), ORCA-ER camera (Hamamatsu Corporation, Bridgewater, NJ), and SlideBook™ 4.2 software (Intelligent Imaging Innovations, Denver, CO). Fluorescence images were taken under the green, red, and blue channels and bright field images provided visualization of full cells and cell-cell borders. To monitor the total amount of FITC-avidin associated to cells, Image-Pro 6.3 (Media Cybernetics, Bethesda, MD) was used to quantify FITC mean intensity per cell, normalized to that of the background in surrounding areas. In addition, to quantify internalization, the percentage of FITC-avidin that did not colocalized with Texas Red dye was quantified, as well as the cell area occupied by this internalized counterpart, as previously described (*38*, *39*). Of note, since FITC fluorescence may be affected by pH, experiments involved cell fixation, which renders the vacuolar H+-ATPase inactive and neutralizes the pH of intracellular compartments. Two independent experiments were analyzed cell-by-cell, for a total sample size of n ≥100 cells. Data were calculated as mean ± standard error of the mean (SEM). Statistical significance for two-way comparisons was determined using Student's *t*-test for both $p<0.05$ and $p<0.5$.

Results and Discussion

PEGylated Polyphosphazenes: PPEGs

PEG remains a gold standard for the steric stabilization of drug carriers by means of creating a highly hydrated shell around them and ensuing prolonged in vivo circulation times (*7*, *9*, *12*, *40*). The molecular design of PEGylated polyphosphazenes (Figure 1) consisted of pendant PEG chains as a neutral hydrophilic component to provide for steric stabilization and carboxylic acid side groups to enable its self-assembly with proteins (*41*, *42*). Polymers were synthesized by first reacting a macromolecular precursor, PDCP with targeted amount of monofunctional PEG (5,000 g/mol) containing primary amino group as an end group, adding ester containing nucleophile, and hydrolyzing the ester functionality to yield carboxylic acid groups (*33*). Figure 1 summarizes compositions, molecular weights (MW), and z-average hydrodynamic diameters (D_z) of PPEGs explored in present paper. All copolymers were fully soluble in water and PBS (pH 7.4).

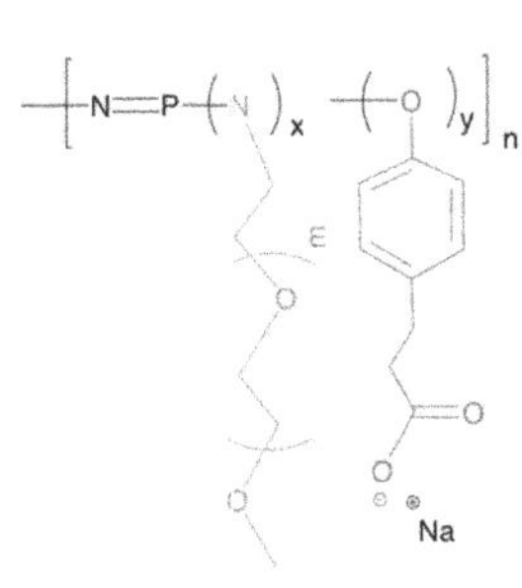

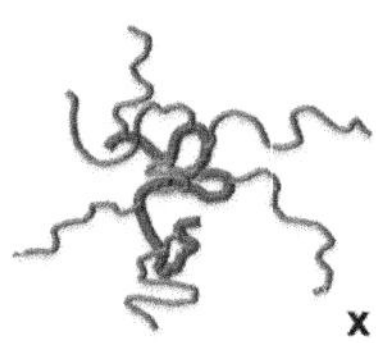

PPEGs

	x %, mol.	y %, mol.	MW, kg/mol	D_z nm
PPEG1	1	99	1,800	67
PPEG5	5	95	720	52
PPEG16	16	84	3,300	75

Figure 1. Schematic presentation of PPEGs and their physico-chemical characteristics (x and y were calculated from ^{1}H NMR measurements, MW -weight average molecular weights (GPC), D_z – z-average hydrodynamic diameters measured by DLS, characterization data from Ref. (33)).

Hydrolytic Degradation of Polyphosphazene Copolymers

PPEGs were first evaluated for their ability to undergo hydrolytic degradation at near physiological and potential storage conditions. Solutions of polyphosphazenes in PBS (pH 7.4) were incubated at various temperatures (4°C, 37°C, 65°C and ambient temperature) and analyzed at various time intervals for molecular mass losses by size-exclusion HPLC. The representative molecular weight loss vs. time dependence at various temperatures is shown for PPEG5 (Figure 2). Hydrolytic degradation of PPEG is evident from the consistent decrease in polymer molecular weight as the incubation proceeded. Notably, PPEG5 underwent rapid degradation at 65°C and somewhat slower breakdown at 37°C. Similar results were obtained for PPEG and PPEG16 with the rate of hydrolysis increasing as the content of PEG groups in polymer rose. A relatively slow degradation rate at 4°C (about 10 % molecular weight loss over 100 days) suggests adequate short-term solution stability of polymers.

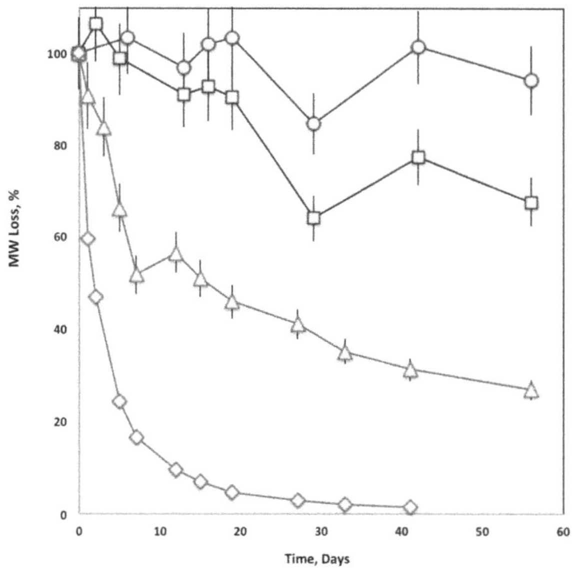

Figure 2. MW loss versus degradation time for PPEG5 at 65 °C (diamonds), 37 °C (triangles), ambient temperature (squares), and 4 °C (spheres) (0.5 mg/mL polymer, GPC, PBS, pH 7.4).

Biocompatibility and pH-Dependent Membrane Disruptive Activity of PPEGs

In vitro evaluation of drug carrier biocompatibility with blood components is an important part of early preclinical screening (*43*). Therefore PPEGs were tested for potential membrane destabilizing behavior in the hemolysis studies. Neither of the polymers showed any hemolytic activity at neutral pH. However, in contrast to other polyphosphazenes with higher content of PEG, PPEG1 exhibited membrane disruptive properties between pH 6 and 7 (Figure 3A), which correspond to the range of early endosomes pH (pH 6.0-6.5) and suggests potential endosomolytic properties of this polymer (*44*).

Environmental Sensitivity of Polyphosphazene Polyacids: The Effect of PEG Side Chains

Polyphosphazene polyacids are generally sensitive to the ionic environment and, in particular, they can form ionically cross-linked hydrogels in the presence of multivalent cations, such as calcium ions (*45–47*). As expected for highly charged anionic polyphosphazene (*46*), PPEG1 demonstrated the formation of large aggregates or microgels (over 1 μm in diameter) upon addition of calcium chloride (Figure 3B). However, higher content of PEG in PPEG5 prevented formation of large aggregates limiting the hydrodynamic diameter of cross-linked PPEG5 particles to approximately 200 nm, which is presumably due to the shielding effect of PEG chains. Furthermore, PPEG16 did not show any detectable increase in peak average size upon addition of calcium chloride. These results emphasize the functional importance of PEG shielding, especially as it relates to preventing intermolecular aggregation and minimizing typical polyelectrolyte solution behavior. Regulating the size of cross-linked PEGylated polyphosphazenes through ionic cross-linking may also open additional opportunities for modulating delivery modalities of polyphosphazene carriers.

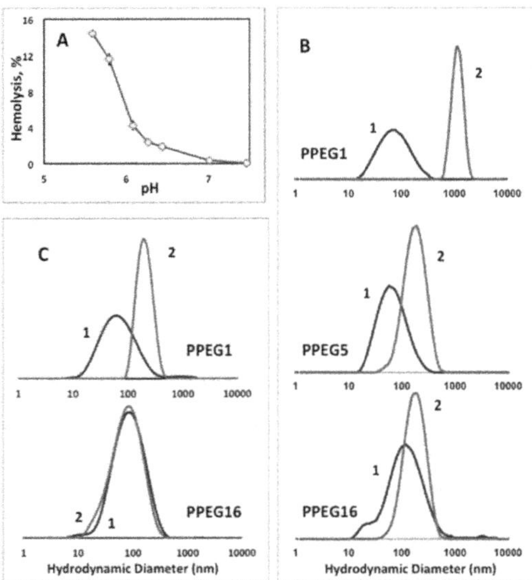

Figure 3. Biocompatibility and environmental sensitivity of PPEGs. (A) Hemolysis; Hemolysis of Red Blood Cells as a function of pH for PPEG1 (0.05 mg/mL polymer, 50 mM PBS); (B) DLS profiles of PPEGs before (1) and after (2) addition of calcium chloride (0.1 mg/mL PPEG, 0.14 mg/mL calcium chloride, 9 mg/mL sodium chloride); (C) DLS profiles of PPEG1 and PPEG16 at (1) pH 8 and (2) pH 5 (0.5 mg/mL PPEG, 9 mg/mL sodium chloride).

Somewhat similar trends were observed upon acid titration of PPEGs (Figure 3C). As seen from the Figure, PPEG1 displayed a significant shift in size upon acidification. The peak average molecular diameter increased from 82 nm at pH 8 to 210 nm at pH 5, which evidences intermolecular aggregation in the system. It has to be noted that nanoparticles formed upon acidification of PPEG1 were sufficiently stable in solution and were characterized by a narrow polydispersity index (0.09 at pH 5 versus 0.37 at pH 8). Furthermore, although PPEG1 aggregates at lower pH, the behavior of this PEGylated polymer is strikingly different from the behavior of the corresponding polyacid homopolymer, which precipitates in acidic solutions (data not shown). A highly PEGylated polymer, PPEG16 does not appear to aggregate under acidic conditions at all (Figure 3C), although a minute decrease in size was observed for the acidified polymer – 90 nm versus 100, which can suggest some intramolecular interactions due to stronger hydrogen bonding at lower pH. Overall, the pH sensitive conformation changes indicate a significant contribution of PEG groups to the polymer solution behavior. They either modulate or completely prevent aggregation and phase separation under acidic conditions. Such steric stabilization of polyelectrolytes against various environmental factors can also be of physiological importance as one of the main reasons for rapid clearance time is flocculation of the carrier leading to protein binding and complement activation (*48*).

Formation of Protein−Polyphosphazene Complexes

PEGylated polyphosphazenes were investigated for their protein carrier capacity. Carboxylic acid functionalities of polymers provide convenient sites for covalent or non-covalent attachment of the biological cargo. Avidin, a glycoprotein characterized by high isoelectric point (pI 10.5) (*49*), was chosen for investigating non-covalent binding characteristics of PPEGs. Figures 4A-C show AF4 profiles of fluorescently labeled avidin in the presence and absence of PPEGs recorded at 495 nm. All spectra of 10:1 (mol/mol) avidin-PPEG formulations (curves 2) demonstrated both broader peak shapes and longer retention times (approximately 6 to 16 minutes increase) compared to those of protein alone (curves 1). Since PPEGs cannot be visualized at this wavelength, their profiles were recorded at 280 nm as a reference and shown as dashed curves (curves 3). As seen from the figure, AF4 profiles of complexes are characterized by peak retention times and shapes similar to those of polymers, but display absorbance at 495 nm, which is only characteristic to the fluorescently labeled protein. This clearly demonstrates the formation of avidin-PPEG complexes for all polymers. Moreover, AF4 profiles of complexes show no absorbance at retention time around seven minutes demonstrating the absence of unbound avidin in polymer-protein formulations. AF4 spectra of unlabeled avidin-PPEG formulations are also characterized by the absence of free protein in the system and demonstrate an increase in peak size at the retention time of PPEG.

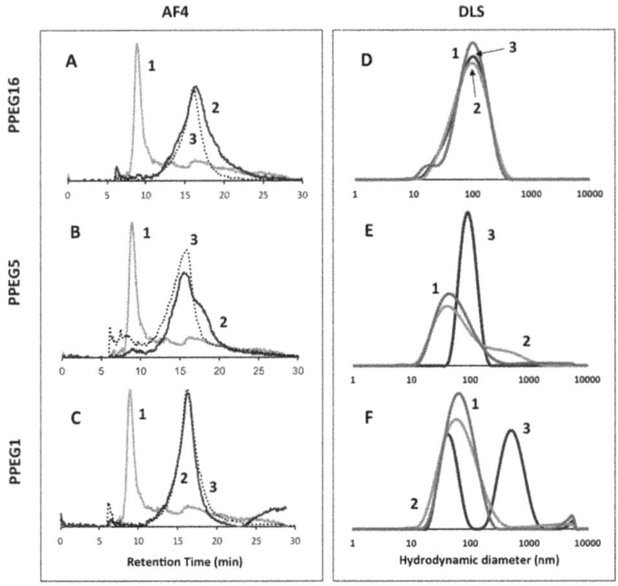

Figure 4. Protein-PPEG binding. (A−C) AF4 profiles of FITC-avidin (1, orange) and FITC-avidin – PPEG complexes (2, blue) recorded at 495 nm for (A) PPEG16, (B) PPEG5, and (C) PPEG1. AF4 profiles for PPEGs (3, dashed lines) recorded at 280 nm are shown for comparison (0.1 mg/mL avidin, 0.1 mg/mL PPEG, complexes: 0.1 mg/mL avidin, polymer concentration adjusted to provide for 10:1 (mol/mol) avidin:PPEG ratio, PBS, pH 7.4); (D−F) DLS profiles of PPEGs (1, brown) and avidin – PPEG complexes at 1:1 (2, green) and 10:1 (3, blue) molar ratios for (D) PPEG16, (E) PPEG5, and (F) PPEG1 (distribution by intensity, 0.1 or 0.01 mg/mL avidin, PBS, pH 7.4).

DLS profiles of avidin-PPEG complexes were affected both by the content of PEG groups in the polymer and the molar ratios of the components (Figures 4D-F). Avidin-PPEG16 complexes showed size distribution, which was practically identical to the polymer and was not sensitive to its composition at least in the 10:1-1:1 (mol/mol) protein to polymer ratio range (Figure 4D). PPEG5 showed similar behavior at the lower protein concentration, but displayed more narrow distribution at 10:1 ratio (Figure 4E), which demonstrate some polymer conformation changes and possible compaction of the ionic core. Finally, complexes of PPEG1 at 1:1 molar ratio showed sizes similar to the polymer. However, some aggregation, indicating intermolecular interactions in the system, was observed at a higher protein concentration (Figure 4F). Similar to the results for calcium cross-linking, the absence of aggregation in avidin formulations with PPEG16 and PPEG5 may be potentially explained by extensive "PEGylation" of their surfaces, which prevents intermolecular interactions that are evident for the polymer with high ionic content, PPEG1.

Maintenance of Protein Conformation and Functional Properties in the Complex

Conformational stability of avidin in non-covalent complexes with PPEGs was evaluated using circular dichroism (CD) spectroscopy. No loss of the native fold of avidin was detected upon addition of polyphosphazene at various concentrations, indicating that the protein maintains its structural integrity in complexes with PPEGs. Representative CD spectra of avidin at various concentrations of PPEG5 are shown in Figure 5A.

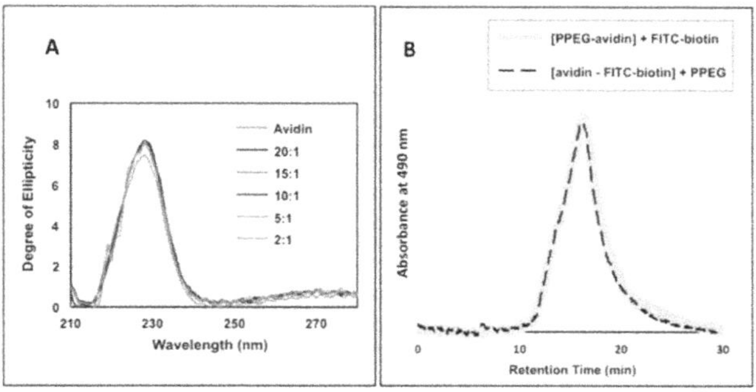

Figure 5. PPEG-avidin complex functionality. (A) Circular dichroism spectra of avidin and avidin-PPEG5 complexes at various protein to polymer molar ratios (0.2 mg/mL avidin, PBS, pH 7.4); (B) AF4 profiles of PPEG16-avidin complexes labeled with FITC-biotin, which was added either directly to the complex (solid blue line) or added first to avidin and then reacted with polymer (dashed black line) (0.51 mg/mL PPEG16, 0.1 mg/mL avidin, 0.004 mg/mL FITC-biotin, absorbance at 495 nm, PBS, pH 7.4).

Maintenance of functional properties of protein was also evaluated by its ability to bind biotin. This was evaluated by reacting avidin-polyphosphazene complex with fluorescently labeled avidin substrate – FITC-biotin. Avidin – FITC-biotin complex, which was prepared first and then mixed with polyphosphazene, was used as a control. Figure 5B shows that AF4 profiles for both formulations using highly PEGylated polyphosphazene, PPEG16, were practically identical. This experiment demonstrates that avidin complexed with polyphosphazene fully maintains its capacity to bind the substrate, which suggests that non-covalent protein-polyphosphazene interactions are gentle enough not to interfere with the biological function of the drug cargo. No dissociation of the complex was detected under these conditions as the peak position remained at the elution time characteristic for avidin-polyphosphazene complex.

Avidin-PPEG Complex Stability in Serum

One of the main factors severely limiting the survival time of polyelectrolyte complexes, such as polyplexes or protein-polyelectrolyte complexes, in the bloodstream is their tendency to undergo polyelectrolyte exchange reactions with endogenous proteins (*48*). Therefore, non-covalent protein-PPEG complexes were evaluated for their stability in the presence of serum proteins. Avidin-PPEG16 complex was incubated in 50% rabbit serum in PBS at 37°C and then evaluated by reacting with fluorescently labeled biotin with subsequent analysis by AF4 at 495 nm. No major changes in AF4 profiles were observed after 24 hours (data not shown). These results demonstrate complex stability in the presence of plasma proteins and the absence of protein exchange reactions in the system.

Hydrodynamic Volumes of the Complexes: The Effect of Protein Content

As was already mentioned, many of therapeutic proteins and peptides have less than optimal pharmacokinetic properties, often because they are smaller than the glomerular filtration cutoff of around 8 nm (*1*, *50*). Figure 6A compares hydrodybanic diameters of avidin:PPEG complexes with avidin at 1:1 and 10:1 molar ratios. A major shift towards larger molecular sizes is evident for avidin-PPEG formulations compared to avidin alone with the hydrodynamic diameters of complexes reaching at least 50 nm (Figure 6B). The molecular diameter of the avidin:PPEG16 complexes was independent of protein content (Figure 6A).

The Effect of PPEGs on Protein Antigenicity

Protein immunogenicity is a significant concern for therapeutic drugs as it can affect both safety and efficacy of the drug (*51*). The consequences of protein immunogenicity vary from no evidence of clinical effect to severe, life-threatening responses and its reduction can be positively reflected in the pharmacokinetic profile of the protein (*12*). The ability of PPEGs to reduce protein immunogenicity was assessed by assaying its antigenicity in vitro by ELISA. The results indicate that all PPEGs were able to mask protein epitopes; however, the effect was polymer dose dependent and was stronger for polymers with higher content of PEG (Figure 6B). The latter suggests that the hydrated PEG cloud was primarily responsible for masking antigenic determinants of the protein. For PPEG16, the maximum reduction in antigenicity relative to free protein was achieved at equimolar complex composition. Polymers with lower PEG content displayed inferior protection against antibodies. Moreover, intermolecular polymer interactions, which are characteristic for PPEG1 and PPEG5 (Figures 3 and 4), apparently contributed to protein masking as the maximum masking effect was achieved when molar excess of these polymers was used (Figure 6B).

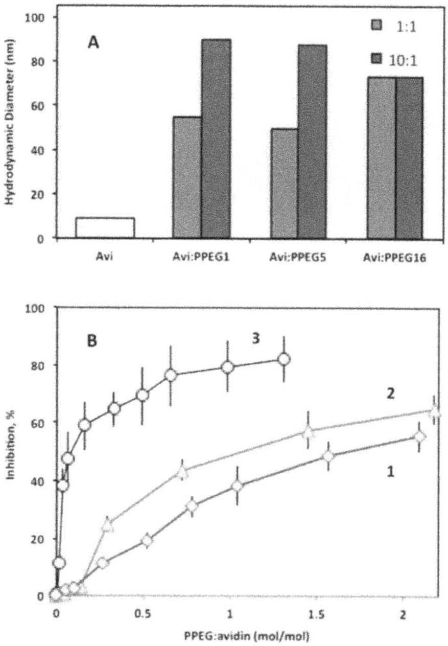

Figure 6. PEGylation effects of PPEGs. (A) Hydrodynamic diameters of avidin and its complexes with PPEGs at 1:1 and 10:1 molar ratios; (B) antigenicity of avidin complexes with (1) PPEG1, (2) PPEG5, and (3) PPEG16 in vitro using ELISA assay (0.75 nmol/L avidin, PBS, pH 7.4).

Binding and Uptake of Protein-"Loaded" Polyphosphazenes by Cells

We next examined whether PPEG16/FITC could facilitate interaction of protein cargo (i.e. avidin) with cells, for which we used oral adenosquamous carcinoma Cal27 line as a model. First, cells were incubated in serum-containing media with PPEG16/FITC-avidin complex or control FITC-avidin alone for 30 minutes, a short incubation time which should mainly show cell-surface interactions (Figure 7A). As shown in fluorescence micrographs and associated image quantification, the polymer markedly enhanced the amount of fluorescent avidin associated per cell from a mean intensity of 283 A.U. to 1,401 A.U., a 5-fold increase in only 30 minutes. In addition, an increase in the number of 100-200 nm diameter vesicular objects was also observed, from 4.5 objects per cell for control FITC-avidin to 73.0 objects/cell for PPEG16/FITC-avidin complex, a 16-fold improvement. This later data suggests that FITC-avidin may begin to internalize within endocytic vesicles by this time when presented as a polymer complex.

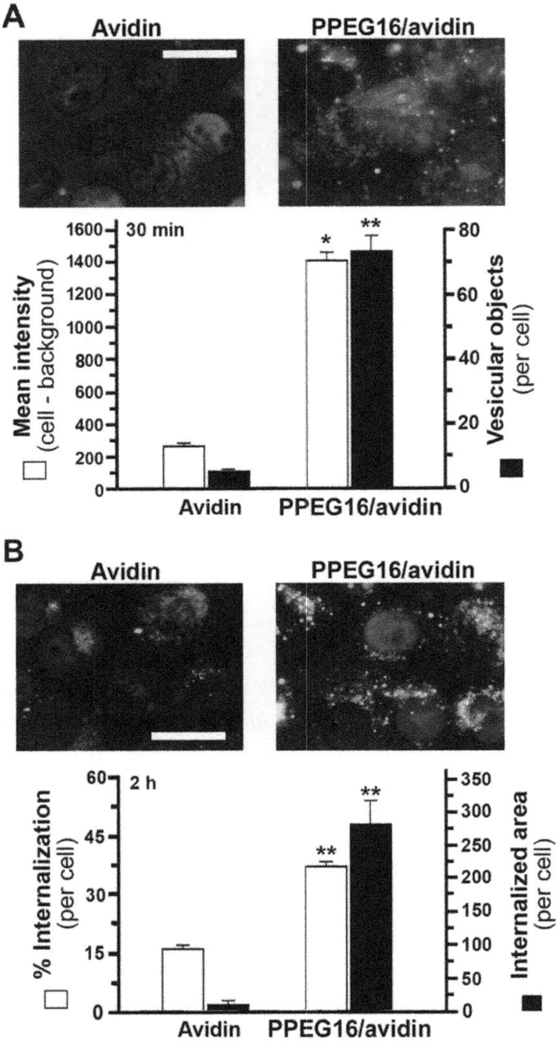

*Figure 7. PPEG16 enhances cell association and internalization of a model protein cargo. (A) Fluorescence microscopy visualization and quantification of background-corrected mean fluorescence intensity (white bars) and the number of 100-200 vesicular objects (black bars) of Cal27 cells after 30 min incubation with control FITC-avidin versus PPEG16/FITC-avidin complex. Green = total FITC-avidin; blue = cell nuclei marked with DAPI; scale bar = 10 μm. (B) Fluorescence microscopy visualization and quantification of the percentage of internalized cargo, and the area occupied by internalized cargo in Cal27 cells, after 2 h incubation with control FITC-avidin versus PPEG16/FITC-avidin complex. Yellow = FITC-avidin counterstained at the cell surface with biotin-IgG + Texas Red-secondary antibody; green = internalized FITC-avidin; blue = cell nuclei (DAPI). (A,B) Data are mean ± standard error of the mean (SEM). * $p<0.5$; ** $p<0.05$, both by Student's t-test.*

To verify this possibility, we next incubated Cal27 cells with either control FITC-avidin or PPEG16/FITC-avidin for 2 hours, to enable potential internalization by cells. This experiment was designed to distinguish cell-surface versus internalized FITC-avidin by using an established method that allows differential visualization of these two pools of cargo (*39*): Briefly, after 2 hour incubation, unbound materials were washed off, cells were fixed, and cell-surface located avidin was counter-labeled using biotin-IgG followed by Texas Red-labeled secondary antibody. As shown in Figure 7B, this renders cell-surface bound avidin as yellow (green + red), while internalized avidin appears green alone. Image quantification showed that the percentage of internalized avidin, calculated from the total avidin associated to cells, was enhanced from 16% in control FITC-avidin to 38% for the polymer conjugate (2.4-fold increase). Additionally, the total surface area occupied by internalized avidin was increased from 11.3 px^2 in control FITC-avidin to 283 px^2 for the polymer conjugate (25-fold enhancement). Therefore, it appears that PPEG16 provides significant enhancement in the cellular interaction and uptake of a model protein cargo.

Conclusions

Non-covalent PEGylation of proteins through their supramolecular assembly with biodegradable carriers containing pendant PEG groups is an appealing approach that can eliminate the complexity of chemical conjugation reactions and associated purification processes. PEGylated polyphosphazenes (PPEGs) offer an interesting pathway, which allows controlled introduction of the PEG steric "shield" and facilitates spontaneous self-assembly with protein through multivalent interactions. Tailorable hydrolytic degradability of PPEGs demonstrated above, which is an important feature of the phosphorus-nitrogen backbone, suggests clinical suitability and potential for modulating pharmacokinetic profiles. Protein carrier capacity of PPEGs is demonstrated by their spontaneous self-assembly with avidin to form complexes in which protein maintains its conformation and function. PEG "shield" of polyphosphazenes imparts stabilizing properties for protein-polymer complexes against aggregation even at high protein loads and in the presence of plasma proteins. Importantly, PPEGs demonstrate their ability to dramatically, up to 10-fold, increase protein size and reduce antigenicity of protein, two important parameters for extending *in vivo* protein half-life. Despite exhibiting protein 'masking' properties, PPEG displayed ability to facilitate intracellular uptake of cargo protein by cancer cells. These results suggest the potential of PPEG as both a biodegradable agent for non-covalent PEGylation and its utility as an intracellular delivery vehicle. It is tempting to speculate that such non-covalent approaches to protein stabilization and delivery can potentially simplify production processes, result in contaminant free formulations, and even broaden the scope of protein drugs to which PEGylation technology can be applied.

Acknowledgments

This work was supported by intramural IBBR funding (A.P.M., S.M. and A.K.A.), MPower Maryland (A.P.M. and A.K.A.), and a National Science Foundation, CBET-1402756 (S.M.). Authors are thankful to Fushimi Pharmaceutical Co. for kindly providing hexachlorocyclotriphosphazene.

References

1. Strohl, W. R. Fusion proteins for half-life extension of biologics as a strategy to make biobetters. *Biodrugs* **2015**, *29*, 215–239.
2. Kaspar, A. A.; Reichert, J. M. Future directions for peptide therapeutics development. *Drug Discovery Today* **2013**, *18*, 807–817.
3. Wang, L.; Ying, T. New directions for half-life extension of protein therapeutics: The rise of antibody Fc domains and fragments. *Curr. Pharm. Biotechnol.* **2016**, *17*, 1348–1352.
4. Vaishya, R.; Khurana, V.; Patel, S.; Mitra, A. K. Long-term delivery of protein therapeutics. *Expert Opin. Drug Delivery* **2015**, *12*, 415–440.
5. Ecker, D. M.; Jones, S. D.; Levine, H. L. In *The Therapeutic Monoclonal Antibody Market*; MAbs, Taylor & Francis, 2015; pp 9−14.
6. Bruno, B. J.; Miller, G. D.; Lim, C. S. Basics and recent advances in peptide and protein drug delivery. *Ther. Del.* **2013**, *4*, 1443–1467.
7. Turecek, P. L.; Bossard, M. J.; Schoetens, F.; Ivens, I. A. PEGylation of biopharmaceuticals: A review of chemistry and nonclinical safety information of approved drugs. *J. Pharm. Sci.* **2016**, *105*, 460–475.
8. Veronese, F. M.; Pasut, G. PEGylation, successful approach to drug delivery. *Drug Discovery Today* **2005**, *10*, 1451–1458.
9. Morar, A. S.; Schrimsher, J. L.; Chavez, M. D. PEGylation of Proteins: A structural approach. *BioPharm Int.* **2006**, *19*.
10. Veronese, F. M.; Harris, J. M. Introduction and overview of peptide and protein pegylation. *Adv Drug Deliv Rev* **2002**, *54*, 453–6.
11. Swierczewska, M.; Lee, K. C.; Lee, S. What is the future of PEGylated therapies? *Expert Opinion on Emerging Drugs* **2015**, *20*, 531–536.
12. Caliceti, P.; Veronese, F. M. Pharmacokinetic and biodistribution properties of poly(ethylene glycol)–protein conjugates. *Adv. Drug Delivery Rev.* **2003**, *55*, 1261–1277.
13. Tabrizi, M. A.; Tseng, C.-M. L.; Roskos, L. K. Elimination mechanisms of therapeutic monoclonal antibodies. *Drug Discovery Today* **2006**, *11*, 81–88.
14. Duncan, R. The dawning era of polymer therapeutics. *Nat. Rev. Drug Discovery* **2003**, *2*, 347–360.
15. Chertok, B.; Webber, M. J.; Succi, M. D.; Langer, R. Drug delivery interfaces in the 21st century: From science fiction ideas to viable technologies. *Mol. Pharmaceutics* **2013**, *10*, 3531–3543.
16. Mero, A.; Ishino, T.; Chaiken, I.; Veronese, F. M.; Pasut, G. Multivalent and flexible PEG-nitrilotriacetic acid derivatives for non-covalent protein pegylation. *Pharm. Res.* **2011**, *28*, 2412.

17. Mueller, C.; Capelle, M. A. H.; Arvinte, T.; Seyrek, E.; Borchard, G. Tryptophan-mPEGs: Novel excipients that stabilize salmon calcitonin against aggregation by non-covalent PEGylation. *Eur. J. Pharm. Biopharm.* **2011**, *79*, 646–657.
18. Fee, C. J.; Van Alstine, J. M. PEG-proteins: Reaction engineering and separation issues. *Chem. Eng. Sci.* **2006**, *61*, 924–939.
19. Mueller, C.; Capelle, M. A. H.; Arvinte, T.; Seyrek, E.; Borchard, G. Noncovalent pegylation by dansyl-poly(ethylene glycol)s as a new means against aggregation of salmon calcitonin. *J. Pharm. Sci.* **2011**, *100*, 1648–1662.
20. Mueller, C.; Capelle, M. A. H.; Seyrek, E.; Martel, S.; Carrupt, P.-A.; Arvinte, T.; Borchard, G. Noncovalent PEGylation: Different effects of dansyl-, l-tryptophan–, phenylbutylamino-, benzyl- and cholesteryl-pegs on the aggregation of salmon calcitonin and lysozyme. *J. Pharm. Sci.* **2012**, *101*, 1995–2008.
21. Webber, M. J.; Appel, E. A.; Vinciguerra, B.; Cortinas, A. B.; Thapa, L. S.; Jhunjhunwala, S.; Isaacs, L.; Langer, R.; Anderson, D. G. Supramolecular PEGylation of biopharmaceuticals. *Proc. Natl. Acad. Sci.* **2016**, *113*, 14189–14194.
22. Khondee, S.; Olsen, C. M.; Zeng, Y.; Middaugh, C. R.; Berkland, C. Noncovalent PEGylation by polyanion complexation as a means to stabilize keratinocyte growth factor-2 (KGF-2). *Biomacromolecules* **2011**, *12*, 3880–3894.
23. Kurinomaru, T.; Shiraki, K. Noncovalent PEGylation of l-asparaginase using PEGylated polyelectrolyte. *J. Pharm. Sci.* **2015**, *104*, 587–592.
24. Kurinomaru, T.; Tomita, S.; Kudo, S.; Ganguli, S.; Nagasaki, Y.; Shiraki, K. Improved Complementary polymer pair system: Switching for enzyme activity by PEGylated polymers. *Langmuir* **2012**, *28*, 4334–4338.
25. Jun, Y. J.; Kim, J. I.; Jun, M. J.; Sohn, Y. S. Selective tumor targeting by enhanced permeability and retention effect. Synthesis and antitumor activity of polyphosphazene–platinum (II) conjugates. *J. Inorg. Biochem.* **2005**, *99*, 1593–1601.
26. Kim, J. I.; Jun, Y. J.; Seong, J.-Y.; Jun, M. J.; Sohn, Y. S. Synthesis and characterization of nanosized poly(organophosphazenes) with methoxy-poly(ethylene glycol) and dipeptide ethyl esters as side groups. *Polymer* **2004**, *45*, 7083–7089.
27. Qiu, L. Y.; Yan, M. Q. Constructing doxorubicin-loaded polymeric micelles through amphiphilic graft polyphosphazenes containing ethyl tryptophan and PEG segments. *Acta Biomater.* **2009**, *5*, 2132–2141.
28. Wilfert, S.; Iturmendi, A.; Schoefberger, W.; Kryeziu, K.; Heffeter, P.; Berger, W.; Brüggemann, O.; Teasdale, I. Water-soluble, biocompatible polyphosphazenes with controllable and pH-promoted degradation behavior. *J. Polym. Sci., Part A: Polym. Chem.* **2014**, *52*, 287–294.
29. Henke, H.; Kryeziu, K.; Banfić, J.; Theiner, S.; Körner, W.; Brüggemann, O.; Berger, W.; Keppler, B. K.; Heffeter, P.; Teasdale, I. Macromolecular Pt(IV) prodrugs from poly(organo)phosphazenes. *Macromol. Biosci.* **2016**, *16*, 1239–1249.

30. Hackl, C. M.; Schoenhacker-Alte, B.; Klose, M. H. M.; Henke, H.; Legina, M. S.; Jakupec, M. A.; Berger, W.; Keppler, B. K.; Bruggemann, O.; Teasdale, I.; Heffeter, P.; Kandioller, W. Synthesis and in vivo anticancer evaluation of poly(organo)phosphazene-based metallodrug conjugates. *Dalton Trans.* **2017**, *46*, 12114–12124.
31. Aichhorn, S.; Linhardt, A.; Halfmann, A.; Nadlinger, M.; Kirchberger, S.; Stadler, M.; Dillinger, B.; Distel, M.; Dohnal, A.; Teasdale, I.; Schöfberger, W. A pH-sensitive macromolecular prodrug as TLR7/8 targeting immune response modifier. *Chem.– Eur. J.* **2017**, *23*, 17721–17726.
32. Tian, Z.; Chen, C.; Allcock, H. R. Injectable and biodegradable supramolecular hydrogels by inclusion complexation between poly (organophosphazenes) and α-cyclodextrin. *Macromolecules* **2013**, *46*, 2715–2724.
33. Andrianov, A. K.; Martinez, A. P.; Weidman, J. L.; Marin, A.; Fuerst, T. R. Biodegradable PEGylated polyion nanocomplexes for protein delivery. *Biomacromolecules* **2018**Submitted.
34. Andrianov, A. K.; Chen, J.; LeGolvan, M. P. Poly(dichlorophosphazene) as a precursor for biologically active polyphosphazenes: Synthesis, characterization, and stabilization. *Macromolecules* **2004**, *37*, 414–420.
35. Yessine, M.-A.; Leroux, J.-C. Membrane-destabilizing polyanions: Interaction with lipid bilayers and endosomal escape of biomacromolecules. *Adv. Drug Delivery Rev.* **2004**, *56*, 999–1021.
36. Lackey, C. A.; Murthy, N.; Press, O. W.; Tirrell, D. A.; Hoffman, A. S.; Stayton, P. S. Hemolytic activity of pH-responsive polymer-streptavidin bioconjugates. *Bioconjugate Chem.* **1999**, *10*, 401–405.
37. Rozema, D. B.; Ekena, K.; Lewis, D. L.; Loomis, A. G.; Wolff, J. A. Endosomolysis by masking of a membrane-active agent (EMMA) for cytoplasmic release of macromolecules. *Bioconjugate Chem.* **2003**, *14*, 51–57.
38. Serrano, D.; Manthe, R. L.; Paul, E.; Chadha, R.; Muro, S. How carrier size and valency modulate receptor-mediated signaling: Understanding the link between binding and endocytosis of ICAM-1-targeted carriers. *Biomacromolecules* **2016**, *17*, 3127–3137.
39. Muro, S.; Muzykantov, V. R.; Murciano, J.-C. Characterization of Endothelial Internalization and Targeting of antibody—Enzyme Conjugates in Cell Cultures and in Laboratory Animals. In *Bioconjugation Protocols: Strategies and Methods*; Niemeyer, C. M., Ed.; Humana Press: Totowa, NJ, 2004; pp 21−36.
40. Gaucher, G.; Dufresne, M.-H.; Sant, V. P.; Kang, N.; Maysinger, D.; Leroux, J.-C. Block copolymer micelles: Preparation, characterization and application in drug delivery. *J. Controlled Release* **2005**, *109*, 169–188.
41. Andrianov, A. K.; Marin, A.; Fuerst, T. R. Molecular-level interactions of polyphosphazene immunoadjuvants and their potential role in antigen presentation and cell stimulation. *Biomacromolecules* **2016**, *17*, 3732–3742.
42. Andrianov, A. K.; Marin, A.; Roberts, B. E. Polyphosphazene polyelectrolytes: A link between the formation of noncovalent complexes

with antigenic proteins and immunostimulating activity. *Biomacromolecules* **2005**, *6*, 1375–1379.
43. Dobrovolskaia, M. A.; Clogston, J. D.; Neun, B. W.; Hall, J. B.; Patri, A. K.; McNeil, S. E. Method for analysis of nanoparticle hemolytic properties in vitro. *Nano Lett.* **2008**, *8*, 2180–2187.
44. Schmaljohann, D. Thermo-and pH-responsive polymers in drug delivery. *Adv. Drug Delivery Rev.* **2006**, *58*, 1655–1670.
45. Allcock, H. R.; Kwon, S. An ionically cross-linkable polyphosphazene: Poly[bis(carboxylatophenoxy)phosphazene] and its hydrogels and membranes. *Macromolecules* **1989**, *22*, 75–79.
46. Andrianov, A. K.; Chen, J.; Payne, L. G. Preparation of hydrogel microspheres by coacervation of aqueous polyphosphazene solutions. *Biomaterials* **1998**, *19*, 109–115.
47. Andrianov, A. K.; Svirkin, Y. Y.; LeGolvan, M. P. Synthesis and biologically relevant properties of polyphosphazene polyacids. *Biomacromolecules* **2004**, *5*, 1999–2006.
48. Seymour, L. W.; Oupicky, D.; Ogris, M. Development of long-circulating polyelectrolyte complexes for systemic delivery of genes. *J. Drug Targeting* **2002**, *10*, 93–98.
49. Bajpayee, A. G.; Wong, C. R.; Bawendi, M. G.; Frank, E. H.; Grodzinsky, A. J. Avidin as a model for charge driven transport into cartilage and drug delivery for treating early stage post-traumatic osteoarthritis. *Biomaterials* **2014**, *35*, 538–549.
50. Longmire, M.; Choyke, P. L.; Kobayashi, H. Clearance Properties of nano-sized particles and molecules as imaging agents: Considerations and caveats. *Nanomedicine (London, England)* **2008**, *3*, 703–717.
51. Shankar, G.; Arkin, S.; Cocea, L.; Devanarayan, V.; Kirshner, S.; Kromminga, A.; Quarmby, V.; Richards, S.; Schneider, C. K.; Subramanyam, M.; Swanson, S.; Verthelyi, D.; Yim, S. Assessment and reporting of the clinical immunogenicity of therapeutic proteins and peptides—Harmonized terminology and tactical recommendations. *AAPS J.* **2014**, *16*, 658–673.

Chapter 7

Applications of Self-Assembled Polyphosphazene Nano-Aggregates in Drug Delivery

Liyan Qiu,* and Jun Fu

Ministry of Education (MOE) Key Laboratory of Macromolecular Synthesis and Functionalization, Department of Polymer Science and Engineering, Zhejiang University, Hangzhou 310027, China
College of Pharmaceutical Sciences, Zhejiang University, Hangzhou 310058, China
***E-mail: lyqiu@zju.edu.cn.**

Polyphosphazenes are a kind of organic-inorganic hybrid polymers with unique chemical structures and powerful functionalities. When the two chlorine atoms attached on polyphosphazene backbone are substituted by hydrophilic chains and hydrophobic groups at optimal graft ratios, amphiphilic polyphosphazenes can be obtained, which can self-assemble into nano-aggregates in different architectures in aqueous solution, such as micelles and polymersomes. In this chapter, we aim to review the large potential of these versatile self-assembled nano-aggregates based on amphiphilic polyphosphazenes as drug carriers for applications in drug delivery.

Introduction

With the development of innovative pharmaceutical therapy, the need for drug carriers becomes increasingly important to deliver the needed medicine to target tissues (*1*). However, materials with the required structural control and high functionality are less common. Currently the most popular polymers under investigation include poly(lactic acid) (*2*), poly(acrylic acid) (*3*), polyethylene glycol (*4*) etc.. Furthermore, studies have demonstrated that precise control of polymer architectures is of critical importance as well as the chemical structure. Polymers provide unique advantages as carriers for drug molecules including evading absorption of serum proteins, improving chemical stability, prolonging circulation time, enhancing concentration at target tissues, minimizing damage to normal tissues and so on. Based on these rigorous and multiple demands, polyphosphazenes, a distinct family of hybrid polymers, generate numerous advantages such as a strong controllable polymerization processes to easily conjugate functional moieties onto the nitrogen and phosphorus atoms of the backbone and endow the resultant polymers with multitudinous functionalities (as shown in Figure 1). The groundbreaking work was explored by Allcock, H. R. and his coworkers in 1979 (*5*). Since then, the literature focused on biodegradable grafted polyphosphazene have captured much attention from researchers and accomplished significant achievements (*6–10*). To date, various kinds of polyphosphazene derivatives have been developed to meet different application requirements. For example, thermosensitive polyphosphazene-based hydrogels displayed great potential for local drug delivery. As Potta, T. and Song, S. C. et al. (*10*) reported in 2010, they designed a novel polyphosphazene bearing α-amino-ω-methoxy-PEG with molecular weight of 550 ($AMPEG_{550}$) as the hydrophilic chain and L-isoleucine ethyl ester (IleOEt) and an amino acid ester as hydrophobic groups. Additionally, Sohn, Y. S. et al. (*11*) attached antitumor drug (diamine)platinum(II) moiety onto the polyphosphazene backbone to synthesize a covalent conjugate. The most important features of this system include enhanced aqueous solubility, reduced renal clearance and thus increased plasma half-life time and a passive tumor accumulation by means of the water-soluble polyphosphazene. Additionally, various polyphosphazene microspheres were prepared by Andrianov, A. K. (*12–14*) for protein delivery, especially for vaccine delivery. These studies have been reviewed in details by Teasdale, I. (*15*). The combination of genes or adjuvants with polyphosphazene microspheres have potentially broadened the stage of immunological research and stimulated the body to produce stronger immune responses. Furthermore, the easily functionalized polyphosphazenes provide the system with longer circulation time and specific transport profile.

Drawing on the recent advances of nano drug delivery systems, the common desired characteristic is to overcome the systematic elimination and enhance circulation time, which in turn maximizes drug activity and reduces side effect. In addition, it is possible to further fine-tune the polymer structure in order to endow them with desired functionality, such as controlled drug release, targeted delivery and stimuli-responsive release. And the achieved nano-vehicles have a wide range of dimensions from 1 to 1,000 nm, mostly falling between 10-250 nm. These factors render nanomedicine an extremely promising trend in this field with a unique combination of properties. Furthermore, the subsequent delivery fate of nano-vehicles need to be excretal, which requires the polymers used as devices to be degradable and the by-products should also be nontoxic. Literature have proved that the protonation of the nitrogen atom facilitates the degradation of polyphosphazenes in acidic media (*16*, *17*). At the same time, varying the substituents also allows control over the polymer degradation rates. And the degradation products of polyphosphazenes (i.e. ammonia, phosphate and alcohol) are nontoxic and could be absorbed or excreted by the body (*18*). Last but most importantly, the unhindered rotation about the backbone bonds around the co-substituted hydrophilic and hydrophobic groups allows the polyphosphazenes self-assemble into nanoscaled micelles, polymersomes or some other specific aggregations. It is worth noting that, building self-assembled nanostructures is a rapidly-growing filed for biomedical applications and nanomedicine over the past decades, whereas micelles and polymersomes are undoubtedly the most eye-catching architectures. On one hand, drug loading into these nanoparticles can be automatically accomplished during the self-assemble process of amphiphilic polymers. On the other hand, the demand of site-specific drug release can be perfectly fulfilled via disaggregation of micelles or polymersomes responsive to various endogenous or external stimuli. In this chapter, therefore, we highlight the design of various amphiphilic polyphosphazenes and attractive applications of polyphosphazene micelles and polymersomes in drug delivery systems.

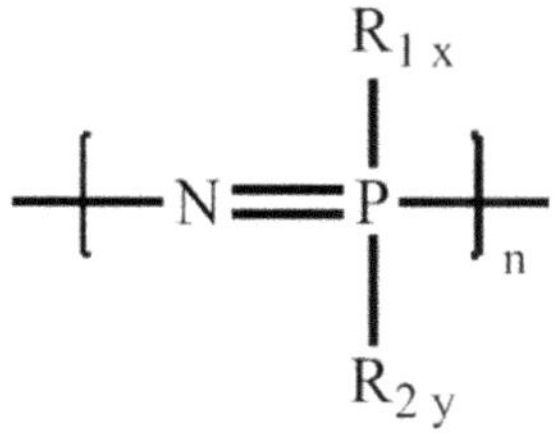

Figure 1. Genetic structure of polyphosphazenes. R_1 and R_2 represent the functional moieties attached onto the backbone, and x and y represent the mole of attached moieties on each unit (x+y=2).

Polyphosphazene-Based Micelles

Since Eisenberg, A. (*19*) raised a novel concept of "block polymer self-assembly" in 1995, self-assembling macromolecules have undergone pervasive exploration and achieved prodigious evolution. Generally, they contained a hydrophilic region and a hydrophobic domain, which would integrate into numerous specific shapes (i.e. nanospheres, nanocapsules, micelles, polymersomes etc.) driven by thermodynamics (*20*). Polymeric micelles have the simple structure and ease of encapsulating and delivering hydrophobic drugs in their water insoluble core, which typically enhances the solubilization of hydrophobic molecules and allows more drugs administered in aqueous solution. And the trend toward developing polyphosphazene-based micelles has motivated numerous research endeavors and has experienced a considerable and sustaining growth.

Several approaches have been explored to prepare polyphosphazene micelles, such as solvent exchange (*21, 22*), o/w emulsion (*23*), solvent evaporation (*24*) and so on. In terms of the first method, polymer is dissolved into an organic solvent owing good miscibility with water for ease of the subsequent self-assembly. The solvent is added dropwise into an aqueous solution under stirring. Stirring is continued at ambient temperature to obtain a homogeneous mixture. After that, the system is transferred into a dialysis bag and dialysis against water until complete removal of the organic solvent. O/w emulsion method is a bit complicated. Firstly, polymer is dissolved in organic solvent followed by pouring into a certain volume of aqueous solution containing surfactant to get a biphasic system. Then the mixture is emulsified by virtue of a mechanical stirrer. Finally, the above emulsion is sonicated using a probe sonicator and the stable nanoparticles formulation is gained once the organic solvent is evaporated by reduced pressure or heating. In this system, surfactant plays an essential role to ensure the emulsion formation. The specific operations for solvent evaporation method are as follows. The polymer is dissolved in a low-boiling organic solvent and then added into a certain volume of water. The mixture is continuously stirred till the organic solvent is all removed from the system. The self-assemblies gradually form along with the solvent evaporation. In the process, the selection of solvent, the volume ratio of water to organic solvent, stirring time and speed will affect the properties of nanoparticles. Hence, the optimal preparation parameters differ for various polymers, which we need to determine for the sake of better drug encapsulation efficiency.

Similar to classic amphiphilic block copolymer polyethylene glycol-poly(lactic acid), amphiphilic block copolymers where polyphosphazenes serve as hydrophobic segments are able to self-assemble into micelles. Contributions of Allcock, H. R. to the field include star-, comb- and linear block copolymers etc.. Take a work produced in 2009 (*25*) as an example, they developed a micellar nanoparticles based on amphiphilic β-cyclodextrin-adamantly substituted polyphosphazene and polystyrene block copolymers (Figure 2). Dynamic light scattering and transmission electron microscope suggested that the host-guest interactions between β-cyclodextrin and adamantly side group changes the polyphosphazenes block from hydrophobic to hydrophilic, which accordingly contributes to the formation of micelles with a diameter around 193 nm. Furthermore, they discovered that controlling the concentration of β-cyclodextrin could ensure the formation of micelles. Even with an accurate command of polymer architecture, the biocompatibility absence of polystyrene block limits the further application for drug delivery. Similar findings are also reported in other literature (*26–30*). In these papers poly[bis(2,2,2-trifluoroethoxy)phosphazene] was chosen as hydrophobic blocks. However, the further drug encapsulation behavior was not investigated.

In addition, a follow-up study by Allcock, H. R. (*31*) reported three novel diaminobutane poly(propyleeimine) dendrimers with polyphosphazene functionalized at the external surface of the dendrimers. Taking pyrene as a model hydrophobic drug, the polymers performed effective incorporation of the active molecule and salt-triggered release properties. The controlled release behavior is attributed to the swelling of the entire polyphosphazene periphery caused by changes in charge density with the increase in the salt concentration. The results indicate that the novel functional dendrimeric system is a promising drug delivery system and its potential applications need to be explored further.

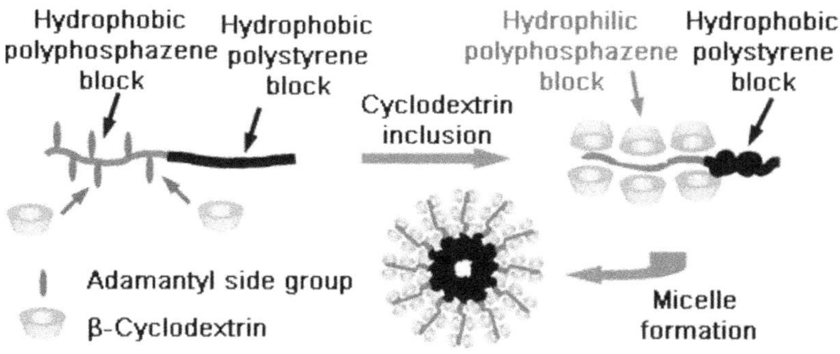

Figure 2. The schematic illustration of micelles formation via inclusion complexes of β-cyclodextrin and adamantly polyphophazene-polystyrene block copolymers (25). (Reproduced with permission from ref. (25). Copyright 2009 American Chemical Society).

Although it has been proven that the formation of block copolymers or dendrimers with polyphosphazenes allows for the fine-tuning of the physical properties through the side group system, the complicated procedure increases the workload and level of difficulty. Therefore, the exploration into using polyphosphazene self-assemblies alone as drug carriers have attracted much attention due to the simple synthesis process. In 2004, Qiu, L.Y. group (*32*) first synthesized a series of amphiphilic graft polyphosphazenes PNIPAm-g-PPPs by sequential nucleophilic substitution actions to link hydrophilic chains (poly(N-isopropylacrylamide) (PNIPAm)) and hydrophobic groups (glycine ethyl ester) onto polymer backbones. Profitting from the reversible hydration-dehydration conversion of PNIPAm sensitive to temperature change, the resultant PNIPAm-g-PPPs displayed thermosensitive behavior with a lower critical solution temperature (LCST) similar to that of PNIPAm at about 32 °C. As shown in Figure 3, PNIPAm-g-PPPs can be dissolved in aqueous solution in form of micelles at room temperature, while the morphology of PNIPAm-g-PPPs self-assembled aggregations was transformed into water insoluble nanoparticles as the temperature increased above the LCST since the PNIPAm chains shrank and collapsed onto the micelle surface. A similar work was reported in 2005 (*33*) that the thermosensitive micelle also formed when substituting glycine ethyl ester for ethyl 4-aminobenzoate as hydrophobic groups in amphiphilic polyphosphazenes with a LCST around 32.6 °C. They further investigated the effect of polymer composition, drug molecular structure and drug feeding amount on drug loading efficiency in polyphosphazene micelles (*21*). The results revealed that the hydrogen bonding interaction between drug and polymer components play critical roles to achieve relatively higher drug loading level. Furthermore, once locally injected, the body temperature induces the dehydration of PNIPAm chains of PNIPAm-g-PPP micelles, resulting in the sustained drug release over a desired period of time. The above work provided a new perspective for researchers to construct a long-acting drug delivery system, which owns a LCST slightly lower than body temperature.

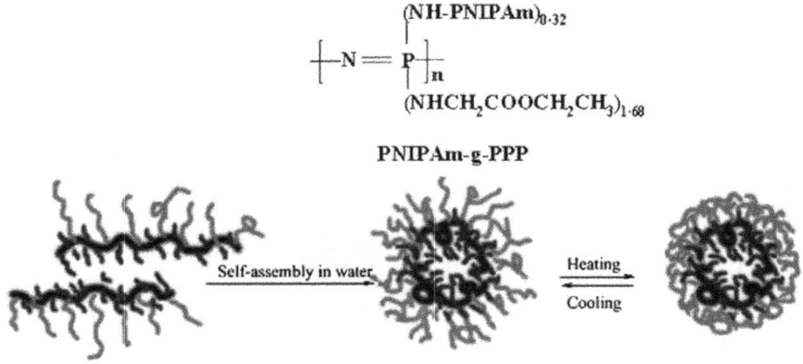

Figure 3. The diagrammatic representation of thermosensitive amphiphilic polyphosphazene micelles formation with PNIPAm and glycine ethyl ester (32). (Reproduced with permission from ref. (32). Copyright 2004 Wiley).

Inspired by the previous work (*32*), Qiu, L. Y. (*23*) and coworkers attempted to develop a novel drug delivery system based on amphiphilic polyphosphazenes which keeps micellar morphology during blood circulation and thus delivers active payloads to tumor via enhanced permeability and retention (EPR) effect. As literature demonstrated, the LCST of PNIPAm can be improved via copolymerization with a more hydrophilic segment (*34*). Furthermore, the ratio of these two monomers in turn accurately regulated the LCST (*35*, *36*). On account of the above research background, they copolymerized NIPAm with *N,N*-dimethylacrylamide (DMAA) to achieve poly(NIPAm-DMAA) with a LCST around 38 °C, which was a bit higher than normal physiological temperature. And then poly(NIPAm-DMAA) was attached onto polyphosphazene backbone with various amounts of hydrophobic ethyl glycinate group to obtain novel thermosensitive polyphosphazene micelles. Furthermore, dialysis and o/w emulsion methods were carried out to prepare doxorubicin-loaded micelles. As expected, the enhanced cytotoxicity of doxorubicin-loaded polyphosphazene micelles was observed compared with free drug solution.

As we know, the majority of polymeric micelles reported to date for cancer treatment selected poly(ethylene glycol) (PEG) as hydrophilic blocks owning to the low interfacial energy and the steric stability provided by the repulsion between PEG molecules (*37–39*). The combination of these two effects makes those polymer aggregations bearing PEG corona become stealth from protein adhesion and macrophage elimination during the systemic circulation, thus successfully deliver drug to tumors. Based on this concept, Zheng, C. et al. (*40*) innovatively constructed a micellar system composed of amphiphilc polyphosphazenes grafted by methoxy-poly(ethylene glycol) as hydrophilic chains and glycine ethyl ester as hydrophobic groups. The results demonstrated that the present polymer could effectively encapsulate hydrophobic antitumor agent with loading content greater than 10% and contributed to improve drug sensitivity of adriamycin-resistant human breast cancer cells by dint of the sustained drug release from micelles in cytoplasma. Similar micelles have been developed to realize a pH-sensitive drug release profile as well due to the protonation of *N,N*-diisopropylethylendiamine (DPA) responsive to acid intracellular environment. To be noticed, DPA-riched polyphosphazene micelles displayed significant cytotoxicity against MCF-7/Adr cancers via faconsiderable blockade of P-glycoprotein activity (*41*).

In addition, Avaji, P. G. et al. (*42*) synthesized a novel polymer-drug coupling micelle using amphiphilic polyphosphazene attached with methoxypoly(ethylene glycol) (M_n=550) and trans-2-diaminocyclohexane-Pt-2-hydroxyethoxydiethymalate. The conjugate micellar prodrugs accumulated at the tumor site via EPR effect, resulting in the higher tumor inhibition rate. Some years later, the other analogical fashion polyphosphazene-drug conjugates were developed (*43*, *44*). It cannot be denied that the polyphosphazene-based micelle is one of the most ideal carriers for drug molecules. Besides the above mentioned explorations in regard to amphiphilic polyphosphazene micelles, there are still so much endeavor devoted into this field. Table 1 displays some recent advances of polyphosphazene-based micelles in the field of drug delivery (*45–53*).

Polyphosphazene-Based Polymersomes

Although the development of micellar carriers is keeping prosperous for targeted drug delivery (*54*), they are generally applicable for hydrophobic drugs. With regard to the hydrophilic or water-soluble molecules, how to deliver them towards target tissue is still a challenge. Fortunately, through unremitting effort, a series of polyphosphazene polymersomes have been constructed and validated their tremendous potentials as drug vehicles. Polymersome, also called polymeric vesicle, is another kind of aggregations which are self-assembled by amphiphilic polymers like micelles. Different from micelles, however, polymersome owns aqueous intra-cavity surrounded by an inter-digitated lamella (*55*) so that polymersome can encapsulate hydrophilic drugs or hydrophobic drugs, or both of them simultaneously. Additionally, polymersome represents similar architecture to liposome, but higher mechanical stability and more tunable modification makes polymersome be promising alternative of liposome.

Table 1. Some recent explorations of polyphosphazene-based micelles in the field of drug delivery

Polymer	*Hydrophilic chain*	*Hydrophobic chain*	*Preparation method*	*Drug*	*Ref.*
poly(β-D-glucose-co-1-dodecyl) phosphazene	β-D-glucose	1-dodecy	crew-cut	-	*(45)*
poly(β-D-glucose-co-1-octyl)phosp-hazenes	β-D-glucose	1-octyl	dialysis	-	*(46)*
poly(4-methylphen oxyphosphazene)-graft-poly(2-methyl-2-oxazoline)	poly(2-methyl-2-oxazoline)	(4-methyl phenoxy-phosphazene	-	pyrene	*(47)*
MEEP/Ph-MEEP-PPP	MEEP	Ph-MEEP	stirring	-	*(48)*
PNIPAm/EAB-PPP	PNIPAm	EAB	dialysis	-	*(49)*
PNIPAm/EtGly-PPP	PNIPAm	Et-Gly	dialysis	indomethacin	*(50)*
PNIPAm/GlyET-PPP	PNIPAm	GlyET	dialysis	diflunisal	*(51)*
PNIPAAm/EAB-PPP	PNIPAAm	EAB	dialysis	indomethacin	*(52)*
PEG/DPA-PPP	PEG	DPA	dialysis	doxorubicin	*(53)*

MEEP represents $[N{=}P(OCH_2CH_2OCH_2CH_2OCH_3)_2]_n$, Ph-MEEP represents $[N{=}PPh(OCH_2CH_2OCH_2CH_2OCH_3)]_m$, PNIPAm represents poly(N-isopropylacrylamide), EAB represents ethyl 4-aminobenzoate, Et-Gly and GlyET represent ethyl glycinate, PEG represents polyethylene glycol, DPA represents N,N-diisopropylethylenediamine.

In 2007, Qiu, L.Y. (*56*) et al. found that by adjusting the feed amount of small molecular compound indomethacin into polyphosphazene solution, the morphology of self-assembled polyphosphazene aggregates changed from micelles into vesicles, mediated by hydrogen bonding interactions between polymers and drugs. On the contrary, the other small molecular drug, medroxyprogesterone acetate, failed to alter the self-assemble morphology due to the absence of strong hydrogen-bonding interaction. This is the first time to report polyphosphazene vesicles. Meanwhile, it also declared the rigorous demands on drug molecules to obtain such polyphosphazene vesicles. In 2009, a follow-up exciting work by Zheng, C. et al. (*57*) broke through the above restrictions, a polyphosphazene-based polymersome was constructed by sequentially substituting methoxy-poly(ethylene glycol) and ethyl-p-aminobenzoate onto the backbone (Figure 4). The polymer showed favorable encapsulation capability of hydrophilic doxorubicin (DOX·HCl). Moreover, DOX·HCl-loaded vesicles enhanced the cytotoxicity of DOX·HCl onto HepG2 cells.

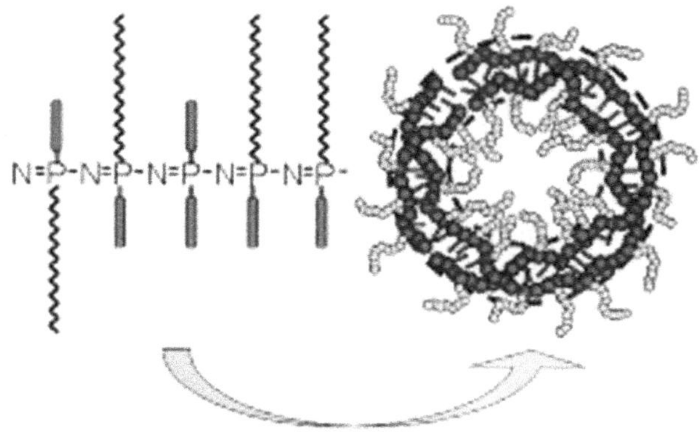

Figure 4. Graphic depiction of a novel polymeric polymersome based on amphiphilic polyphosphazene attached with ethyl-p-benzoate and methoxy-poly(ethylene glycol) to encapsulate DOX·HCl (57). (Reproduced with permission from ref. (57). Copyright 2009 Elsevier).

According to the literature on self-assemble behavior of amphiphilic block copolymers, the mass or volume fraction of hydrophilic segment (*f*) determines to great extent whether polymers self-assemble into polymersomes or not (*58*). As amphiphilic block polymers have a high interaction parameter between hydrophobic segment and water (*58*), the current uniform rule for achieving polymersomes structures in water is a phospholipid-like ratio of hydrophilic block to total mass of polymer: $f \approx 35 \pm 10\%$ (*59*, *60*). The most remarkable exploration about the self-assembly rule of polyphosphazenes was conducted by Qiu's group (Figure 5) (*61*). They meticulously inspected the influence of the volume fraction of hydrophilic chain on the copolymer nano-aggregations. Similar to block copolymers, grafted polyphosphazenes possess the parallel trend but with a bit higher critical value. Specifically, polyphosphazenes can exist as polymersomes with *f* ranging from 0.67~0.79, while the traditional value for amphiphilic block copolymers is 35%±10% (*62*). This roughly regularity of structure transitions provides a rational direction for the accurate design of novel polymersome carriers. Consequently, several polyphosphazene vesicles with different functions were explored and extensively studied in the last few years for drug delivery.

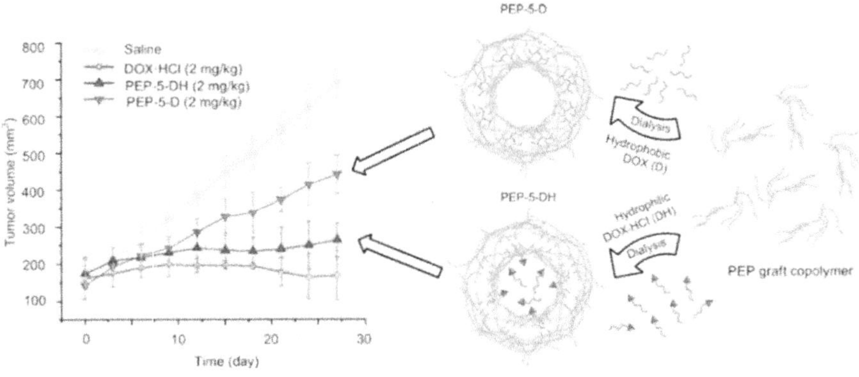

Figure 5. Vesicles based on polyphosphazene attached with ethyl-p-benzoate and polyethylene glycol to encapsulate hydrophilic or hydrophobic doxorubicin (61). (Reproduced with permission from ref. (61). Copyright 2014 Elsevier).

For instance, Zheng, C. et al. (*63*) constructed a polyphosphazene polymersome containing hydrophilic methoxy-poly(ethylene glycol) and hydrophobic DPA, which possessed pH-responsive protonation to endow the vesicle with pH-sensitive property. And the PEG mass fraction of this new grafted polymer is 0.57, which falls into the responding range to form vesicles. Furthermore, hydrophobic drug doxorubicin loaded into polymersomes showed burst release in weak acidic environment leading to the higher toxicity to MCF-7/Adr cells. This excellent performance attributed to the perforation formation on the surface of vesicles induced by the protonation of tertiary amino groups in DPA (Figure 6). Moreover, Zhu, X. M. and her colleagues (*64*) gained another polyphosphazene vesicle according to the above results, which could efficiently load anti-tumor drug carboplatin via the hydrogen-bonding interaction between therapeutic molecules and the hydrophobic groups of polyphosphazene. After integrated with polymeric vesicles, the CT-26 colon adenocarcinoma tumor growth was significantly inhibited accompanied with the alleviative carboplatin-caused myelotoxicity.

As known (*65*), multiple drug resistance is a major obstacle in cancer treatment, then the idea of a combination treatment rises in response to the proper time and conditions. In the study of Xu, J. et al. (*66*), an amphiphilic polyphosphazene polymersome containing ethyl-*p*-aminobenzoate as hydrophobic chain and amino-terminal poly(ethylene glycol) as hydrophilic group was prepared to co-deliver doxorubicin and chloroquine (Figure 7). The results indicated a prominent increase toxicity of this dual drug-loaded polymersome against doxorubicin-resistant breast cancer cell line and leukemia cell line when compared with free doxorubicin, free chloroquine as well as the single drug loaded ones. It has been established that the reason of doxorubicin-resistance in cancer cells is closely related to P-glycoprotein overexpression, which effluxes the already entered doxorubicin from cells resulting in a low concentration level of intracellular drug. However, with the endocytosis of co-loaded polymersomes, chloroquine could prohibit doxorubicin exocytosis as a competitive substrate of P-glycoprotein (*67–69*) and ultimately inhibit drug efflux more effectively. As a consequence, the co-loaded polymersomes displayed more potent cytotoxicity on breast cancer cells with IC_{50} 7.45-fold lower than that of free doxorubicin in P-glycoprotein overexpressed zebrafish model.

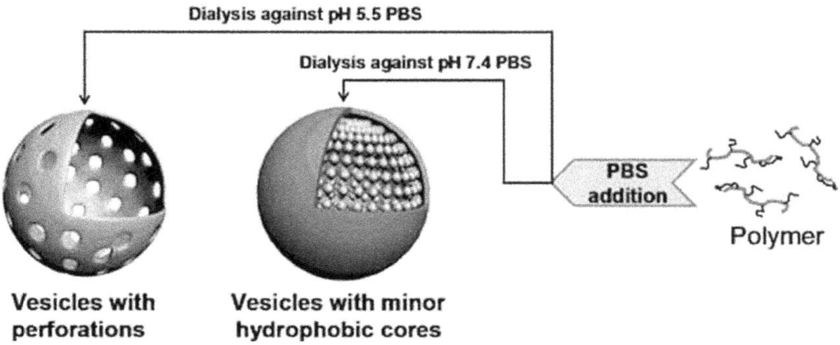

Figure 6. Graphic illustration of possible route for pH-responsive amphiphilic polyphosphazene vesicle formation (63). (Reproduced with permission from ref. (63). Copyright 2011 Wiley).

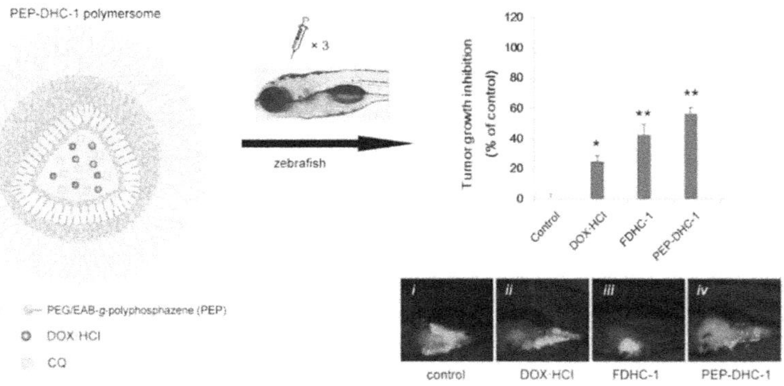

Figure 7. A polyphosphazene polymersomes co-encapsulating hydrophilic molecules, i.e. doxorubicin hydrochloride and chloroquine to reverse drug-resistance (66). FDHC-1 represents the mixture of free doxorubicin hydrochloride and chloroquine phosphate and PEP-DHC-1 means doxorubicin hydrochloride and chloroquine phosphate co-loaded PEP vesicles. (Reproduced with permission from ref. (66). Copyright 2016 Elsevier).

A successful drug carrier is designed to deliver an active agent to its targeted site at the right time and effective concentration. While in the quest to develop the optimal drug carriers, there still exist challenges to drugs about lack of action specificity and severe side effects (*70*). Over the past decades, the primary attention of targeted drug delivery is focused on cancer therapies, and some significant advances have been made by means of the subtle variation in tumor tissue, such as temperature (*71*), pH (*63*) and so on. In addition, various active-targeting moieties attached onto the surface of polymersomes are able to guide tumor-specific delivery of payload. Therefore, the popular targeting molecules, such as folic acid and RGD-containing peptides, are extensively tried to modify nanoparticles (*72*, *73*). There is, however, a small number of reports highlighting on the decoration of active-targeting moieties on polymersomes. Li, X. et al. (*74*) synthesized cholesterol-linked aptamer AS1411 to modify a polyphosphazene polymersome through hydrophobic supermolecular interactions, which acquired active target function to nucleolin expressed in the plasma membranes of MCF-7 cells (Figure 8). After loading hydrophilic doxorubicin with high efficiency, the aptamer anchored polymersomes performed higher assimilation property by breast cancer cells. After 1 h of exposure, the intracellular geometric mean of fluorescence incubated with aptamer-modified polymersomes was 1.73-fold greater than that of aptamer free polymersomes. Also the antiproliferative activity was more than a 50% stronger after 48 h in comparison with the aptamer absence polymersomes. Furthermore, the in vivo distribution studies ulteriorly demonstrated the fate of the polymersomes after intravenous administration. Note that nanovesicles with aptamer tended to gather at the tumor site during the entire experimental period and the amount of fluorescence in tumor tissue was 1.75-fold higher than that of the literalness polymersomes. In conclusion, the aptamer-functionalized polymersomes present superior cellular drug uptake, higher cytotoxicity and better in vivo performance.

As the research further developed, Qiu's group discovered a common issue existed in polymersomes, that is, when they highly encapsulated molecules, especially for hydrophilic drugs, the drug leakage from vesicles during the circulation process is obvious (*75–77*). As is known to all, this undesired performance will lead to the decrease of therapeutic efficiency and the enhancement of side effects. Therefore, Fu, J. and Qiu, L.Y. (*78*) prepared a neoteric polyphosphazene polymersome with tertiary amino groups in the hydrophilic chain. Once chloroauric acid ions adding into the system, the electrostatic adsorption between tertiary amino groups and ions leads to the uniform distribution of them among the lamella. With the aid of sodium borohydride, these ions in situ turned into gold nanoparticles to obtain an organic-inorganic hybrid vesicles. And then gold nanoparticles were introduced into the hydrophobic lamella and distributed uniformly via the chelation between gold nanoparticles and tertiary amino groups to obtain an organic-inorganic hybrid vesicle. As shown in Figure 9, the joined gold nanoparticles facilitate to enhance the compactness of hydrophobic membrane and obstruct the loaded doxorubicin hydrochloride (DOX·HCl) escape under physiological conditions. Furthermore, the introduced amount of gold nanoparticles into vesicles was investigated to screen the optimal ratio. On the other hand, the protonation of tertiary amino groups conduced to the holes formation on the surface of vesicles as the previous work reported (*63*), resulting in the burst drug release in tumor site which own the weak acidic circumstance. Finally, the accumulative release ratio was only 24% in pH 7.4, but 64.5% in pH 5.5 in the first 10 h. *In vivo* experiments of the hybrid polymersome exhibited better anti-tumor efficiency and slight cardiotoxicity which might induced by DOX·HCl simultaneously. Apart from the above strategy, Liang, L. N. et al. (*79*) incorporated cholesteryl hemisuccinate into polyphosphazene vesicles via the hydrogen bonding interaction. The lamella permeability was significantly decreased to retard the cargo leakage during the systemic circulation. Furthermore, the pH-sensitive benzimidazole contributed to the intracellular drug release. These two features worked synergistically to realize the high efficiency and low toxicity of medicines *in vivo*.

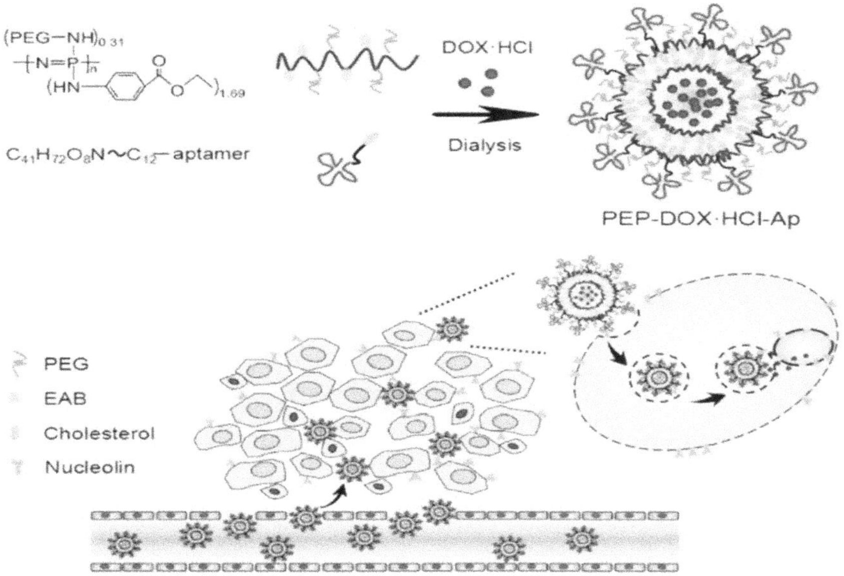

Figure 8. Polyphosphazene polymersomes targeted tumor cells via aptamer to selective accumulate at the specific site (74). (Reproduced with permission from ref. (74). Copyright 2016 Elsevier).

Except for the application of small drug molecules delivery, amphiphilic polyphosphazene vesicles have been built for gene and protein administration as well. Nowadays, the traditional way for gene or protein delivery via non-virus vehicles is to form polyplexes with polyphosphazenes by electrostatic interactions directly. In addition, the more examples about polyplexes based on polyphosphazenes and genes have been reviewed by Teasdale, I (*15*). Other than the aforementioned works, Peng, Y. et al. (*80*) set up a marvelous composite vesicles consisting of two amphiphilic polyphosphazenes, i. e. $[NP(PEG)_{0.3}(EAB)_{1.7}]_n$ (PEEP) and weakly positive $[NP(PEG)_{0.5}(DPA)_{1.5}]_n$ (PEDP) (Figure 10). The successful loading of MiR-200c ascribed to the synergistic effect of physical encapsulation and electrostatic interaction. Taking advantage of the almost neutral potential on the surface, the system was much safer than conventional positive charged carriers. And the subsequent tumor growth inhibition in vivo reached 68%, which was even higher than that of Lipo2000.

Another research achievement made by Gao, M. H. and her workmates (*81*) succeeded to package recombinant murine interleukin 12 plasmid (pmIL-12) into polyphosphazene polymersomes (Figure 11). The hydrophilic cavity and the weakly cationic N,N-diisopropylethylenediamine jointly facilitated the embarkation of pmIL-12. This system avoided the absorption of plasmid by serum protein and the destruction by deoxyribonuclease effectively in the blood circulation. In comparison with naked pmIL-12, plasmid-loaded vesicles accumulated at tumor site through enhanced permeation and retention effect and evoked the immune response. Furthermore, IL-12 generation was beneficial for the production of interferon-γ which could in turn stimulate phagocytes and dendritic cells to secret more IL-12. This positive feedback was conducive to reinforce the immune response and inhibit the tumor growth. It is worth mentioning that the severe systemic toxicity of IL-12 plasmid was circumvented and extended the administration mode from local delivery to systematic dose.

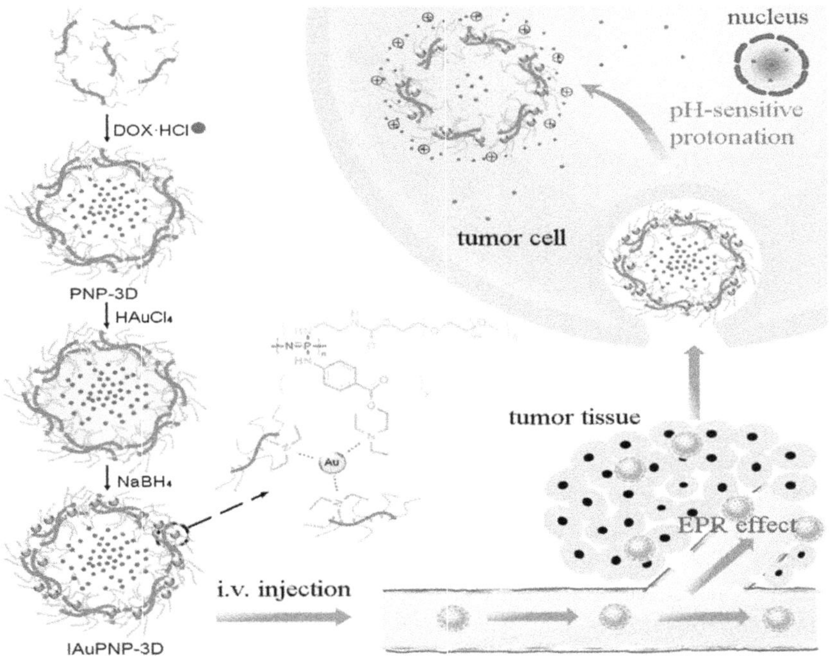

Figure 9. Preparation of PNP polyphosphazene vesicles with in-situ generated gold nanoparticles and the pH-responsive intracellular release (78). (Reproduced with permission from ref. (78). Copyright 2017 Wiley).

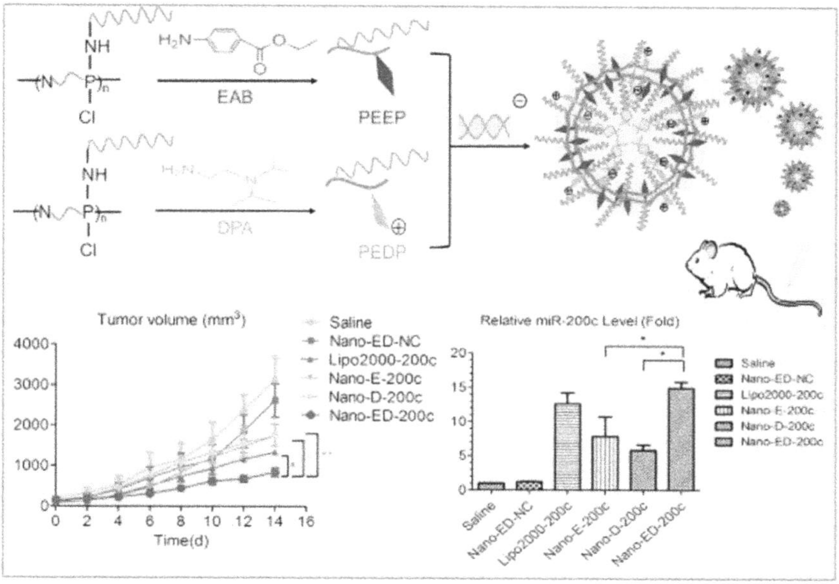

Figure 10. MiR-200c-loaded PEEP/PEDP vesicles achieved the improved in vivo anti-tumor efficiency (80). (Reproduced with permission from ref. (80). Copyright 2016 Elsevier).

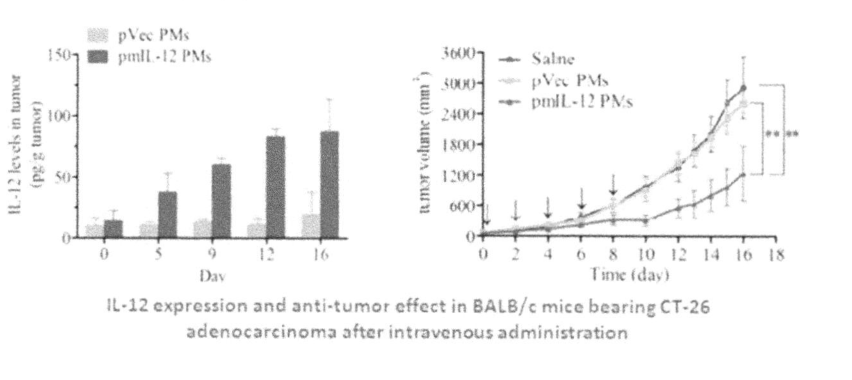

Figure 11. Preparation of pmIL-12-loaded polyphosphazene vesicles and the in vivo anti-tumor efficiency (81). (Reproduced with permission from ref. (81). Copyright 2016 American Chemical Society).

In short, thanks to the endeavor of researchers, more and more polymersomes based on polyphosphazenes have been explored alongside innumerable great achievements. Overall, the development of polyphosphazenes vesicles have changed the scientific landscape and provided a promising choice to drug carriers.

Conclusion

To sum up, over the past decade, a considerable and unremitting efforts have been made on the development of polyphosphazenes from micelle to polymersome. Literature studies have validated the concept that amphiphilic grafted polyphosphazenes have considerable advantages over conventional polymers owing to their unique structure. We believe that the comprehensive understanding of the relationship between polymer and treatment will facilitate the exploitation of the next generation polyphosphazene with stronger functions as drug devices. It is with this in mind that many research groups have devoted a considerable effort to the development of polyphosphazenes self-assembled nano-aggregates. The role played by polyphosphazenes in the field of drug vehicles will keep growing in importance in the future in order to deal with the harsh environment after administration. And there is no doubt that more intelligent polyphosphazene nanocarriers will emerge in the coming years and they will show extraordinary talents on the stage of drug delivery system. We eagerly awaited the arrival of this great era and report on these in due course.

References

1. Onaca, O.; Enea, R.; Hughes, D. W.; Meier, W. Stimuli-Responsive Polymersomes as Nanocarriers for Drug and Gene Delivery. *Macromol. Biosci.* **2009**, *9*, 129–139.
2. Hu, M.; Shen, Y.; Zhang, L.; Qiu, L. Polymersomes Via Self-Assembly of Amphiphilic Beta-Cyclodextrin-Centered Triarm Star Polymers for Enhanced Oral Bioavailability of Water-Soluble Chemotherapeutics. *Biomacromolecules* **2016**, *17*, 1026–1039.
3. Chun, M. K.; Cho, C. S.; Choi, H. K. Mucoadhesive Drug Carrier Based on Interpolymer Complex of Poly(Vinyl Pyrrolidone) and Poly(Acrylic Acid) Prepared by Template Polymerization. *J. Control. Release* **2002**, *81*, 327–334.
4. Zhang, J.; Rana, S.; Srivastava, R. S.; Misra, R. D. On the Chemical Synthesis and Drug Delivery Response of Folate Receptor-Activated, Polyethylene Glycol-Functionalized Magnetite Nanoparticles. *Acta Biomater.* **2008**, *4*, 40–48.
5. Allcock, H. R. Small-Molecule Phosphazene Rings as Models for High Polymeric Chains. *Acc. Chem. Res.* **1979**, *12*, 351–358.
6. Potta, T.; Chun, C.; Song, S. C. Rapid Photocrosslinkable Thermoresponsive Injectable Polyphosphazene Hydrogels. *Macromol. Rapid Commun.* **2010**, *31*, 2133–2139.

7. Potta, T.; Chun, C.; Song, S. C. Injectable, Dual Cross-Linkable Polyphosphazene Blend Hydrogels. *Biomaterials* **2010**, *31*, 8107–8120.
8. Lee, B. H.; Lee, Y. M.; Sohn, Y. S.; Song, S.-C. A Thermosensitive Poly (Organophosphazene) Gel. *Macromolecules* **2002**, *35*, 3876–3879.
9. Lee, B. H.; Song, S.-C. Synthesis and Characterization of Biodegradable Thermosensitive Poly (Organophosphazene) Gels. *Macromolecules* **2004**, *37*, 4533–4537.
10. Potta, T.; Chun, C.; Song, S. C. Injectable, Dual Cross-Linkable Polyphosphazene Blend Hydrogels. *Biomaterials* **2010**, *31*, 8107–8120.
11. Sohn, Y. S.; Baek, H.; Cho, Y. H.; Lee, Y.-A.; Jung, O.-S.; Lee, C. O.; Kim, Y. S. Synthesis and Antitumor Activity of Novel Polyphosphazene-(Diamine) Platinum (Ii) Conjugates. *Int. J. Pharm.* **1997**, *153*, 79–91.
12. Andrianov, A. K.; Chen, J.; Payne, L. G. Preparation of Hydrogel Microspheres by Coacervation of Aqueous Polyphosphazene Solutions. *Biomaterials* **1998**, *19*, 109–115.
13. Andrianov, A. K.; Payne, L. G.; Visscher, K. B.; Allcock, H. R.; Langer, R. Hydrolytic Degradation of Ionically Cross-Linked Polyphosphazene Microspheres. *J. Appl. Polym. Sci.* **1994**, *53*, 1573–1578.
14. Andrianov, A. K.; Chen, J. Polyphosphazene Microspheres: Preparation by Ionic Complexation of Phosphazene Polyacids with Spermine. *J. Appl. Polym. Sci.* **2006**, *101*, 414–419.
15. Teasdale, I. *Polyphosphazenes for Medical Applications*; Smithers Rapra: Shrewsbury, U.K., 2014.
16. Allcock, H. R.; Fuller, T. J.; Matsumura, K. Hydrolysis Pathways for Aminophosphazenes. *Inorg. Chem.* **1982**, *21*, 515–521.
17. Allcock, H. R.; Morozowich, N. L. Bioerodible Polyphosphazenes and Their Medical Potential. *Polym. Chem.* **2012**, *3*, 578–590.
18. Al-Abd, A. M.; Hong, K. Y.; Song, S. C.; Kuh, H. J. Pharmacokinetics of Doxorubicin after Intratumoral Injection Using a Thermosensitive Hydrogel in Tumor-Bearing Mice. *J. Control. Release* **2010**, *142*, 101–107.
19. Zhang, L.; Eisenberg, A. Multiple Morphologies of "Crew-Cut" Aggregates of Polystyrene-B-Poly(Acrylic Acid) Block Copolymers. *Science* **1995**, *268*, 1728–1731.
20. Antonietti, M. Self-Organization of Functional Polymers. *Nat. Mater.* **2003**, *2*, 9–10.
21. Zhang, J. X.; Li, X. J.; Qiu, L. Y.; Li, X. H.; Yan, M. Q.; Yi, J.; Zhu, K. J. Indomethacin-Loaded Polymeric Nanocarriers Based on Amphiphilic Polyphosphazenes with Poly (N-Isopropylacrylamide) and Ethyl Tryptophan as Side Groups: Preparation, in Vitro and in Vivo Evaluation. *J. Control. Release* **2006**, *116*, 322–329.
22. Chen, C.; Qian, Y. C.; Sun, C. B.; Huang, X. J. Self-Assembly and Morphological Transitions of Random Amphiphilic Poly(Beta-D-Glucose-Co-1-Octyl) Phosphazenes. *Soft Matter* **2015**, *11*, 6266–6274.
23. Qiu, L. Y.; Wu, X. L.; Jin, Y. Doxorubicin-Loaded Polymeric Micelles Based on Amphiphilic Polyphosphazenes with Poly(N-Isopropylacrylamide-Co-N,N-Dimethylacrylamide) and Ethyl Glycinate as Side Groups: Synthesis, Preparation and in Vitro Evaluation. *Pharm. Res.* **2009**, *26*, 946–957.

24. Qiu, L. Y.; Yan, M. Q. Constructing Doxorubicin-Loaded Polymeric Micelles through Amphiphilic Graft Polyphosphazenes Containing Ethyl Tryptophan and Peg Segments. *Acta Biomater.* **2009**, *5*, 2132–2141.
25. Cho, S. Y.; Allcock, H. R. Synthesis of Adamantyl Polyphosphazene−Polystyrene Block Copolymers, and B-Cyclodextrin-Adamantyl Side Group Complexation. *Macromolecules* **2009**, *42*, 4484–4490.
26. Chang, Y.; Prange, R.; Allcock, H. R.; Lee, S. C.; Kim, C. Amphiphilic Poly [Bis (Trifluoroethoxy) Phosphazene]−Poly (Ethylene Oxide) Block Copolymers: Synthesis and Micellar Characteristics. *Macromolecules* **2002**, *35*, 8556–8559.
27. Chang, Y.; Powell, E. S.; Allcock, H. R. Environmentally Responsive Micelles from Polystyrene–Poly [Bis (Potassium Carboxylatophenoxy) Phosphazene] Block Copolymers. *J. Polym. Sci. Part A: Polym. Chem.* **2005**, *43*, 2912–2920.
28. Allcock, H. R.; Cho, S. Y.; Steely, L. B. New Amphiphilic Poly [Bis (2 2,2-Trifluoroethoxy) Phosphazene]/Poly (Propylene Glycol) Triblock Copolymers: Synthesis and Micellar Characteristics. *Macromolecules* **2006**, *39*, 8334–8338.
29. Tian, Z.; Liu, X.; Chen, C.; Allcock, H. R. Synthesis and Micellar Behavior of Novel Amphiphilic Poly [Bis (Trifluoroethoxy) Phosphazene]-Co-Poly [(Dimethylamino) Ethyl Methacrylate] Block Copolymers. *Macromolecules* **2012**, *45*, 2502–2508.
30. Chang, Y.; Cho, S. Y.; Steely, L. B.; Allcock, H. R. Synthesis and Self-Association Behavior of Poly [Bis (2-(2-Methoxyethoxy) Ethoxy) Phosphazene]-B-Poly (Propyleneglycol) Triblock Copolymers. *J. Polym. Sci. Part A: Polym. Chem.* **2009**, *47*, 692–699.
31. Cho, S. Y.; Allcock, H. R. Dendrimers Derived from Polyphosphazene−Poly (Propyleneimine) Systems: Encapsulation and Triggered Release of Hydrophobic Guest Molecules. *Macromolecules* **2007**, *40*, 3115–3121.
32. Zhang, J. X.; Qiu, L. Y.; Zhu, K. J.; Jin, Y. Thermosensitive Micelles Self-Assembled by Novel N-Isopropylacrylamide Oligomer Grafted Polyphosphazene. *Macromol. Rapid Commun.* **2004**, *25*, 1563–1567.
33. Zhang, J. X.; Qiu, L. Y.; Jin, Y.; Zhu, K. J. Physicochemical Characterization of Polymeric Micelles Constructed from Novel Amphiphilic Polyphosphazene with Poly(N-Isopropylacrylamide) and Ethyl 4-Aminobenzoate as Side Groups. *Colloids Surf., B* **2005**, *43*, 123–130.
34. Liu, H.; Zhu, X. Lower Critical Solution Temperatures of N-Substituted Acrylamide Copolymers in Aqueous Solutions. *Polymer* **1999**, *40*, 6985–6990.
35. Nakayama, M.; Okano, T.; Miyazaki, T.; Kohori, F.; Sakai, K.; Yokoyama, M. Molecular Design of Biodegradable Polymeric Micelles for Temperature-Responsive Drug Release. *J. Control. Release* **2006**, *115*, 46–56.
36. Barker, I.; Cowie, J.; Huckerby, T.; Shaw, D.; Soutar, I.; Swanson, L. Studies of the "Smart" Thermoresponsive Behavior of Copolymers of

N-Isopropylacrylamide and N, N-Dimethylacrylamide in Dilute Aqueous Solution. *Macromolecules* **2003**, *36*, 7765–7770.
37. Meng, F.; Engbers, G. H.; Gessner, A.; Muller, R. H.; Feijen, J. Pegylated Polystyrene Particles as a Model System for Artificial Cells. *J. Biomed. Mater. Res. A* **2004**, *70*, 97–106.
38. Jeon, S. I.; Lee, J. H.; Andrade, J. D.; De Gennes, P. G. Protein—Surface Interactions in the Presence of Polyethylene Oxide: I. Simplified Theory. *J. Colloid Interface Sci.* **1991**, *142*, 149–158.
39. Jeon, S. I.; Andrade, J. D. Protein—Surface Interactions in the Presence of Polyethylene Oxide: Ii. Effect of Protein Size. *J. Colloid Interface Sci.* **1991**, *142*, 159–166.
40. Zheng, C.; Qiu, L. Y.; Yao, X. P.; Zhu, K. J. Novel Micelles from Graft Polyphosphazenes as Potential Anti-Cancer Drug Delivery Systems: Drug Encapsulation and in Vitro Evaluation. *Int. J. Pharm.* **2009**, *373*, 133–140.
41. Qiu, L.; Zheng, C.; Zhao, Q. Mechanisms of Drug Resistance Reversal in Dox-Resistant Mcf-7 Cells by Ph-Responsive Amphiphilic Polyphosphazene Containing Diisopropylamino Side Groups. *Mol. Pharm.* **2012**, *9*, 1109–1117.
42. Avaji, P. G.; Joo, H. I.; Park, J. H.; Park, K. S.; Jun, Y. J.; Lee, H. J.; Sohn, Y. S. Synthesis and Properties of a New Micellar Polyphosphazene-Platinum(Ii) Conjugate Drug. *J. Inorg. Biochem.* **2014**, *140*, 45–52.
43. Avaji, P. G.; Park, J. H.; Lee, H. J.; Jun, Y. J.; Park, K. S.; Lee, K. E.; Choi, S. J.; Sohn, Y. S. Design of a Novel Theranostic Nanomedicine: Synthesis and Physicochemical Properties of a Biocompatible Polyphosphazene-Platinum(Ii) Conjugate. *Int. J. Nanomed.* **2016**, *11*, 837–851.
44. Jun, Y. J.; Park, J. H.; Avaji, P. G.; Park, K. S.; Lee, K. E.; Lee, H. J.; Sohn, Y. S. Design of Theranostic Nanomedicine (Ii): Synthesis and Physicochemical Properties of a Biocompatible Polyphosphazene-Docetaxel Conjugate. *Int. J. Nanomed.* **2017**, *12*, 5373–5386.
45. Chen, C.; Huang, X.-J.; Liu, Y.; Qian, Y.-C.; Xu, Z.-K. Synthesis and Self-Assembly of Amphiphilic Polyphosphazene with Controllable Composition Via Two Step Thiol-Ene Click Reaction. *Polymer* **2014**, *55*, 833–839.
46. Chen, C.; Qian, Y.-C.; Sun, C.-b.; Huang, X.-J. Self-Assembly and Morphological Transitions of Random Amphiphilic Poly (B-D-Glucose-Co-1-Octyl) Phosphazenes. *Soft Matter* **2015**, *11*, 6266–6274.
47. Chang, J. Y.; Park, P. J.; Han, M. J. Synthesis of Poly (4-Methylphenoxyphosphazene)-Graft-Poly (2-Methyl-2-Oxazoline) Copolymers and Their Micelle Formation in Water. *Macromolecules* **2000**, *33*, 321–325.
48. Chang, Y.; Lee, S. C.; Kim, K. T.; Kim, C.; Reeves, S. D.; Allcock, H. R. Synthesis and Micellar Characterization of an Amphiphilic Diblock Copolyphosphazene. *Macromolecules* **2001**, *34*, 269–274.
49. Zhang, J. X.; Qiu, L. Y.; Jin, Y.; Zhu, K. J. Multimorphological Self-Assemblies of Amphiphilic Graft Polyphosphazenes with Oligopoly (N-Isopropylacrylamide) and Ethyl 4-Aminobenzoate as Side Groups. *Macromolecules* **2006**, *39*, 451–455.

50. Zhang, J. X.; Li, S. H.; Li, X. H.; Qiu, L. Y.; Li, X. D.; Li, X. J.; Jin, Y.; Zhu, K. J. Physicochemical Characterization, in Vitro, and in Vivo Evaluation of Indomethacin-Loaded Nanocarriers Self-Assembled by Amphiphilic Polyphosphazene. *J. Biomed. Mater. Res. A* **2008**, *86*, 914–925.
51. Zhang, J. X.; Qiu, L. Y.; Jin, Y.; Zhu, K. J. Thermally Responsive Polymeric Micelles Self-Assembled by Amphiphilic Polyphosphazene with Poly (N-Isopropylacrylamide) and Ethyl Glycinate as Side Groups: Polymer Synthesis, Characterization, and in Vitro Drug Release Study. *J. Biomed. Mater. Res. A* **2006**, *76*, 773–780.
52. Zhang, J. X.; Yan, M. Q.; Li, X. H.; Qiu, L. Y.; Li, X. D.; Li, X. J.; Jin, Y.; Zhu, K. J. Local Delivery of Indomethacin to Arthritis-Bearing Rats through Polymeric Micelles Based on Amphiphilic Polyphosphazenes. *Pharm. Res.* **2007**, *24*, 1944–1953.
53. Zheng, C.; Xu, J.; Yao, X.; Qiu, L. Polyphosphazene Nanoparticles for Cytoplasmic Release of Doxorubicin with Improved Cytotoxicity against Dox-Resistant Tumor Cells. *J. Colloid Interface Sci.* **2011**, *355*, 374–382.
54. Trivedi, R.; Kompella, U. B. Nanomicellar Formulations for Sustained Drug Delivery: Strategies and Underlying Principles. *Nanomedicine (London)* **2010**, *5*, 485–505.
55. Ahmed, F.; Discher, D. E. Self-Porating Polymersomes of Peg-Pla and Peg-Pcl: Hydrolysis-Triggered Controlled Release Vesicles. *J. Control. Release* **2004**, *96*, 37–53.
56. Zhang, J. X.; Li, X. D.; Yan, M. Q.; Qiu, L. Y.; Jin, Y.; Zhu, K. J. Hydrogen Bonding-Induced Transformation of Network Aggregates into Vesicles-a Potential Method for the Preparation of Composite Vesicles. *Macromol. Rapid Commun.* **2007**, *28*, 710–717.
57. Zheng, C.; Qiu, L.; Zhu, K. Novel Polymersomes Based on Amphiphilic Graft Polyphosphazenes and Their Encapsulation of Water-Soluble Anti-Cancer Drug. *Polymer* **2009**, *50*, 1173–1177.
58. Lee, J. S.; Feijen, J. Polymersomes for Drug Delivery: Design, Formation and Characterization. *J. Control. Release* **2012**, *161*, 473–483.
59. Discher, D. E.; Ahmed, F. Polymersomes. *Annu. Rev. Biomed. Eng.* **2006**, *8*, 323–341.
60. Discher, B. M.; Won, Y. Y.; Ege, D. S.; Lee, J. C.; Bates, F. S.; Discher, D. E.; Hammer, D. A. Polymersomes: Tough Vesicles Made from Diblock Copolymers. *Science* **1999**, *284*, 1143–1146.
61. Xu, J.; Zhao, Q. H.; Jin, Y. M.; Qiu, L. Y. High Loading of Hydrophilic/Hydrophobic Doxorubicin into Polyphosphazene Polymersome for Breast Cancer Therapy. *Nanomed. Nanotechnol.* **2014**, *10*, 349–358.
62. Discher, D. E.; Ortiz, V.; Srinivas, G.; Klein, M. L.; Kim, Y.; Christian, D.; Cai, S.; Photos, P.; Ahmed, F. Emerging Applications of Polymersomes in Delivery: From Molecular Dynamics to Shrinkage of Tumors. *Prog. Polym. Sci.* **2007**, *32*, 838–857.
63. Zheng, C.; Yao, X.; Qiu, L. Novel Polymeric Vesicles with Ph-Induced Deformation Character for Advanced Drug Delivery. *Macromol. Biosci.* **2011**, *11*, 338–343.

64. Zhu, X.; Peng, Y.; Qiu, L. Amino-Functionalized Nano-Vesicles for Enhanced Anticancer Efficacy and Reduced Myelotoxicity of Carboplatin. *Colloid Surface. B* **2017**, *157*, 56–64.
65. Wilson, T. R.; Johnston, P. G.; Longley, D. B. Anti-Apoptotic Mechanisms of Drug Resistance in Cancer. *Curr. Cancer Drug Targets* **2009**, *9*, 307–319.
66. Xu, J.; Zhu, X. M.; Qiu, L. Y. Polyphosphazene Vesicles for Co-Delivery of Doxorubicin and Chloroquine with Enhanced Anticancer Efficacy by Drug Resistance Reversal. *Int. J. Pharm.* **2016**, *498*, 70–81.
67. Fu, L. W.; Zhang, Y. M.; Liang, Y. J.; Yang, X. P.; Pan, Q. C. The Multidrug Resistance of Tumour Cells Was Reversed by Tetrandrine in Vitro and in Xenografts Derived from Human Breast Adenocarcinoma Mcf-7/Adr Cells. *Eur. J. Cancer* **2002**, *38*, 418–426.
68. Xia, C. Q.; Smith, P. G. Drug Efflux Transporters and Multidrug Resistance in Acute Leukemia: Therapeutic Impact and Novel Approaches to Mediation. *Mol. Pharmacol.* **2012**, *82*, 1008–1021.
69. Varma, M. V.; Ashokraj, Y.; Dey, C. S.; Panchagnula, R. P-Glycoprotein Inhibitors and Their Screening: A Perspective from Bioavailability Enhancement. *Pharmacol. Res.* **2003**, *48*, 347–359.
70. Hussein, A. S.; Abdullah, N.; Ahmadun, F. l.-R. Polymeric Micro and Nanoparticles as Promising Drug Delivery Vehicles: A Review. *Int. J. Nanoparticles* **2012**, *5*, 316–337.
71. Couffin-Hoarau, A. C.; Leroux, J. C. Report on the Use of Poly(Organophosphazenes) for the Design of Stimuli-Responsive Vesicles. *Biomacromolecules* **2004**, *5*, 2082–2087.
72. Lale, S. V.; Kumar, A.; Prasad, S.; Bharti, A. C.; Koul, V. Folic Acid and Trastuzumab Functionalized Redox Responsive Polymersomes for Intracellular Doxorubicin Delivery in Breast Cancer. *Biomacromolecules* **2015**, *16*, 1736–1752.
73. Das, S.; Sharma, D. K.; Chakrabarty, S.; Chowdhury, A.; Sen Gupta, S. Bioactive Polymersomes Self-Assembled from Amphiphilic Ppo-Glycopolypeptides: Synthesis, Characterization, and Dual-Dye Encapsulation. *Langmuir* **2015**, *31*, 3402–3412.
74. Li, X.; Zhu, X.; Qiu, L. Constructing Aptamer Anchored Nanovesicles for Enhanced Tumor Penetration and Cellular Uptake of Water Soluble Chemotherapeutics. *Acta Biomater.* **2016**, *35*, 269–279.
75. Liu, G. J.; Ma, S. B.; Li, S. K.; Cheng, R.; Meng, F. H.; Liu, H. Y.; Zhong, Z. Y. The Highly Efficient Delivery of Exogenous Proteins into Cells Mediated by Biodegradable Chimaeric Polymersomes. *Biomaterials* **2010**, *31*, 7575–7585.
76. Meng, F. H.; Engbers, G. H. M.; Feijen, J. Biodegradable Polymersomes as a Basis for Artificial Cells: Encapsulation, Release and Targeting. *J. Control. Release* **2005**, *101*, 187–198.
77. Upadhyay, K. K.; Bhatt, A. N.; Mishra, A. K.; Dwarakanath, B. S.; Jain, S.; Schatz, C.; Le Meins, J. F.; Farooque, A.; Chandraiah, G.; Jain, A. K.; Misra, A.; Lecommandoux, S. The Intracellular Drug Delivery and Anti Tumor Activity of Doxorubicin Loaded Poly(Gamma-Benzyl

L-Glutamate)-B-Hyaluronan Polymersomes. *Biomaterials* **2010**, *31*, 2882–2892.

78. Fu, J.; Liang, L.; Qiu, L. In Situ Generated Gold Nanoparticle Hybrid Polymersomes for Water-Soluble Chemotherapeutics: Inhibited Leakage and Ph-Responsive Intracellular Release. *Adv Funct. Mater.* **2017**, *27*, 1604981–1604993.
79. Liang, L.; Fu, J.; Qiu, L. Design of Ph-Sensitive Nanovesicles Via Cholesterol Analogue Incorporation for Improving in Vivo Delivery of Chemotherapeutics. *ACS Appl. Mater. Interfaces* **2018**, *10*, 5213–5226.
80. Peng, Y.; Zhu, X.; Qiu, L. Electroneutral Composite Polymersomes Self-Assembled by Amphiphilic Polyphosphazenes for Effective Mir-200c in Vivo Delivery to Inhibit Drug Resistant Lung Cancer. *Biomaterials* **2016**, *106*, 1–12.
81. Gao, M.; Zhu, X.; Wu, L.; Qiu, L. Cationic Polyphosphazene Vesicles for Cancer Immunotherapy by Efficient in Vivo Cytokine Il-12 Plasmid Delivery. *Biomacromolecules* **2016**, *17*, 2199–2209.

Fundamentals: Synthesis, Degradation Pathways, and Molecular Modeling

Chapter 8

Synthesis and Modification of Poly(alkyl/arylphosphazenes)

Patty Wisian-Neilson*,[1] and Robert H. Neilson[2]

[1]Department of Chemistry, Southern Methodist University, Dallas, Texas 75275, United States
[2]Department of Chemistry and Biochemistry, Texas Christian University, Fort Worth, Texas 76129, United States
***E-mail: pwisian@smu.edu.**

Polyphosphazenes with all substituents attached by direct phosphorus–carbon bonds are accessible via the thermal condensation-polymerization reaction of Si–N–P compounds. The general scope of this approach, modification reactions of the simplest polymers, and some of the basic properties of these materials are reviewed.

Although the best-known process for making polyphosphazenes remains the ring-opening-substitution approach, condensation-polymerization methods are gaining ground in the effort to make this unique class of polymers more accessible in large, reproducible quantities that allow for unique applications of this versatile polymer system. The first polyphosphazene made by condensation-polymerization was poly(bistrifluoroethoxyphosphazene), reported by Flindt and Rose in 1977 (*1*), about a dozen years after Allcock and Kugel (*2*) first made soluble poly(aminophosphazenes) and poly(alkoxyphosphazenes) by ring-opening of hexachlorocyclotriphosphazene, followed by nucleophilic substitution of chlorine in the reactive polymer, $[Cl_2PN]_n$. In 1980, as part of our ongoing studies of main group element compounds with Si–N–P linkages containing P(V), we observed the loss of fluorotrimethylsilane in our attempts to prepare (disilylamino)phosphoranes (*3*). Not only did we see elimination of Me_3SiF producing N-silylphosphoranimines, we also observed that further elimination of Me_3SiF occurred in some cases. Based on the Flindt and Rose work, we changed the halogen leaving groups to trifluoroethoxy groups and soon found that thermolysis produced essentially quantitative amounts of $Me_3SiOCH_2CF_3$ as well as the then elusive (*4*), fully methylated polymer, poly(dimethylphosphazene) (*5*), which has the distinction of being the isoelectronic analog of the well-studied silicone polymer, $[Me_2SiO]_n$. Since then, the scope of the condensation-polymerization has been significantly expanded in both reaction scale and diversity of substituents attached to the phosphazene backbone by P–C bonds (*6–11*). In addition, condensation processes involving loss of $P(O)Cl_3$ (*12*) and N_2 (*13*) have also been reported and several catalytic processes have been studied for polymerization of Si–N–P compounds (*14–19*).

Precursor Synthesis

An important aspect of the condensation-polymerization route to polyphosphazenes was the development of the straightforward, Wilburn "one-pot" synthesis of silylaminophosphines containing alkyl and aryl groups (*20*). This method (a) uses the bulky $(Me_3Si)_2N$ group to block reactivity of P–Cl bonds in commercially available phosphorus reagents such as PCl_3 and $PhPCl_2$, (b) can be conducted on a 1 to 3 mole scale (*21*), (c) avoids problems of earlier methods which used difficult to handle or prepare materials such as Me_2PCl or Ph(Me)PCl, and (d) is useful for preparing both symmetrical and unsymmetrical phosphines, $(Me_3Si)_2NPRR'$ (*10, 11, 20, 21*). (Scheme 1) In general, these silylaminophosphines can be obtained in high yields and are easily purified by vacuum distillation. Somewhat lower yields are obtained when sequential treatment with RMgCl/MeMgBr is employed because of formation of the disubstituted products with less sterically hindering R groups.

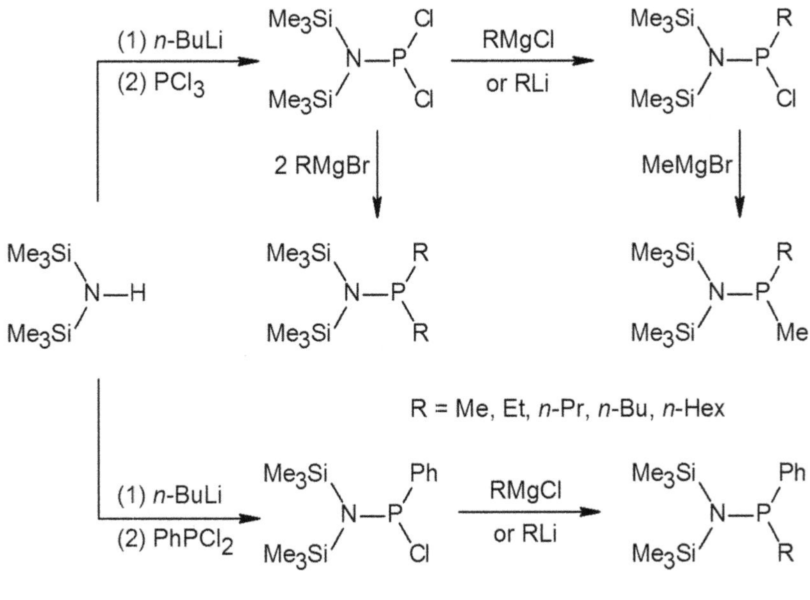

Scheme 1. Synthesis of (Disilylamino)phosphines

This step builds in the P–C bonded alkyl and aryl groups that cannot be attached via nucleophilic substitution of chlorine in poly(dichloro-phosphazene) (*22*), To achieve the P(V) oxidation state of polyphosphazenes, however, it is necessary to oxidize the P(III) center in the silylaminophosphines. Although there are many methods to do this, the role of the reactive Si–N bond and the nature of the oxidizing agent are ultimately crucial for the formation of polymerizable P(V) phosphoranimines. For example, oxidation of $(Me_3Si)_2N–PMe_2$ with Me_3SiN_3 (*23*) or Me_3SiOO*t*-Bu (*24*) results in N-silylphosphoranimines, $Me_3SiN{=}P(ESiMe_3)Me_2$ (where E = $NSiMe_3$ or O, respectively), but these types of P(V) compounds do not easily undergo condensation-polymerization. On the other hand, the labile Si–N bond responsible for the shift of the Me_3Si group in formation of these products results in the elimination of halotrimethylsilanes, Me_3SiX, when (disilylamino)phosphines, $(Me_3Si)_2NPRR'$ are oxidized by Br_2 (*25*) or C_2Cl_6 (*8, 9, 26*) (Scheme 2). In many cases, these P-halophosphoranimines have been isolated by vacuum distillation, but as noted below, this is not always necessary before conversion to more useful polyphosphazene precursors.

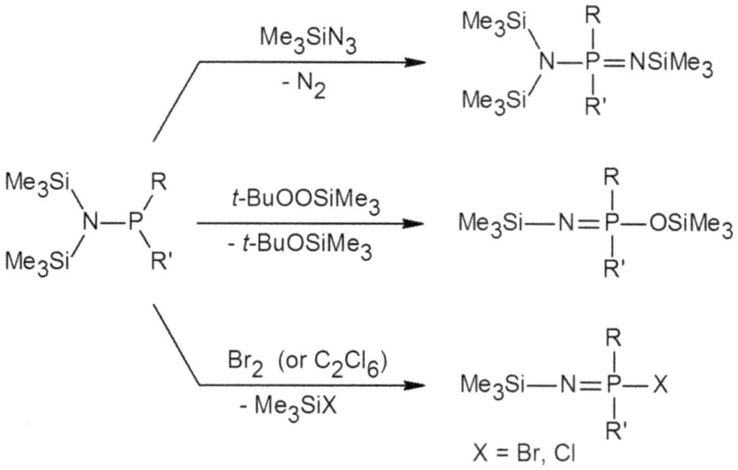

Scheme 2. Oxidation Reactions of (Disilylamino)phosphines

Both the Si–N and P–X bonds in these P-halophosphoranimines are useful reactive sites as we have discussed in a previous review (*27*). Changing the leaving group at phosphorus effectively alters the thermal stability of the phosphoranimine. For example, $Me_3SiN{=}P(X)Me_2$, where X = NMe_2 (*25*), OMe (*28*), or $OSiMe_3$ (*24*), showed no sign of decomposition even on heating above 200 °C for extended periods, but where X was Br, OCH_2CF_3 (*25*), or OPh (*26*), elimination of Me_3SiX occurred under milder conditions. Simple nucleophilic substitution reactions on P-halophosphoranimines (Scheme 3) were developed to produce large, mole scale yields of distillable phosphoranimines with trifluoroethoxy or phenoxy leaving groups. It is most noteworthy that, in many cases, including the preparation of $Me_3SiN{=}P(OPh)(Me)(Ph)$, the precursor to the easily modified polymer $[Me(Ph)PN]_n$ discussed in the last section of this review, all three steps can be done on a one-mole scale, in one day, in a single reaction flask, with optimal yields. (Scheme 4)

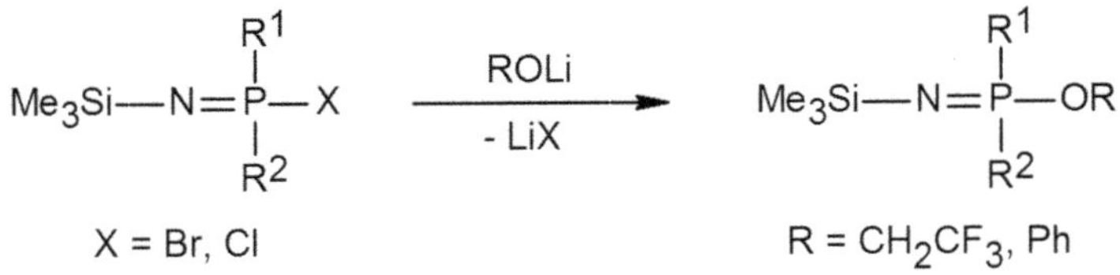

Scheme 3. Synthesis of Precursors to Poly(alkyl/arylphosphazenes)

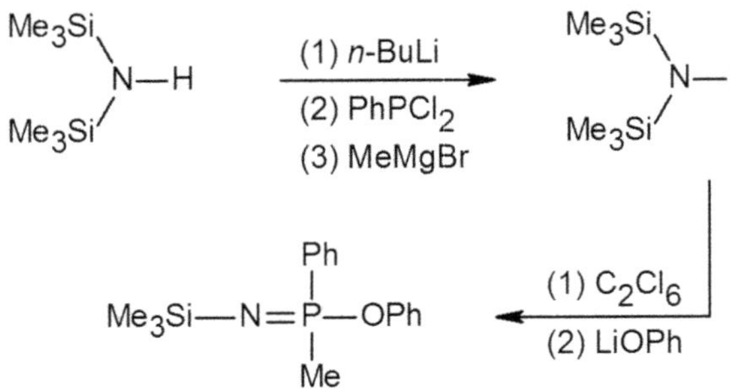

Scheme 4. One-Pot Synthesis of Precursor to $[Me(Ph)PN]_n$

Synthesis of Poly(alkyl/arylphosphazenes)

The extensive utility of the P-trifluoroethoxy and P-phenoxy-N-silyl-phosphoranimines is demonstrated by their uncatalyzed thermolysis reactions (Scheme 5) which produce essentially quantitative yields of a large variety of poly(dialkylphosphazenes), **A**, and poly(alkyl/arylphosphazenes), **B**, along with the easily removed, volatile siloxane, $Me_3SiOCH_2CF_3$ or Me_3SiOPh byproducts (*5–7*, *10*, *11*, *21*, *26*). These condensation-polymerization reactions are typically done on multigram to mole scales simply by heating the phosphor-animines in either sealed glass ampules or stainless steel reaction bombs, or by inert atmosphere reflux at temperatures between 150 to 180 °C. Similarly, many different random copolymers containing dimethyl/dialkyl, **C**, dimethyl/methyl-alkyl, **D**, methylphenyl/dialkyl, **E**, and methylphenyl/methylalkyl, **F**, groups have been prepared (*10*, *11*). Thus, in broad scope, this condensation-polymerization pathway to polyphosphazenes with P–C bonded side groups complements the ring-opening-substitution approach which is most useful for the preparation of polymers with P–N- and P–O-linked side groups.

As expected, both the P–C linkage and the different combinations of substituents in both the homo- and copolymers impart a wide range of properties (*11*). This is most obvious in the solubility differences. For example, semicrystalline poly(dimethylphosphazene), $[Me_2PN]_n$, is only soluble in chlorinated hydrocarbons such as $CHCl_3$ and CH_2Cl_2, while amorphous poly(methylphenylphosphazene), $[Me(Ph)PN]_n$, is also soluble in THF. Neither is soluble in hexanes, diethyl ether, acetone, or water, but $[Me_2PN]_n$ will dissolve in a 50:50 mixture of THF:H_2O or in water containing small amounts of a weak acid (*7*). By contrast, the other symmetrical poly(dialkylphosphazenes), $[R_2PN]_n$, are insoluble in all solvents, but disruption of their microcrystallinity by partial protonation of the backbone enhances solubility in $CHCl_3$ and CH_2Cl_2 containing small amounts of weak acids (*11*). This greatly facilitates spectroscopic characterization of these polymers.

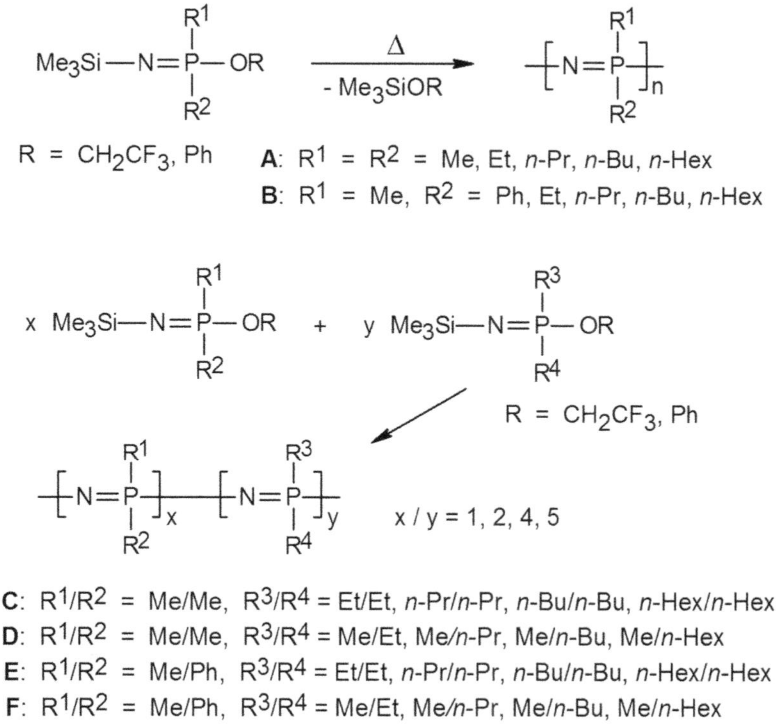

Scheme 5. Synthesis of Poly(alkyl/arylphosphazenes)

The significant differences in glass transition temperatures of these polymers also reflect the scope of properties for this class of polyphosphazenes. (Table 1) The T_g values range from above room temperature (37 °C) for $[Me(Ph)PN]_n$ (*7*) to ca. −70 °C for polymers with methylhexyl components (*11*), thus maintaining the well-established torsional flexibility of the PN backbone. We have also shown that larger aromatic groups significantly affect the glass transition temperatures. For example, poly(methylbiphenylphosphazene) has a T_g of ca. 100 °C (*29*). (Scheme 6)

As expected, the surface hydrophobicity is also significantly affected by the nature of the alkyl and aryl side groups in these polymers. Despite the low polarity of phenyl and methyl groups, the water contact angle θ for $[Me(Ph)PN]_n$ is low (73°) relative to poly(diphenoxyphosphazene), $[(PhO)_2PN]_n$ (92°) (*30*) and $[Et(Ph)PN]_n$ (*7*). However, with only alkyl groups on each phosphorus atom, the contact angles change drastically. For example, the contact angle for $[Me(n\text{-}Hex)PN]_n$ is only 22° and $[Me(n\text{-}Bu)PN]_n$ is so hydrophilic that it absorbed 15 mole % water from a saturated atmosphere in only a week (*9*). Clearly, the basicity of the backbone nitrogen atom is enhanced by the electron releasing alkyl groups. This is further demonstrated by studies of coordination of monovalent metals and protons to $[Me(Ph)PN]_n$ and $[Me_2PN]_n$ (*31*) and by the stabilization of gold nanoparticles by $[Me(Ph)PN]_n$ (*32*).

Table 1. Molecular weight and thermal analysis data for some alkyl/aryl-phosphazene polymers, $[R^1(R^2)PN]_n$, and their 1:1 copolymers, $\{[R^1(R^2)PN]-[R^3(R^4)PN]\}_n$

R^1/R^2	R^3/R^4	M_w (pdi)[a]	T_g (°C)	T_m (°C)
Me/Me		108,300 (1.7)	−46	143
Et/Et[b]			−8	217
n-Pr/*n*-Pr[b]				231
n-Bu/*n*-Bu[b]				190
n-Hex/*n*-Hex[b]			−29	129
Me/Ph		136,000 (1.7)	37	
Me/*n*-Pr		68,400 (2.3)	−55	41
Me/*n*-Bu		62,100 (4.5)	−51	
Me/*n*-Hex		108,000 (2.5)	−69	
Me/Me	Me/Ph	93,900 (1.6)	−3	
Me/Me	Me/Et	74,100 (3.4)	−54	103
Me/Me	Me/*n*-Pr	56,000 (1.7)	−59	
Me/Me	Me/*n*-Bu	170,000 (3.6)	−56	
Me/Me	Me/*n*-Hex	101,000 (2.4)	−71	
Me/Ph	Me/*n*-Pr	76,400 (3.5)	−47	
Me/Ph	Me/*n*-Bu	107,000 (2.7)	−24	
Me/Ph	Me/*n*-Hex	98,900 (1.9)	−35	

[a] Polydispersity index (M_w/M_n). [b] Insufficient solubility for SEC analysis.

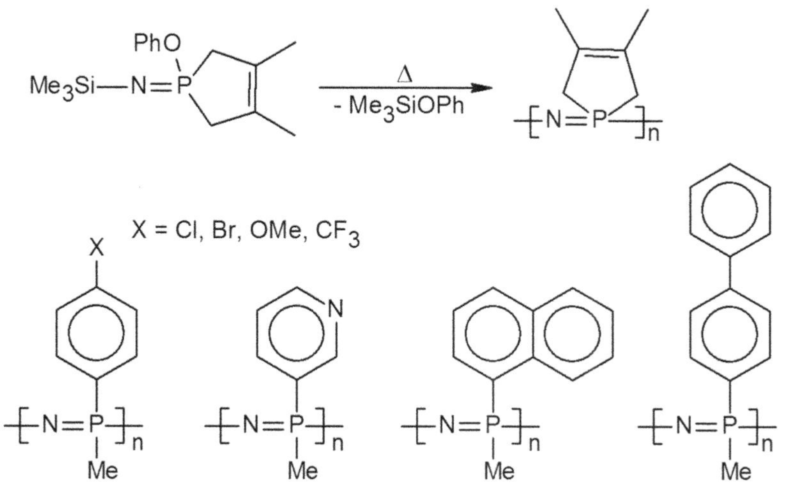

Scheme 6. Other Polyphosphazenes from Condensation-Polymerization

Generally, the poly(alkyl/arylphosphazenes) are also thermally stable with onsets of decompositions well above 300 °C. Thermal studies of $[Me_2PN]_n$ and $[Me(Ph)PN]_n$ showed that degradation occurs by simple unzipping to cyclic trimers and tetramers upon heating for extended periods (*33*). One of the most noteworthy features of the polymers is their hydrolytic stability. Unlike polyphosphazenes with P–N and P–O bonded groups which hydrolyze to phosphates, ammonia, and amine or alcohols, these polymers with P–C bonds are stable in water and even in strong bases. Moreover, simple cytotoxicity agarose overlay and MEM (minimum essential medium) elution tests of $[Me_2PN]_n$ and $[Me(Ph)PN]_n$ indicated no toxicity as well as nonhemolytic properties (*34*). Hence, the poly(alkyl/arylphosphazenes) may be useful biostable materials for long term biomedical applications.

Since the initial report of the preparation of poly(alkyl/arylphosphazenes) by condensation-polymerization, the P-alkyl-P-aryl-N-silylphosphoranimines have been used to make a number of other unusual polyphosphazenes (*8*, *26*, *29*, *35*) (Scheme 6), as well as many analogous cyclic phosphazenes (*36*, *37*). Furthermore, condensation-polymerization of N-silylphosphoranimines has been extended to include the preparation of some polyphosphazenes previously prepared by ring-opening substitution, including living polymerizations catalyzed by PCl_5 (*15*, *16*), $P(OMe)_3$ (*17*, *18*), F^- (*14*), and, most recently, by H_2O/imidazole (*19*). Two other condensation-polymerization processes involving loss of $P(O)Cl_3$ (*12*) or N_2 (*13*), rather than Me_3SiX, have also been reported.

Modification of Poly(alkyl/arylphosphazenes)

Despite the scope of the condensation-polymerization reaction for the preparation of many polyphosphazenes with P–C bonded side-groups, those side groups must not contain functional groups that interact with the reactive Si–N or P–O bonds in the N-silylphosphoranimine precursors and must withstand either thermolysis or catalytic conditions used in the polymerization process.

Post-polymerization reactions of simple preformed poly(alkyl/arylphosphazenes), however, significantly extend the functionality and properties of this class of phosphazene polymers. Poly(methylphenylphosphazene), **PMPP**, has generally been used as the parent polymer for functionalization reactions, because, unlike the poly(dialkylphosphazenes), $[Ph(Me)PN]_n$ is soluble in THF, a solvent suitable for deprotonation of the P-methyl group with *n*-butyllithium. (Scheme 7) (*38*) The intermediate anion has been treated with a variety of electrophiles to produce several subfamilies of functionalized polymers (Scheme 8) (*39–43*).

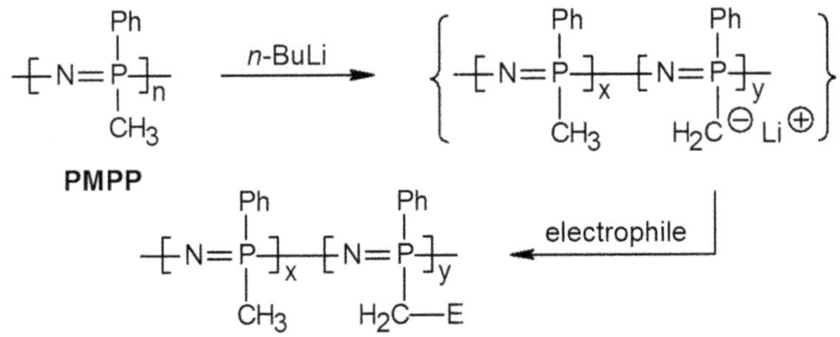

Scheme 7. Deprotonation-Substitution of Poly(methylphenylphosphazene)

There are several important aspects of the deprotonation-substitution method, including the following. (a) Up to 95 to 100 % of the methyl groups can be deprotonated, but bulky electrophiles limit the final degree of substitution (*44*). (b) Copolymers $[Me_2PN]_x[Ph(Me)PN]_x$ which are also soluble in THF, can be similarly modified (*45, 46*). (c) Polymers with methyl groups and a different alkyl group, $[alkyl(Me)PN]_n$, can be selectively modified at the methyl group (*9*). (d) Cosubstitution can be achieved with certain electrophiles (*47*). (e) Properties can be changed, and ultimately tuned, through these modifications.

The polymers in Scheme 8 demonstrate the tunability of properties of the parent polymer, $[Ph(Me)PN]_n$. The most obvious and first observed differences are the solubility changes that the new side groups impart to the polymer system, especially when carboxylate groups are added. Indeed, with only 50% of the methyl groups converted to $CH_2COO^-Na^+$, the polymer is water soluble (*41*). As expected, hydrophobicity is enhanced through incorporation of long alkyl and fluoroalkyl groups in either the silyl (*47*) or alcohol derivatives (*48*). Glass transition temperatures are somewhat higher in the alcohol derivatives as expected with hydrogen bonding (*40*), but the length of the alkyl group also has a significant, ultimately tunable, effect on the T_g values of these polymers (*49*). Thermal stability is enhanced when sulfide, SR, groups are oxidized to sulfones (*43*) and redox activity is imparted by addition of ferrocene groups via ferrocenyl aldehyde and ketone electrophiles (*50*). Gas permeability and mechanical integrity is tuned by varying amounts and combinations of alkyl and fluoroalkylsilyl side-groups (*48*). The feasibility of these one-step modifications was extended to both copolymers and poly(methylalkylphosphazenes) to further alter properties (*45, 46*); e.g., the polymer is water soluble when only ca. 25% of the methyl groups on the one-to-one copolymer, $\{[Me_2PN][Ph(Me)PN]\}_n$, are converted to $CH_2COO^-Na^+$ groups. Surface modification of the parent polymer and several derivatives was also examined (*51*). Most importantly, the functional groups add reactive sites to the polymer system. For example, the alcohol derivatives have been used to prepare esters (*52*), including one suitable for initiation of free radical polymerization reactions (discussed below).

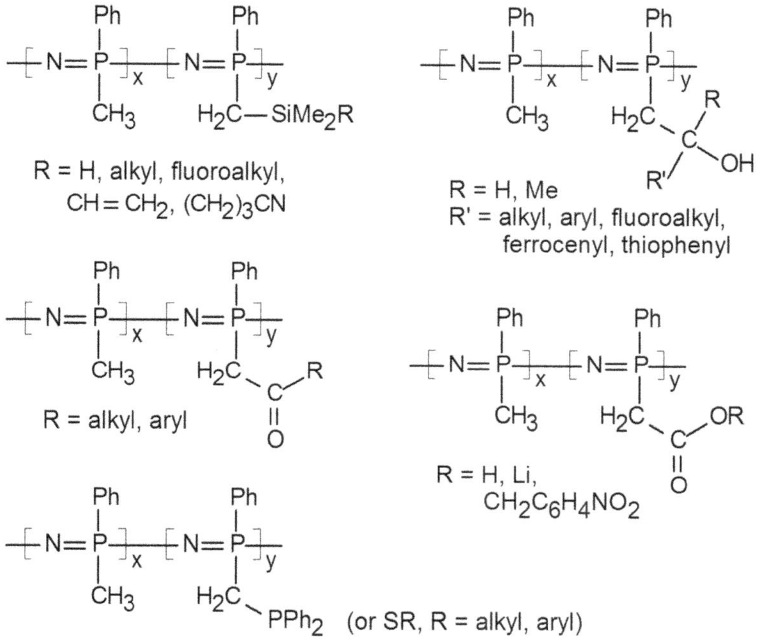

Scheme 8. Some Functionalized Poly(alkyl/arylphosphazenes) from Modification of P-Methyl Groups in [Ph(Me)PN]$_n$

The ability to combine the useful features of organic polymer systems with the poly(alkyl/arylphosphazenes) through grafting was established by anionic polymerization initiated by the polymer anion intermediate shown in Scheme 7. In this manner, both poly(methyl methacrylate) (*53*) and poly(styrene) (*54*) were grafted onto the parent polymer [Ph(Me)PN]$_n$. (Scheme 9) Anionic ring-opening polymerization of the trisiloxane ring produced one of the first examples of this isoelectronic inorganic-inorganic graft system (*55*). As noted above, a suitable site for the initiation of controlled radical polymerization via atom transfer radical polymerization (ATRP) was achieved via esterification of a simple alcohol derivative. (Scheme 10) (*56*) More recently, a slight variation of the ATRP process was studied in our labs for the preparation of hybrid polyphosphazene-*graft*-polyethylene glycol systems (*57*).

The phenyl group has been modified by nitration reactions and subsequent conversion to amides, but this process requires rigorous conditions, leading to partial chain-degradation (*58*). Other methods for modification of aromatic groups have been conducted in our labs using the substituted aromatic phenyl groups shown in Scheme 6. For example, bromophenyl groups were used for palladium-catalyzed, cross-coupling reactions to build new phenyl–C and phenyl–N functional groups into the poly(alkyl/arylphosphazene) system (*59*).

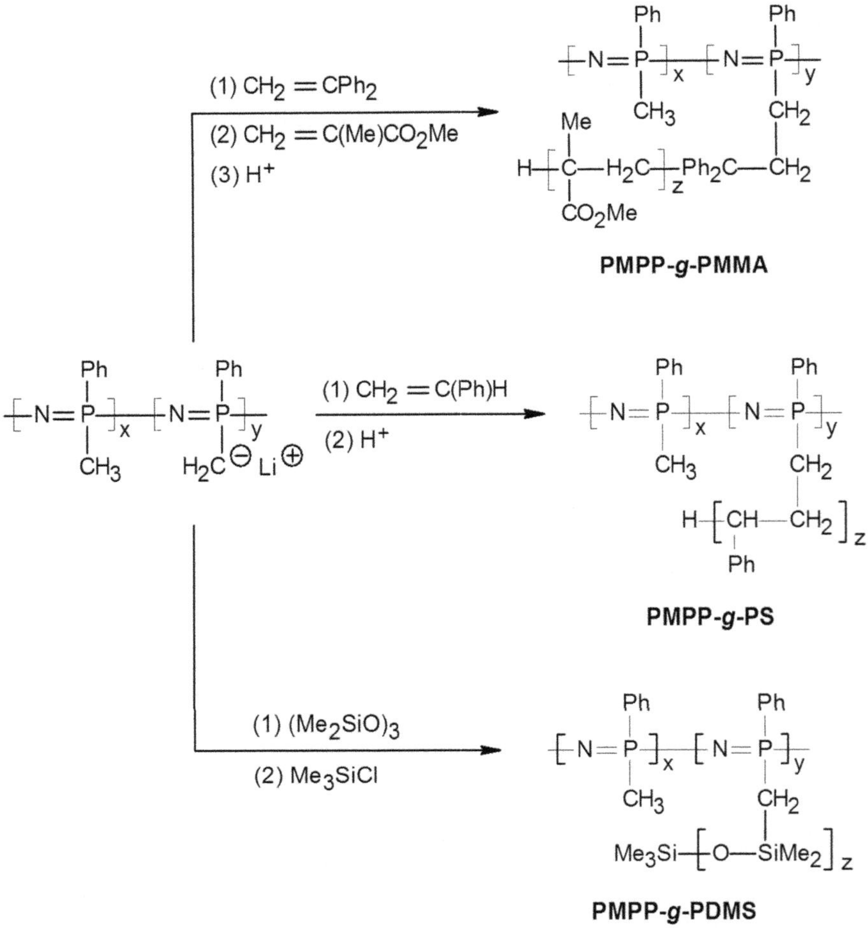

Scheme 9. Graft Copolymers from Anionic Polymerization

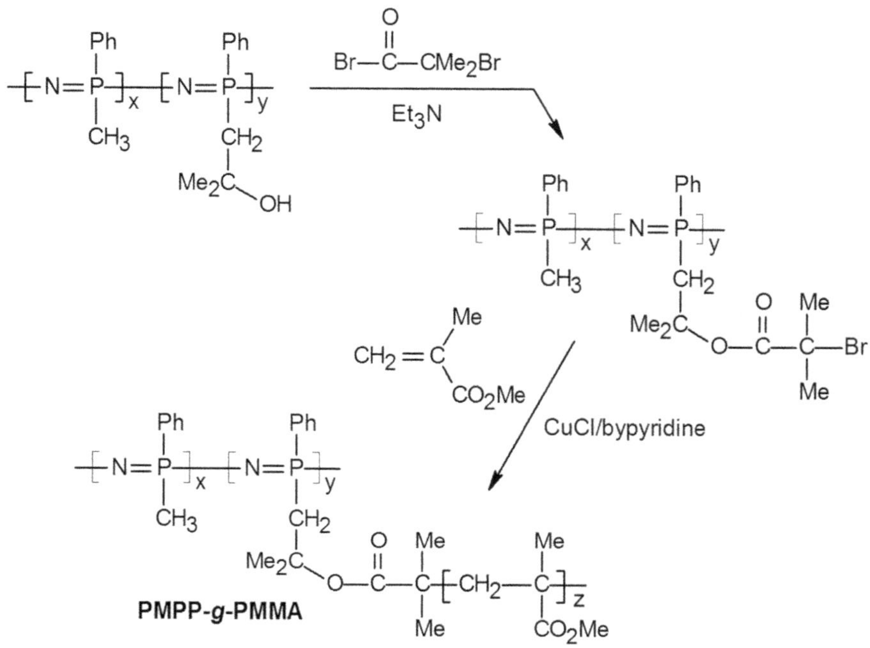

Scheme 10. Graft Copolymers via Atom Transfer Radical Polymerization (ATRP)

Summary

Since the initial report of poly(dimethylphosphazene) thirty-eight years ago, the scope of the synthesis of polyphosphazenes by condensation-polymerization has expanded extensively. Not only have over 40 to 50 poly(alkyl/aryl-phosphazenes) and poly(dialkylphosphazenes) been prepared by this method, condensation-polymerization of N-silylphosphoranimines has been used to make several polyphosphazenes with P–O bonded groups and several different types of catalysts/initiators have been reported for this process. The poly(alkyl/aryl-phosphazenes) possess most of the advantages of the polyphosphazene family, including the flexible PN backbone and property tunability, but these P–C bonded polymers do not hydrolyze readily as do those with P–N and P–C substituents. Hence, the poly(alkyl/arylphosphazenes) have complementary properties that may render them useful for hydrolytically, chemically stable, and bioinert coatings, structural materials, membranes, etc. In addition, this chemical stability may make them recoverable and recyclable. It can be advantageous that broad tuning of the properties of poly(alkyl/arylphosphazenes) can be done at the small molecule stage where purity and reaction efficiency is better controlled. However, the diversity of properties of a single, easily prepared polymer, $[Ph(Me)PN]_n$, can be significantly altered and many functional groups for attaching cleavable or stable moieties have already been attached to this simple polymer system by deprotonation-substitution reactions. In addition, studies of reactions at aromatic groups show promise as one-step processes to functionalize phosphazene the polymers.

References

1. Flindt, E. P.; Rose, H. Trivalent-pentavalente Phosphorverbindungen/ Phosphazene. I. Eine neue Methode zur Darstellung von Poly[bis(trifluoräthoxy)phosphazene]. *Anorg. Allg. Chem.* **1977**, *428*, 204–208.
2. Allcock, H. R.; Kugel, R. L. Synthesis of High Polymeric Alkoxy- and Aryloxyphosphonitriles. *J. Am. Chem. Soc.* **1965**, *87*, 4216–4217.
3. Wisian-Neilson, P.; Neilson, R. H.; Cowley, A. H. The Reactions of Amines and Pyridine with Tri- and Tetrafluorophosphoranes: Formation of HPF_5 and H_2PF_4 Anions. *Inorg. Chem.* **1977**, *16*, 1460–1463.
4. Allcock, H. R.; Patterson, D. B. Ring-Ring and Ring-Chain Equilibration of Dimethylphosphazenes. Relationship to Phosphazene Polymerization. *Inorg. Chem.* **1977**, *16*, 197–200.
5. Wisian-Neilson, P.; Neilson, R. H. Poly(dimethylphosphazene), $(Me_2PN)_n$. *J. Am. Chem. Soc.* **1980**, *102*, 2848–2849.
6. Neilson, R. H.; Wisian-Neilson, P. Synthesis of Poly(dialkylphosphazenes) from N-Silylphosphinimines. *J. Macromol. Sci., Chem.* **1981**, *A16*, 425–439.
7. Neilson, R. H.; Hani, R.; Wisian-Neilson, P.; Meister, J. J.; Roy, A. K.; Hagnauer, G. L. Synthesis and Characterization of Poly(alkyl/aryl-phosphazenes). *Macromolecules* **1987**, *20*, 910–916.
8. Gruneich, J. A.; Wisian-Neilson, P. A New Phospholenazene Polymer. *Macromolecules* **1996**, *29*, 5511–5512.
9. Wisian-Neilson, P.; Koch, K. A.; Zhang, C. Deprotonation-Substitution Reactions of Poly(dialkylphosphazenes) and Their Phosphoranimine Precursors. *Macromolecules* **1998**, *31*, 1808–1813.
10. Hani, R.; Jinkerson, D. L.; Wood, C.; Neilson, R. H. Poly(alkyl/ arylphosphazene) Chains and Rings via Condensation Reactions of Si–N–P Precursors. *Phosphorus, Sulfur, Silicon* **1989**, *41*, 159–164.
11. Jinkerson, J. L. *Alky/Aryl Phosphazenes: New Polymers, Cyclic Analogs, and Precursors.* Ph.D. Dissertation, Texas Christian University, August, 1989.
12. D'Hallum, G.; De Jaeger, R.; Chambrette, J. P.; Potin, P. Synthesis of Poly(dichlorophosphazenes) from $Cl_3P{=}NP(O)Cl_2$. *Macromolecules* **1992**, *25*, 1254–1258.
13. Franz, U.; Nuyken, O.; Matyjaszewski, K. Synthesis of Poly(phenyl-*o*-tolylphosphazene) via Reaction of Trimethylsilyl Azide with 2,2,2-Trifluoroethyl Phenyl-*o*-tolylphosphinite. *Macromolecules* **1993**, *26*, 3723–3725.
14. Montague, R. A.; Matyjaszewski, K. Synthesis of Poly[bis(trifluoro-ethoxy)phosphazene] under Mild Conditions Using a Fluoride Initiator. *J. Am. Chem. Soc.* **1990**, *112*, 6721–6123.
15. Honeyman, C. H.; Manners, I.; Morrissey, C. T.; Allcock, H. R. Ambient Temperature Synthesis of Poly(dichlorophosphazene) with Molecular Weight Control. *J. Am. Chem. Soc.* **1995**, *117*, 7035–7036.

16. Allcock, H. R.; Nelson, J. M.; Reeves, S. D.; Honeyman, C. H.; Manners, I. Ambient-Temperature Direct Synthesis of Poly(organophosphazenes) via the "Living" Cationic Polymerization of Organo-Substituted Phosphoranimines. *Macromolecules* **1997**, *30*, 50–56.
17. Huynh, K.; Lough, A. J.; Manners, I. Reaction of P-Donor Ligands with N-Silyl(halogeno)organophosphoranimines: Formation of Cations with P–P Coordination Bonds and Poly(alkyl/aryl)phosphazenes at Ambient Temperature. *J. Am. Chem. Soc.* **2006**, *128*, 14002–14003.
18. Taylor, T. J.; Presa Soto, A.; Huynh, K.; Lough, A. J; Swain, A. C.; Norman, N. C.; Russell, C. A.; Manners, I. Synthesis of Poly(alkyl/arylphosphazenes) via the Ambient Temperature Phosphite-Mediated Chain-Growth Polycondensation of (N-Silyl)bromo-phosphoranimines. *Macromolecules* **2010**, *43*, 7446–7452.
19. Steinke, J. H. G.; Greenland, B. W.; Johns, S.; Parker, M. P; Atkinson, R. C. J.; Cade, I. A.; Golding, P.; Trussell, S. J. Robust and Operationally Simple Synthesis of Poly(bis(2,2,2-trifluoroethoxy)phosphazene) with Controlled Molecular Weight, Low PDI, and High Conversion. *ACS Macro Lett.* **2014**, *3*, 548–551.
20. Neilson, R. H.; Wisian-Neilson, P. Synthesis of (Silylamino)phosphines by the Wilburn Method. *Inorg. Chem.* **1982**, *21*, 3568–3569.
21. Wisian-Neilson, P.; Neilson, R. H. Poly(dimethylphosphazene) and Poly(methylphenyl-phosphazene). *Inorg. Synth.* **1989**, *25*, 69–74.
22. Allcock, H. R.; Evans, T. L.; Patterson, D. B. Reaction of Poly(difluorophosphazene) with Phenyllithium. *Macromolecules* **1980**, *13*, 201–207.
23. Wilburn, J. C.; Wisian-Neilson, P.; Neilson, R. H. Synthesis and Structural Isomerism of Some Silylaminophosphinimines. *Inorg. Chem.* **1979**, *18*, 1429–1433.
24. Neilson, R. H.; Wisian-Neilson, P.; Wilburn, J. C. Synthesis and Structural Isomerism of Some (Silylamino)phosphine Oxides. *Inorg. Chem.* **1980**, *19*, 413–416.
25. Wisian-Neilson, P.; Neilson, R. H. Synthesis of New N-Silylphosphinimines: Phosphazene Precursors. *Inorg. Chem.* **1980**, *19*, 1875–1878.
26. Neilson, R. H.; Jinkerson, J. L.; Kucera, W. R.; Longlet, J. J.; Samuel, R. C.; Wood, C. E. Synthesis of Poly(alkyl/arylphosphazenes) and Their Precursors. *ACS Symposium Series* **1994**, *572*, 232–245.
27. Neilson, R. H.; Wisian-Neilson, P. Poly(alkyl/arylphosphazenes) and Their Precursors. *Chem. Rev.* **1988**, *88*, 541–562.
28. Wisian-Neilson, P.; Neilson, R. H., unpublished results.
29. Jung, J.-H.; Zhang, C.; Wisian-Neilson, P., Southern Methodist University, unpublished work.
30. Allcock, H. R.; Steely, L. B.; Singh, A. Hydrophobic and Superhydrophobic Surfaces from Polyphosphazenes. *Polym. Int.* **2006**, *55*, 621–625.
31. Wisian-Neilson, P.; García-Alonso, F. J. Coordination of Poly(methylphenylphosphazene) and Poly(dimethylphosphazene). *Macromolecules* **1993**, *26*, 7156–7160.

32. Walker, C. H.; St. John, J. V.; Wisian-Neilson, P. Synthesis and Size Control of Gold Nanoparticles Stabilized by Poly(methylphenyl-phosphazene). *J. Am. Chem. Soc.* **2001,** *123*, 3846–3847.
33. Allcock, H. R.; McDonnell, G. S.; Riding, G. H.; Manners, I. Influence of Different Organic Side Groups on the Thermal Behavior of Polyphosphazenes: Random Chain Cleavage, Depolymerization and Pyrolytic Cross-Linking. *Chem. Mat.* **1990**, *2*, 425–432.
34. Wisian-Neilson, P. Polyphosphazenes from Condensation Polymerization. In *Biomedical Applications of Polyphosphazenes*; Andrianov, A. K. Ed.; Wiley: New Jersey, 2009; Chapter 10, pp 155–167.
35. Wang, B.; Klaehn, J. R.; Neilson, R. H. New Cyclic and Polymeric Phosphazenes Derived from N-Silylphosphoranimines. *Phosphorus, Sulfur, Silicon* **2004**, *179*, 821–825.
36. Wisian-Neilson, P.; Johnson, R. S.; Zhang, H.; Jung, J.-H.; Neilson, R. H.; Ji, J.; Watson, W. H.; Krawiec, M. Reactions of N-Silylphosphoranimines with Alcohols: Synthesis and Structure of Cyclotriphosphazenes with Nongeminal Methyl and Phenyl Substituents. *Inorg. Chem.* **2002**, *41*, 4775–4779.
37. Kumas, C.; Jung, J. H.; Potluri, S. K.; Wang, X.; Mauldin, C. E.; Zhang, H.; Devulapalli, P.; Wisian-Neilson, P. *Synthesis and Structural Characterization of Nongeminal P-Aryl-P-Methyl Cyclophosphazenes Prepared from N-Silylphosphoranimines*, to be submitted for publication.
38. Wisian-Neilson, P.; Ford, R. R.; Neilson, R. H.; Roy, A. K. Silylated Derivatives of Poly[(methyl)(phenyl)phosphazene]. *Macromolecules* **1986**, *19*, 2089–2091.
39. Wisian-Neilson, P.; Ford, R. R. Ferrocenyl Substituted Poly(alkyl/aryl-phosphazenes). *Organometallics* **1987**, *6*, 2258–2259.
40. Wisian-Neilson, P.; Ford, R. R. Alcohol Derivatives of Poly(methylphenylphosphazene). *Macromolecules* **1989**, *22*, 72–75.
41. Wisian-Neilson, P.; Islam, M. S.; Ganapathiappan, S.; Scott, D. L.; Raghuveer, K. S.; Ford, R. R. Carboxylic Acid, Ester, and Lithium Carboxylate Derivatives of Poly(methylphenylphosphazene). *Macromolecules* **1989**, *22*, 4382–4384.
42. Wisian-Neilson; Zhang, C. Ketone Derivatives of Poly(methylphenyl-phosphazene): Incorporation of an Electrophilic Site into Poly(alkyl/aryl-phosphazenes). *Macromolecules* **1998**, *31*, 9084–9086.
43. Jung, J.-H.; Kmecko, T.; Claypool, C. L.; Zhang, H.; Wisian-Neilson, P. Synthesis and Characterization of Sulfur-Containing Nongeminal Cyclic and Polymeric Alkylarylphosphazenes. *Macromolecules* **2005**, *38*, 2122–2130.
44. Wisian-Neilson, P.; Claypool, C. L.; Bahadur, M. Formation of $[(Li^{+}CH_2^{-})(Ph)PN]_n$ at Room Temperature. *Macromolecules* **1994**, *27*, 7494–7495.
45. Kraiter, D. C.; Wisian-Neilson, P.; Zhang, C.; Crumbliss, A. L. Synthesis and Redox Characterization of Phosphazene Terpolymers with Pendant Ferrocene Groups. *Macromolecules* **2012**, *45*, 3658–3668.
46. Crane, R. A.; Wisian-Neilson, P., Southern Methodist University, unpublished work.

47. Wisian-Neilson, P.; Xu, G.-F. Gas Permeation Studies of Silylated Derivatives of Poly(methylphenylphosphazene). *Macromolecules* **1996**, *29*, 3457–3461.
48. Wisian-Neilson, P.; Xu, G.-F.; Wang, T. Fluorinated Polyphosphazenes Derived from Poly(methylphenylphosphazene) and Fluorinated Aldehydes and Ketones. *Macromolecules* **1995**, *28*, 8657–8661.
49. Wisian-Neilson, P.; Bailey, L.; Bahadur, M. Modification of Poly(methylphenylphosphazene) to Increase Hydrophilicity. *Macromolecules* **1994**, *27*, 7713–7717.
50. Crumbliss, A. L.; Cooke, D.; Castillo, J.; Wisian-Neilson, P. Redox Properties of Phosphazene Polymers with Pendant Ferrocene Groups. *Inorg. Chem.* **1993**, *32*, 6088–6094.
51. Kmecko, T.; Wang, X.; Wisian-Neilson, P. Surface Chemical Modification of Poly(methylphenylphosphazene). *J. Inorganic and Organometallic Polymers and Materials* **2007**, *17*, 413–321.
52. Wisian-Neilson, P.; Huang, L.; Islam, M. Q.; Crane, R. A. Synthesis of Poly(alkyl/arylphosphazenes) with Ester Substituents. *Polymer* **1994**, *35*, 4985–4989.
53. Garcia-Alonso, F. J.; Wisian-Neilson, P., Southern Methodist University, unpublished work.
54. Wisian-Neilson, P.; Schaefer, M. A. Synthesis and Characterization of Poly(methylphenylphosphazene)-*graft*-polystyrene Copolymers. *Macro-molecules* **1989**, *22*, 2003–2007.
55. Wisian-Neilson, P.; Islam, M. S. Poly(methylphenylphosphazene)-*graft*-poly(dimethylsiloxane). *Macromolecules* **1989**, *22*, 2026–2028.
56. Cambre, J. N.; Wisian-Neilson, P. Poly(methylphenylphosphazene)-*graft*-Poly(methyl methacrylate) Copolymers via Atom Transfer Radical Polymerization. *J. Inorg. Organomet. Polym. Mater.* **2006**, *16*, 311–318.
57. Zugell, D. A.; Wisian-Neilson, P.; Tsarevsky, N. V. *Hybrid Polyphosphazene-graft-PEG Derivatives for Use as Drug Delivery Vessels*, to be submitted for publication.
58. Wisian-Neilson, P.; Bahadur, M.; Iriarte, J. M.; Wood, C. E.; Ford, R. R. Nitro, Amino, and N-Amido Derivatives of Poly(methylphenylphosphazene). *Macromolecules* **1994**, *27*, 4471–4476.
59. Kumas, C. *Phosphazenes as Scaffolds for the Synthesis of New Molecules by Palladium and Copper Catalyzed Reactions*. Ph.D. Dissertation, Southern Methodist University, May, 2012.

Chapter 9

Water Soluble (Bio)degradable Poly(organo)phosphazenes

Aitziber Iturmendi and Ian Teasdale*

Institute of Polymer Chemistry, Johannes Kepler University, Altenberger Straße 69, 4040 Linz, Austria
***E-mail: ian.teasdale@jku.at.**

Water soluble and degradable poly(organo)phosphazenes are promising polymers for a wide range of applications, especially in the biomedical field. This chapter aims to summarize the variety of water soluble polyphosphazenes available ranging from ionic to amphiphilic, and thermoresponsive polymers to hydrogels. The most recent developments in the design and synthesis of water soluble poly(organo)phosphazenes are presented. Furthermore, we focus on the degradation pathways of the polymers with special attention paid to their stimulated degradation mechanisms.

Degradable Water Soluble Polymers

Water soluble synthetic polymers are widely used in many applications including in the food and pharmaceutical industry, water treatment and electrical applications (*1*). In addition to this, many bioerodible polymers are used, which by definition become water soluble under physiological conditions (*2*). Due to their miscibility with water they are removed from the local site. However, while they "disappear" to the naked eye, being classified as bioerodible does not necessarily imply they are (chemically) degradable to small molecules. For environmental reasons (*3*, *4*) as well as medical applications (*5*–*7*) the degradability of high molecular weight polymers is also an important issue which has to be considered carefully. Degradation is a chemical process based on the cleavage of covalent bonds. Among the possible degradation mechanisms, such as oxidative, photodegradative, or enzymatic (*8*–*10*), hydrolysis is the most common chemical process in which water is responsible for the covalent bond cleavage.

Although the terms are used interchangeably in the scientific literature, the term biodegradation is ideally reserved for processes in which degradation is caused by biological agents, such as enzymes, cells, or microorganisms. Furthermore, since it can be assumed that all polymers degrade somehow, it is only a matter of time, we follow herein the definition of degradable polymers suggested by Göpferich, as polymers that "degrade within the timescale of the application and specific conditions" (*11*).

The environmental impact of non-degradable synthetic polymers has become one of global concern in recent years (*12*, *13*). Recently, a number of polymers such as polylactic acid (PLA), polyhydroxybutyrate (PHB), poly(lactic-co-glycolic) acid (PLGA) and polycaprolactone (PCL), which are considered to be, and marketed as, biodegradable polymers, have been studied recently in fresh water and sea water (*14*). These polymers were stored for one year in vials with holes in their caps for oxygen exchange and immersed in 3 mL fresh water as well as artificial sea water. Thus, under fluorescent light and constant temperature (25°C) the results showed that only PLGA degraded totally in less than one year, whereas PCL, PLA and PET did not degrade at all. In the same timeframe PHB showed only ~8% degradation. Therefore, it is important to define the specific requirements of the materials in order to classify them as degradable or non-degradable polymers. In this sense, many common water-soluble polymers such as polyethylene glycol (PEG), poly(vinyl alcohol) (PVA), and poly[N-(2-hydroxypropyl)methacrylamide] (poly(HPMA)), do not have tendency of the backbone to breakdown so they are regarded as non-degradable polymers (*15*). Semi-degradable polymers that degrade to smaller polymer fragments, can be prepared *via* copolymerization of persistent blocks with degradable units, as has been shown for example with HPMA copolymers (*16–18*). Some polymers, such as, poly(L-glutamic acid), some polyphosphazenes and poly(phosphoesters) (Figure 1), are known to fully degrade to small molecules, an essential feature for some applications (*6*, *19*, *20*).

Of the many medical applications, drug delivery systems represent one of the most widely researched applications for water-soluble polymers (*21*, *22*). Numerous types of polymer-drug conjugates, nanocarriers and polymer therapeutics are under development or entering clinics to improve the effectiveness and biodistribution of small-molecule drugs, as well as biomacromolecular drugs (*21*, *22*). Any polymeric carriers which enter the blood-stream for such therapeutic applications should fulfill not only the necessary water-solubility and biocompatibility requirements, but also be proven to undergo complete clearance from the body in order to avoid deleterious long term side effects (*23–25*). Since the essence of macromolecular carriers (long retention times, enhanced permeation and retention effect etc.) is in their high molecular weights, it follows that polymers which degrade in a timely manner to small molecules are essential.

Phosphorous-based polymers are an exceptional possibility for such applications due to the inherent tendency of phosphorus to undergo hydrolysis and hence when incorporated into macromolecules, main chain polymer degradation (*26*). In combination with their facile functionalization this importantly allows the degradation rates to be tailored across a wide range spectrum to suit the specified application.

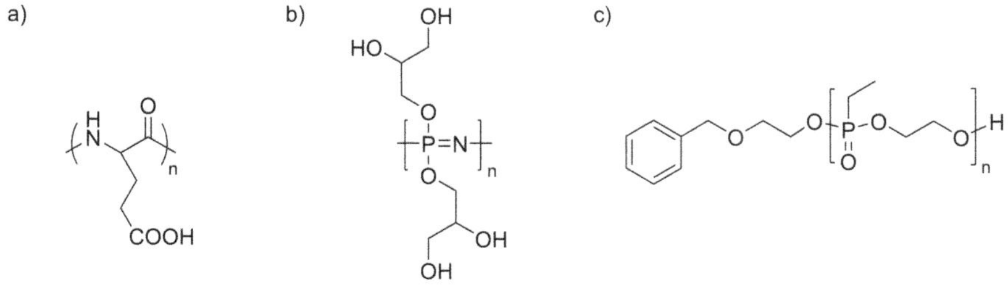

Figure 1. Some selected examples of degradable water-soluble synthetic polymers: a) poly(L-glutamic acid), b) poly(diglycerophosphazene) and c) poly(ethylene methylphosphonate).

Polyphosphazenes

Poly(organo)phosphazenes are polymers with an inorganic backbone based on alternating nitrogen and phosphorous atoms (*27*). They are unique due to their flexible, high thermal stability and intrinsic hydrolytic backbone. Side groups, organic and covalently linked to the phosphorous atom by macromolecular substitution (*27*), can tune the properties of the polymer (including e.g. hydrolysis rates, glass transition temperatures) (*28*). Polyphosphazenes are used in a wide variety of applications that range from elastomers, lithium ion conductive polymers to biomedical materials (*29, 30*).

Of the various routes to poly(organo)phosphazenes (*31, 32*), the mostly used is *via* the inorganic macromolecular precursor poly(dichloro)phosphazene, $[NPCl_2]_n$. There are different routes available for the synthesis of $[NPCl_2]_n$ (*27, 31*), but the most used methods are ring-opening polymerization (ROP) of hexachlorophosphazene and living cationic polymerization. ROP (Figure 2a) is the most widely used route to obtain polymers with high molecular weight and broad polydispersities, but high purity of monomer, for reproducible results, and usually high temperatures (250°C) are required (*27*)

The alternative method is *via* living cationic polymerization of trichloro(trimethylsilyl)phosphoranimine ($Cl_3PNSi(CH_3)_3$) monomer at room temperature using phosphorus pentachloride (PCl_5) as initiator (Figure 2b) (*33*). Such procedure gives shorter polymer chains ($n<75$) (*34*) with narrow molecular weight distribution but with living chain ends (*35*). More recently, polyphosphazenes with defined chain ends have been synthesized utilizing chlorinated tertiary phosphines, such as triphenylphosphine dichloride (Ph_3PCl_2), as initiators (Figure 2c) (*36, 37*). In this case, it is ensured the monodirectional growth of the polymer enduring the controlled chain growth and narrow polydispersities. Furthermore, it opens the opportunity to achieve block copolymers as only one chain end remains active (*38*).

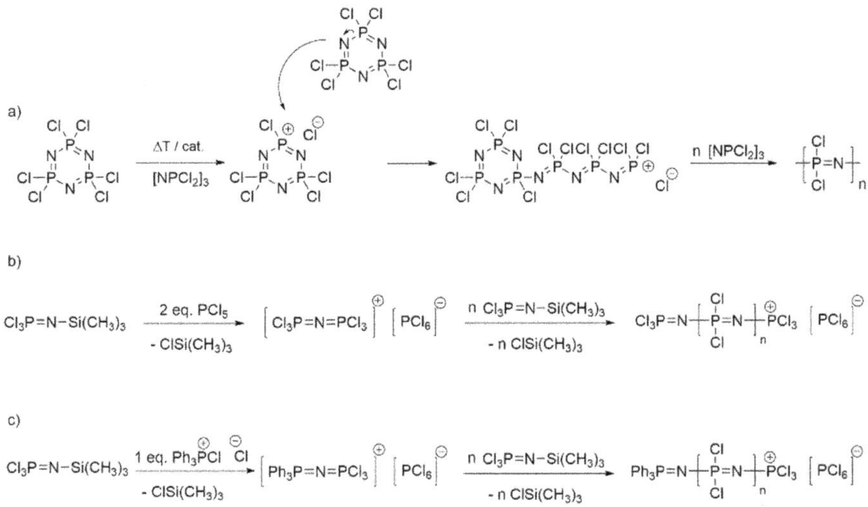

Figure 2. Some synthetic routes for the synthesis of poly(dichloro)phosphazene $[NPCl_2]_n$. a) Ring-opening polymerization of hexachlorophosphazene. Living-cationic polymerization of trichloro(trimethylsilyl)phosphoranimine ($Cl_3PNSi(CH_3)_3$) with b) phosphorus pentachloride (PCl_5) and c) triphenylphosphine dichloride (Ph_3PCl_2) as initiators.

An useful property of poly(organo)phosphazenes is the facile synthesis of a wide variety of polymers (*29, 39–41*). Once the backbone is assembled different organic nucleophiles can be introduced through macromolecular substitution of the chlorine atoms. Therefore, it makes it easier compared to many other methods of polymer synthesis where the side groups are incorporated into the starting monomer.

Water-Soluble Poly(organo)phosphazenes

In the following sections a selection of some water-soluble poly(organo)phosphazenes will be presented to denote the potential of these polymers for biomedical uses.

Anionic Poly(organo)phosphazenes

Acidic/anionic polyphosphazene electrolytes are the most extensively studied water-soluble polyphosphazenes in nanomedicine. Poly[di(carboxylatophenoxy)phosphazene] (PCPP), first synthesized by Kwon and Allcock (*42*) and later demonstrated by Andrianov and co-workers its outstanding ability to function as an immunoadjuvant (*43, 44*), has been extensively investigated. While the free acid is not soluble in neutral or acidic water, it shows excellent solubility in its sodium salt form (Figure 3a) (*42*). Once the backbone is assembled, the chlorine atoms from the $[NPCl_2]_n$ precursor are replaced with the sodium salt of propyl p-hydroxybenzoate (sodium paraben) nucleophile. The use of protecting groups is required as the free acid groups lead to the breakdown of polymer backbone (*45*).

Different polyphosphazene polyelectrolytes have been developed with some variations, but among them poly[di(sodiumcarboxylatoethylphenoxy)phosphazene] (PCEP) (Figure 3b) have already shown improved immunoadjuvant activity compared to PCPP (*46, 47*). In this case, the polymer backbone is substituted with methyl 3-(4-hydroxyphenyl)propionate groups (*48*). There is also an alternative route to afford anionic poly(organo)phosphazenes *via* aromatic sulfonation (*49–51*). While at higher temperatures considerable polymer degradation occurs, poly[bis(phenoxyethoxy)phosphazene] could be sulfonated using concentrated H_2SO_4 at room temperature and treated with sodium carbonate solution to afford a water-soluble poly(organo)phosphazene (Figure 3c) (*50*). It is also possible to prepare water-soluble polythionylphosphazene polyelectrolytes (*32*). After ring-opening polymerization of the cyclic thionylphosphazene, subsequent macromolecular substitution with an amine bearing a protected carboxylic acid and finally upon removal of the protective silyl groups under mild conditions (e.g. tetrabutylammonium fluoride), the anionic polymer is obtained. The triethylammonium counterion was exchanged to sodium and a readily water soluble polyelectrolyte was obtained.

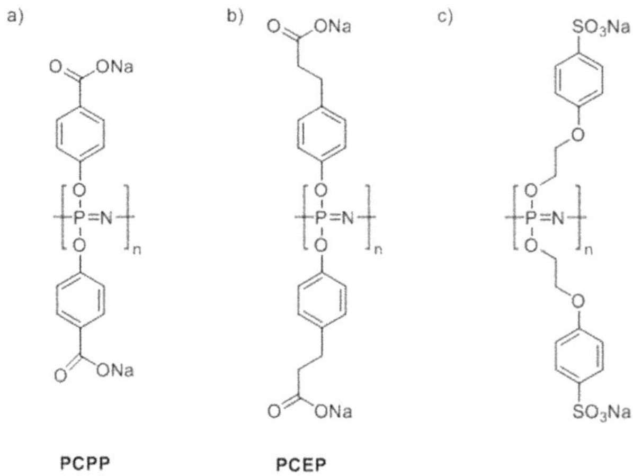

Figure 3. Chemical structures of some anionic poly(organo)phosphazenes: a) Poly[di(sodiumcarboxylatophenoxy)phosphazene] (PCPP), b) Poly[di(sodiumcarboxylatoethylphenoxy)phosphazene] (PCEP), and c) Sulfonated poly[bis(phenoxyethoxy)phosphazene].

Cationic Poly(organo)phosphazenes

Cationic polymers can also be prepared. Poly[(2-dimethylaminoethylamino)phosphazene] (pDMAEA-ppz) (Figure 4a) is one of the first cationic polyphosphazenes reported by Luten et al (*52*) for its in vitro gene delivery characteristics. It is a polymer bearing a tertiary amine which is positively charged at physiological pH. Furthermore, through the addition of imidazole side groups the degradation rates could be tailored (Figure 4b) (*53*). More recently, cationic polyphosphazenes have been reported through cosubstitution of imidazole groups with cyclic polyamine (*54*). Among the polymers under study, Im-PPZ-cyclen (Figure 4c) showed promising results due to the tetraamine structure in which the positive charge is delocalized on the heterocyclic ring. Quaternization is also an alternative method to prepare cationic poly(organo)phosphazenes. In this context, some organic-inorganic hybrid brush polymers have been synthesized by grafting [2-(dimethylamino)ethyl methacrylate] into the linear poly(organo)phosphazene *via* atom transfer radical polymerization (ATRP) and subsequent quaternization with different alkyl chain length iodides (*55*). In a different study, allyl bromide was incorporated into the tertiary amine of pDMAEA-ppz obtaining the cationic polymer which could lead also to cross-linking reactions to obtain different hydrogels (See Hydrogels section) (*56*).

Recent synthesis of water soluble amino acid ester polyphosphazenes with free amino functional groups has been also investigated (*57*). Boc-serine ethyl ester was reacted with the precursor poly(dichloro)phosphazene affording an amino-protected O-linked polymer. After deprotection in mild acidic conditions (using trifluoroacetic acid at room temperature) the free pendent hydrophilic amino groups contributed to the water solubility of the poly(organo)phosphazene.

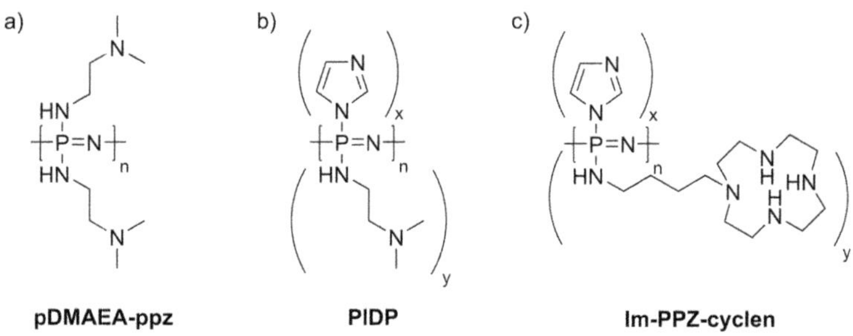

Figure 4. Chemical structure of some cationic poly(organo)phosphazenes: a) Poly[(2-dimethylaminoethylamino)phosphazene] (pDMAEA-ppz), b) Poly(imidazole/dimethylaminoethylamino)phosphazene (PIDP), and c) poly(imidazole/1,4,7,10-tetraazyclodocane)phosphazene (Im-PPZ-cyclen).

Non-Charged Water Soluble Poly(organo)phosphazenes

Polyphosphazenes that bear hydroxyl units show significant features due to the functionality. Poly(diglycerophosphazene) (Figure 1b) has excellent water solubility and complete hydrolytic degradation in 720h at 37°C (*58*). However, investigations into this polymer have been rare, probably, due to the tedious protection and deprotection reactions required, in order to avoid crosslinking during polymer synthesis. In the last few years, the interest in hyperbranched or dendritic polyglycerols has increased significantly as their potential use in biomedical applications (*59–61*) and very recently degradable, dendritic polyols on a branched poly(organo)phosphazene backbone have been reported. Applying a recently developed method to synthesize poly(organo)phosphazenes with star dendritic type structures (*62*), fully degradable hydroxyl-functionalized polymers with exceptionally high functionality and controlled structures have been prepared (Figure 5) (*63*). Furthermore, the use of photochemical thiol-yne addition of the glyceryl moieties circumvents the requirement for protection group chemistry. Indeed this type of chemistry has proven highly practical for the preparation of such highly functionalized, hydrophilic polyphosphazenes (see also glycopolymers below), not just due to the orthogonality towards a high number of functional groups, but also the high efficiency of the reaction and also the doubling of the number of functional groups per phosphorus. Importantly the sterically unhindered and reactive propargyl amine allows facile, complete substitution of the $[NPCl_2]_n$ backbone, before adding bulkier functional groups (Figure 6). This is a key feature since the presence of residual Cl atoms is well-known to have a decisive effect on the hydrolytic stability of polyphosphazenes (*64*).

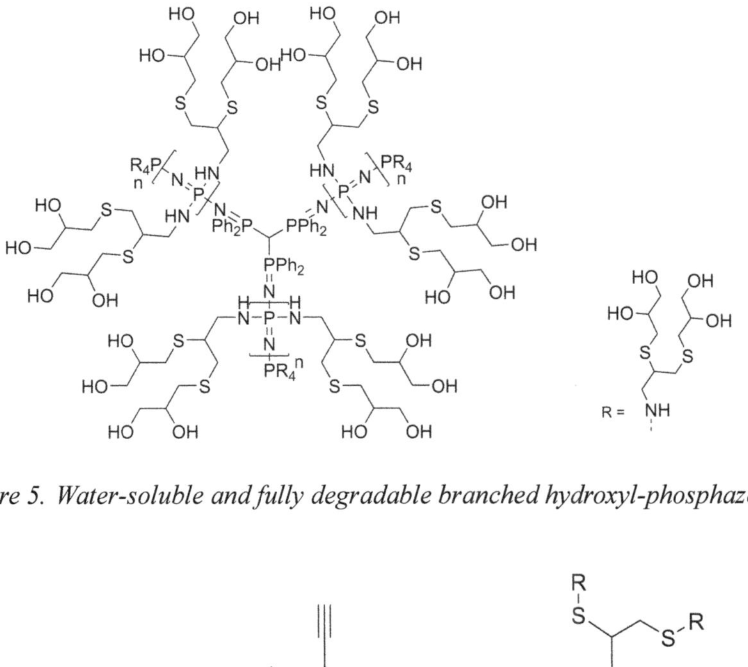

Figure 5. Water-soluble and fully degradable branched hydroxyl-phosphazene.

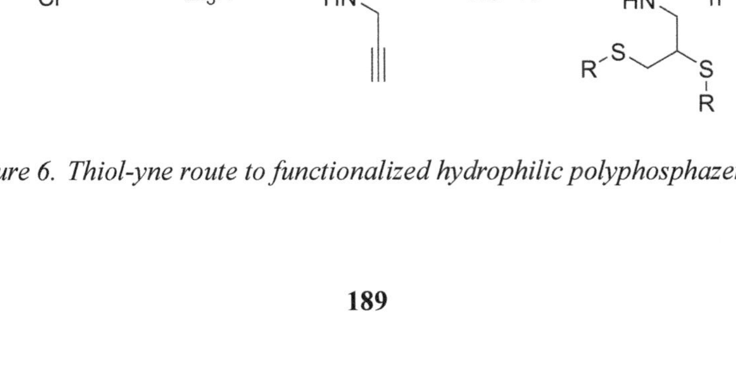

Figure 6. Thiol-yne route to functionalized hydrophilic polyphosphazenes.

Polyvinylpyrrolidone (PVP) and PEG polymers are among the most widely used synthetic water-soluble polymers. However, as they are non-degradable their use in some medical applications, such as intravenous application, is restricted. Thus synthesis of new degradable polymers with the well-known pharmaceutical carriers is of great interest. One route to overcome this limitation is the preparation of hybrid polyphosphazenes with a similar chemical functionality to PVP. Poly(di[2-(2-oxo-1-pyrrolidinyl)ethoxy]phosphazene) (PYRP) (Figure 7a) was synthesized by Andrianov and co-workers (*65*) using 1-(2-hydroxyethyl)-2-pyrrolidone as substituent. In this way the biological relevant properties of the neutral *N*-ethylpyrrolidone group were combined with the advantage of the backbone degradability. It was also demonstrated that *N*-ethylpyrrolidone could modulate the degradation rates of mixed-substituent poly(organo)phosphazenes) (*65*, *66*). PYRP has been used as a replacement for non-degradable PVP, for example in the solubilization of the non water-soluble photosensitizer hypericin, providing comparable results to the known hypericin-PVP conjugate but with a degradable polymer (*67*).

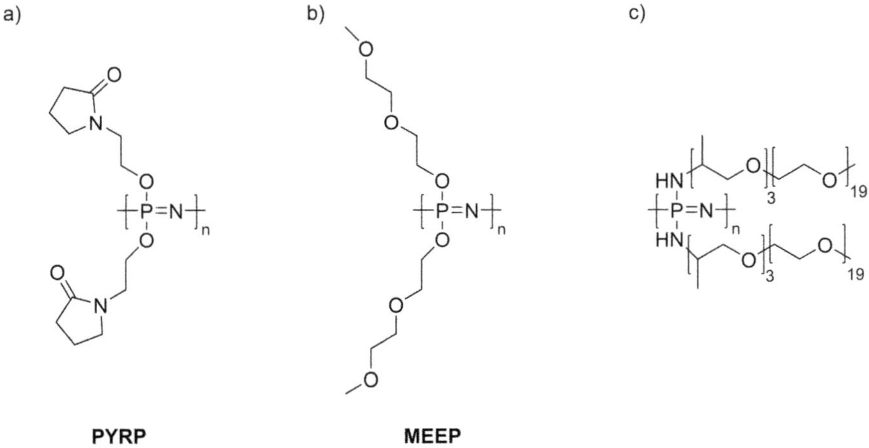

Figure 7. Chemical structure of some non-charged poly(organo)phosphazenes: a) Poly(di[2-(2-oxo-1-pyrrolidinyl)ethoxy]phosphazene) (PYRP), b) Poly[bis(2-(2-methoxyethoxy)ethoxy)phosphazene] (MEEP), and c) Jeffamine M-1000-substituted polyphosphazene.

Poly[bis(2-(2-methoxyethoxy)ethoxy)phosphazene] (MEEP) (Figure 7b) synthesized by Allcock and co-workers (*68*) showed also to be water-soluble and used as solid polymer electrolytes (*68–70*). Indeed water soluble polymers with a wide range of side-chain lengths and various branched architectures are readily accessible widely commercially available PEGs (*71*). However, such PEGylated polyphosphazenes tend to show highly stable towards hydrolytic degradation. Cosubstitution with hydrolysis sensitizing moieties can be used to prepare degradable variants (*72*). Alternatively degradable water-soluble poly(organo)phosphazenes have been prepared using the hydrophilic Jeffamine M-1000 oligomer (Figure 7c) (*10*). The higher lability of the P-N bond compared to the P-O (as in MEEP) increases the susceptibility of the polymer to hydrolytic degradation. Water-soluble poly(organo)phosphazenes with controlled dimensions can be synthesized with varied lengths of Jeffamine polyetheramine grafted to the polymer backbone (*34*). The branching density of the polymer could be increased by incorporation of two more arms per repeat unit. By incorporation of alkyne-functionalized groups into the polymer backbone and subsequent thiol-yne reaction, two thiol groups could be coupled onto a single alkyne group. Furthermore, multi-arm star poly(organo)phosphazenes have been synthesized (*62*). as well as dendritic molecular brush structures with up to 30 000 functional end groups, water solubility, narrow polydispersities and hydrodynamic diameters ranging from 10 to 70 nm.

Post-polymerization functionalization also allows incorporation of diverse saccharides on a poly(organo)phosphazene backbone to give water soluble glycopolymers (*27*). A direct route has also been developed *via* the conjugation of lactobionic acid onto a primary amine-functionalized poly(organo)phosphazene (*73*). A alternative route is provided by photochemical thiol-yne addition to a propargylamine bearing mixed-poly(organo)phosphazene. The high reactivity towards thioglucopyranose allows incorporation of two saccharide groups per triple bond (*74*), with the thiol-yne reaction again proving useful to prepare highly functionalized, hydrophilic polyphosphazenes without the need for protecting group chemistry.

Amphiphilic Poly(organo)phosphazenes

Amphiphilic poly(organo)phosphazenes consisting of hydrophilic and hydrophobic substituents are of great interest. A unique feature compared to the conventional block-polymers usually utilized for self-assembly, is that even a randomly substituted polyphosphazene backbone allows the micellar-like structure formation. This is presumably due to the high rotational freedom of the backbone allowing agglomeration of hydrophobic moieties This behavior for example can be observed in water-soluble polyphosphazenes loaded with hydrophobic drugs (*75*), as well as in peptide based hybrid poly(organo)phosphazenes (Figure 8) (*76*). These polymers were synthesized by addition of the Gly-Phe-Leu-Gly (GFLG) sequence in combination with the hydrophilic Jeffamine M-1000 oligomer.

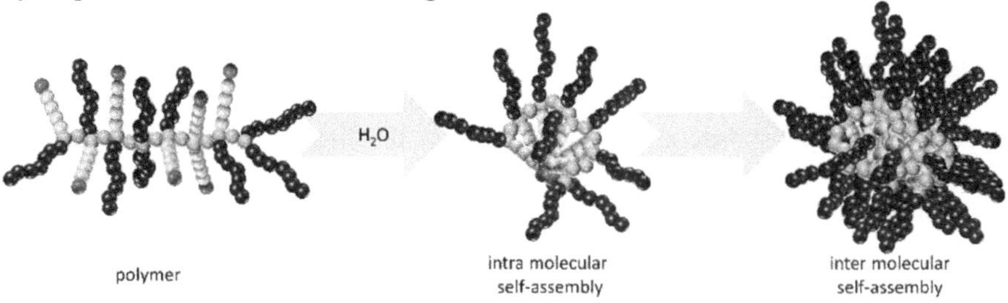

Figure 8. Graphic representation of intra- and intermolecular self-assembly of the amphiphilic poly(organo)phosphazene in aqueous solution. (Reproduced with permission from reference (76). 2016 MDPI.)

A series of amphiphilic graft poly(organo)phosphazenes, poly(NIPAm-co-DMAA)/Ethylglycinate-polyphosphazene (PNDGPs), have been synthetized with ethyl glycinate as hydrophobic side group and a copolymer bearing *N*-isopropylacrylamide (NIPAm) and *N,N*-dimethylacrylamide (DMAA) as hydrophilic side groups (*77*). By variation of the ratio of hydrophilic to hydrophobic side groups some amphiphilic poly(organo)phosphazenes were investigated with tunable critical micelle concentration (CMC). In this case, the hydrophobic drug doxorubicin (DOX) was trapped in the core of the micelle. Cosubstitution of poly(organo)phosphazenes with hydrophilic poly(ethylene glycol) and hydrophobic ethyl-*p*-aminobenzoate also lead to amphiphilic polymers with tunable supramolecular structures due to the variation of the hydrophilic ratio (*78*). Furthermore, by exchange of the hydrophobic ethyl-*p*-aminobenzoate to the pH-sensitive *N,N*-diisopropylethylenediamine (DPA) group some pH-responsive amphiphilic poly(organo)phosphazenes has been investigated, upon the protonation of the tertiary amine at lower pH values (*79*).

Thermoresponsive Polyphosphazenes

Thermoresponsive, sometimes referred as thermosensitive polymers, are polymers that demonstrate changes in their physical properties with temperature, for example a number of amphiphilic polymers show a lower critical solution temperature (LCST) in aqueous solutions, whereas above of this temperature the polymer chains collapse with a consequent precipitation or gelation of the polymer. Indeed some of the amphiphilic polymers described in the previous section also demonstrate thermoresponsive behavior. For example, PNDGPs shows an LCST around 39°C and its temperature sensitivity is utilized to release the encapsulated material (*77*). Thermoresponsive polymers with an LCST around or below body temperature are of particular interest for medical applications.

Branched and linear alkylene ether based poly(organo)phosphazenes also present LCST behavior in the range of 40°C-80°C (*60*). More recently, amine-capped Jeffamine oligomers with different ratios of ethylene oxide (EO) to propylene oxide (PO) have been used as substituents. These polymers have LCST values in the range of 18°C to 90°C (Figure 9a) (*80*). Interestingly, these polymers were observed to agglomerate into colloids in the 100 nm range above their LCST, presumably due the high flexibility of the poly(organo)phosphazenes (Figure 9b).

Another commonly employed method to prepare thermoresponsive poly(organo)phosphazenes is through mixed substitution of hydrophobic amino acid esters with oligomeric alkoxy PEGs (Figure 10b). The LCST values can be tailored through varying the hydrophilic/hydrophobic composition. For example, the polymer bearing glycine ethyl ester and 0.31 mol of PEG_{350}/unit has an LCST of 35°C while 1.12 mol/unit shows LCST of 83.7°C. Longer PEG chains also increase the LCST as it is the case of 1.09 mol PEG_{750}/unit in which the LCST is 98.5°C (*81*). In contrast, the more hydrophobic amino acid ester group contributes to lower LCST values. In a similar way, thermoresponsive poly(organo)phosphazene bearing L-isoleucine ethyl ester, as hydrophobic substituent, and hydrophilic amino-methoxy-PEG oligomers has been synthesized (*82*). Again, the LCST values also increase with the ratio of PEG. Variations on the ratio between PEG and L-isoleucine ethyl ester allow tailoring the degradation rate and mechanical properties of the polymer.

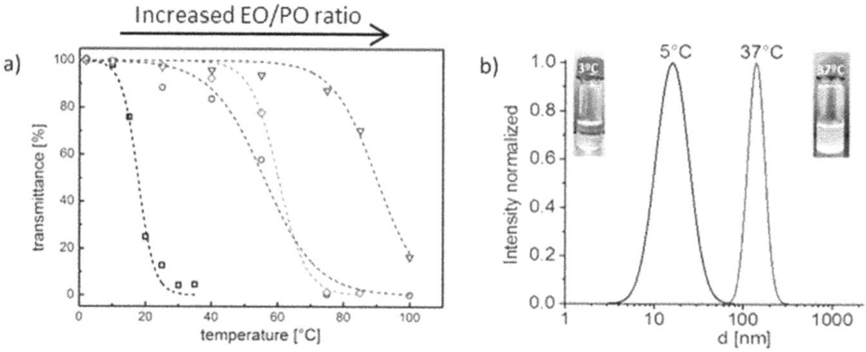

Figure 9. a) Jeffamine substituted polyphosphazenes, with increasing hydrophilicity show higher LCST ranges and b) temperature driven self-assembly of amphiphilic polyphosphazenes. (Adapted with permission from reference (80). Copyright 2014 Wiley.)

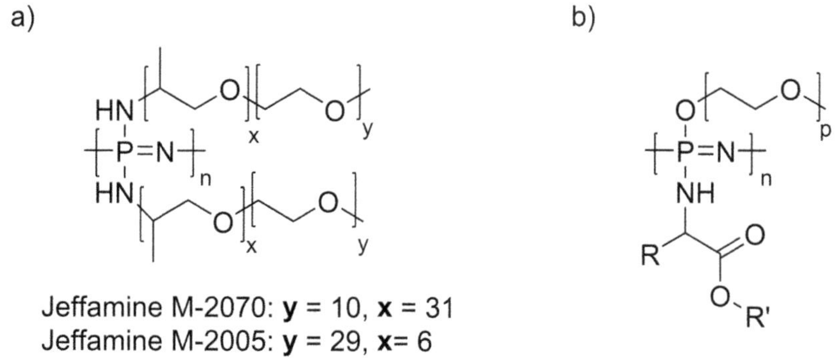

Figure 10. Structure of some typical thermoresponsive poly(organo)phosphazenes: a) Jeffamine-substituted polyphosphazenes, and b) Mixed substituted polyphosphazene with hydrophobic amino acid ester and oligomeric alkoxy PEG.

Hydrogels

The ability of thermoresponsive polymers to afford physical and chemical changes as a function of temperature opens the opportunity to synthesize thermoresponsive hydrogels. A significant amount of work related to poly(organo)phosphazene-based injectable hydrogels for drug delivery systems has been developed by S.C. Song and co-workers in which its synthesis, sol-gel properties and degradability have been studied (*81–83*). The gelation properties of the polymers, based on amino acid esters and amino-PEG, can be tailored by variations of PEG chain length, composition of substituents, amino acid esters and pH (*83*). Solution temperatures influence not only sol-gel transition, but also degradation rates (*84*).

Fine-tuning of the systems, including incorporation carboxylic moieties facilitated the dual interaction of polymeric nanoparticles (D-NP) (*85*). The study showed the effectiveness of the injectable bone morphogenetic protein-2(BMP-2)/D-NP nanocomplex hydrogel system, in which the aqueous solution becomes a hydrogel at body temperature, and the sustained protein release system besides its degradation in vivo and in vitro (Figure 11). The effectiveness of the hydrogel to regenerate a vertical bone at the peri-implant site has also been demonstrated (*86*). Moreover, varying the chain-lengths and amounts of the carboxylic acid side group the physical properties, such as dissolution/degradation behavior, water uptake, or BMP-2 release rate could be tailored (*87*).

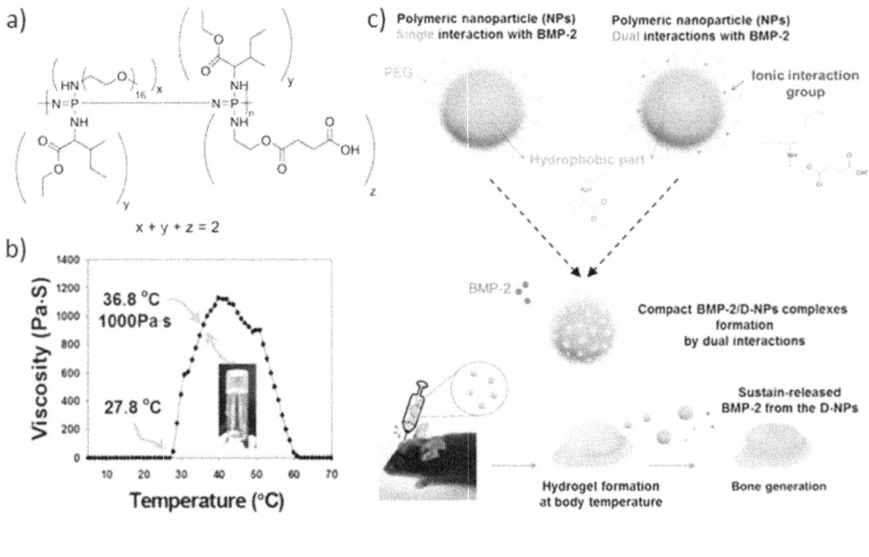

Figure 11. a) Structure of the poly(organo)phosphazene for D-NPs. b) Temperature-dependent viscosity changes of D-NPs in aqueous solution. c) Schematic illustration of the nanocomplexes formation between BMP-2 and D-NPs due to the ionic and hydrophobic interactions, followed by hydrogel formation after injection and its sustained release to assist the bone generation. (Adapted with permission from reference (85). Copyright 2015 Elsevier.)

Supramolecular-structured hydrogels can be also prepared from water-soluble polyphosphazenes, for example by inclusion complexation between mPEG grafted poly(organo)phosphazenes and cyclodextrin in aqueous solution. Through the addition of cyclodextrin physical cross-linking occurs with mPEG while unreacted mPEG chains assist the hydrophilicity (Figure 12). Gelation times, stability and release of a model protein could be tailored with variation of polymer structure and concentration (*88*). Moreover, ionic hydrogels, in which anionic polymer chains form "bridges" with cations, have been synthesized for encapsulation (*89*). For example PCPP polyelectrolytes in presence of di- or trivalent cations, such as calcium, copper, or aluminum, can be crosslinked to form ionic hydrogels (*42*). The salts form "bridges" between the polymer chains leading to cross-linking that can be broken in presence of monovalent cations.

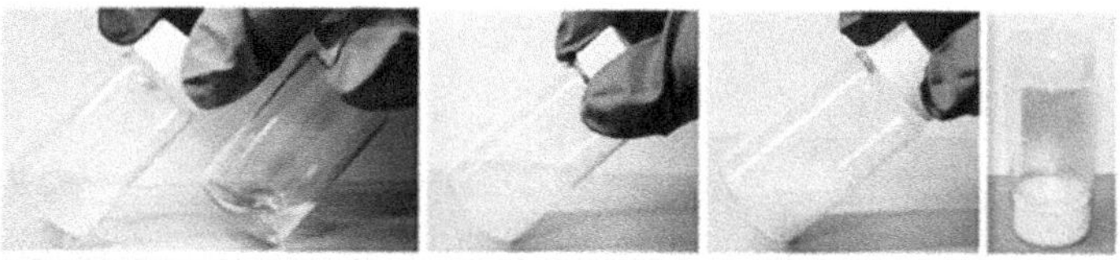

Figure 12. Combination of poly(ethylene glycol) methyl ether-based poly(organo)phosphazene (PmPEG2000) with α-cyclodextrin (α-CD) lead to a rapid supramolecular hydrogel formation. (Adapted with permission from reference (88). Copyright 2013 American Chemical Society.)

Of the hydrogels described above, all of them are formed by physical gelation in absence of covalent interactions. Chemical cross-linkers have also been employed in gelation, primarily with the goal of improving the mechanical properties of the gel. Cosubstitution of poly(organo)phosphazenes with methacrylate groups has also been used, leading to photo-crosslinked polymers upon irradiation with UV light (*90*). After physical cross-linking at body temperature, the gel is irradiated producing a cross-linked system with improved mechanical properties. Varying the amount of methacrylate the mechanical gel property and degradation rate could be controlled. However, UV light exposure is not appropriate for use in biological tissues due to its damaging nature and low penetration. In order to overcome these limitations, systems have been developed in which the polymer is UV-pretreated and cross-linked only after injection (*91*). Interestingly, the polymer is exposed to UV light in aqueous solution at low temperatures and it crosslinked only after injection into the body due to the rise in temperature. Allyl groups could be incorporated *via* quaternization method into a tertiary amine to assist the thiol-ene reaction. After UV light irradiation, hydrogels were easily obtained through the cross-linking reaction between PEG-dithiol and the allyl-functionalized polyphosphazene (Figure 13) (*56*). This cationic hydrogel has been studied for its binding and release effects on biomacromolecules.

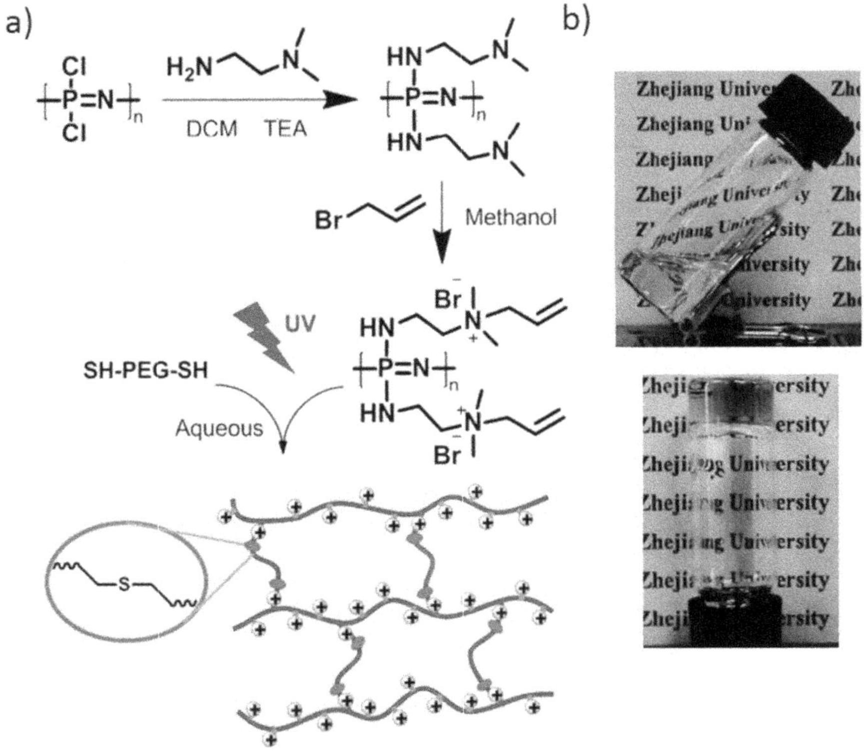

Figure 13. a) Schematic illustration of the quaternization of the polyphosphazene by allyl bromide followed by thiol-ene reaction with PEG-dithiol in aqueous solution. b) Polymer solution before (top) and after (bottom) UV irradiation. Adapted with permission from reference (56). Copyright 2015 The Royal Society of Chemistry.

Degradation Mechanisms

The complete backbone degradation of some poly(organo)phosphazenes is of considerable interest in numerous biomedical applications. Poly(organo)phosphazene degradation occurs *via* hydrolytic cleavage of the organic substituents leading to the instable hydroxyphosphazenes and phosphazenes species triggering further degradation to phosphate and ammonium ions (*31*, *72*, *92*) (Figure 14 route (a)). Importantly, the mixture of phosphates and ammonium salts is well-known to be benign (*72*) so in combination with a suitable choice of the organic substituents, materials with non-toxic degradation products can be synthesized.

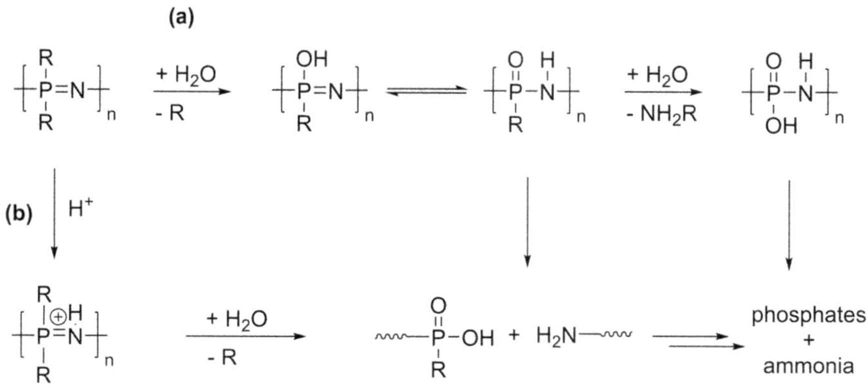

R = NHR' or OR'

Figure 14. Proposed hydrolytic degradation mechanism of poly(organo)phosphazenes in a) neutral and b) acidic media.

The degradation rate is tailored by the characteristics of the organic substituents, which also directly affect a number of different parameters such as mechanical stability and physicochemical properties, which also must be taken into account when the organic substituents are selected. Another aspect to consider is the complete replacement of chlorine atoms in the backbone by the nucleophiles as remaining chlorine atoms influence on the degradation mechanisms (*45*, *48*). Uncompleted macromolecular substitution shows accelerated hydrolytic breakdown and hence must be carefully assessed when preparing polymers with controlled degradation profiles.

Poly(organo)phosphazenes bearing amino acid derivatives have attracted much attention due to their biocompatibility and availability and hence intense research efforts have been carried out into the hydrolysis of different poly[(amino acid ester)phosphazenes] in the solid state (*64*, *92–94*) with a simultaneous decrease in polymer mass, molecular weight and the release of low molecular degradation products. Although these polymers are generally hydrophobic, their hydrolytic sensitivity has also been applied *via* cosubstitution to a wide variety of water-soluble polymers. One example could be by incorporation of amino acid moieties as spacer between the polymer backbone and the hydrophilic Jeffamine M-1000 oligomer (*10*). In this way enhanced degradation could be observed compared to the polyphosphazenes without amino acid linkers (Figure 7c). The polymer without a spacer in pH 5 solution at 37°C showed 14 % of the backbone degraded to phosphate within 4 weeks, while the polymer bearing valine spacer showed 50 % of the phosphate formation. Using a less bulky amino acid, such as glycine, the polymer underwent almost complete degradation under the same conditions.

Despite the fact that P-N bond is more hydrolytically sensitive compared to the stronger P-O bond, there are some water soluble polyphosphazenes with P-O linkages that show hydrolytic instability with acceptable degradation times. One example is the water soluble PYRP polymer (Figure 7a) which showed a decrease in the molecular weight and release of 1-(2-hydroxyethly)-2-pyrrolidone confirming the hydrolytic cleavage of the side group (*65*). The results also showed hydrolytic degradation of the polymer in aqueous solutions at different pH values, where degradation rates increased in the acidic environment (Figure 15). PCPP polymer (Figure 3a) also showed reduction in the molecular weight over time (*95*) but in this case promoted also by the carboxyl group which can catalyze the degradation (see discussion of self-catalyzed degradation in Stimuli responsive degradation mechanisms).

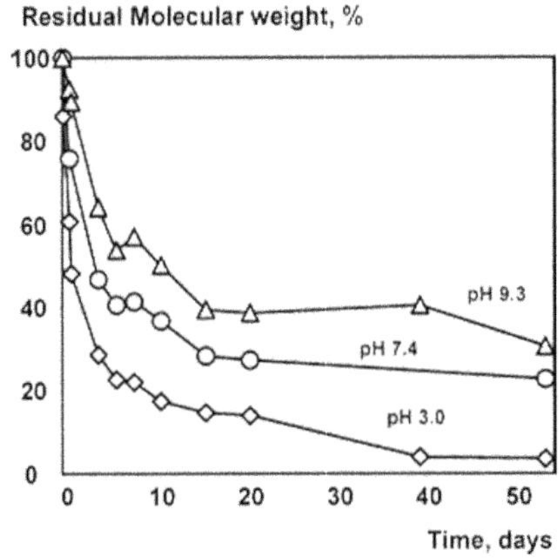

Figure 15. Hydrolytic degradation of PYRP at 55°C and different pH. Representation of molecular weight loss by gel permeation chromatography (GPC) with poly(ethylene oxide) PEO standards. (Adapted with permission from reference (65). Copyright 2005 American Chemical Society.)

Although the degradation rates of different poly(organo)phosphazenes are to some degree predictable from the chemical structures (see Allcock et al. (*72*)), a direct comparison of the degradation kinetics of most reported polymers is difficult due to the differences in the experimental conditions (e.g. temperature, solid or solution state and deionized water or buffered systems) between the investigated systems. The PYRP polymer (Figure 7a) for example shows ~80 % of M_w loss in ~ 20 days (pH 7.4 tris buffer at 55°C), while the PCPP polymer (Figure 3a) in the same conditions shows only ~30 % of M_w loss (*65*). DMAEA-ppz (Figure 4a) with P-N bond has a half-life of 24 days at 37°C and neutral pH (pH 7.5 tris-HCl buffer), and enhanced degradation at pH 5 with a half-life of 5 days (*52*). By addition of some imidazole groups (Figure 4b) the degradation of the polymer has shown to be slightly faster shortening 2 days respectively (*53*).

Stimuli Responsive Degradation Mechanisms

Degradable water soluble polymers discussed until now show also pH dependence in which the acidic conditions enhance the degradation kinetics of many poly(organo)phosphazenes (*10*, *64*, *65*). At lower pH values the protonation of the nitrogen atom occurs followed by a migration of the positive charge to the backbone and therefore, the nucleophilic attack of water is assisted (*96*)

(Figure 14 route (b)). This pH triggered-response is an interesting mechanism for biomedical applications, as it opens the possibility to design systems which respond to changes in pH, for example for targeting of intracellular environments or diseased or cancerous tissue (*97*).

It is also reported that the presence of acidic groups close to the polymer backbone enhances the hydrolysis rates due to intra-molecular catalysis (*93*, *98*). The nitrogen atom is protonated due to the proximity of the acid group and further reaction with water leads to side group cleavage. In this context, poly(organo)phosphazenes bearing a tetrapeptide sequence Gly-Phe-Leu-Gly presented by showed enzymatically triggered degradation by which after enzymatic cleavage of a peptide bond, the free carboxyl group promoted the backbone degradation (Figure 16) (*76*).

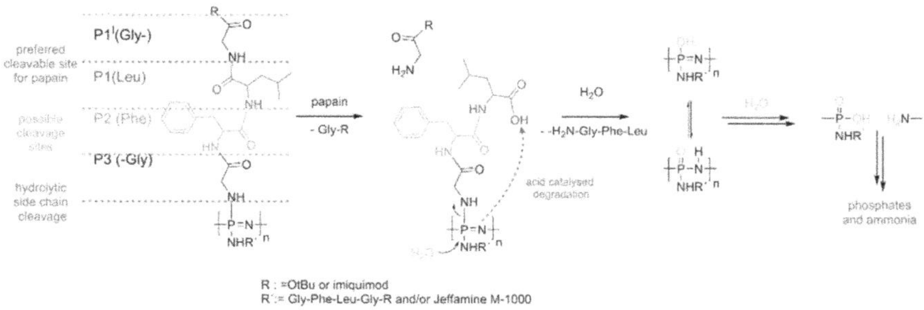

Figure 16. Schematic illustration of the enzymatically triggered and acid-catalyzed degradation of GFLG-peptide based poly(organo)phosphazenes. (Reproduced with permission from reference (76). 2016 MDPI.)

In early works by Allcock and coworkers, poly(glycine)phosphazene was reported to be difficult to isolate due to its rapid hydrolysis (*98*), presumably due to the proximity of the carboxylic acid moiety to the phosphazene backbone. In a recent study, the inherently instable poly(glycine)phosphazene, has hence been protected (caged) by phenylboronate moieties, enhancing it hydrolytic stability (*99*). A mixed-substituent polymer was then prepared, based on a glycinate arylboronic acid pinacol ester with Jeffamine M-1000 to provide the water solubility to the system. The arylboronate pinacol ester group oxidizes to phenol in presence of the reactive oxygen species (ROS) H_2O_2 (*100*, *101*). ROS are important in a wide range of biological functions but it may damage biomolecules and therefore contribute to several diseases when it is overproduced. The polymer showed selective degradation in presence of 10 mM aqueous solution of H_2O_2 due to the oxidation of the caging group to phenol and thus, its self-immolation to uncage the unstable glycine-substituted polyphosphazene which catalyzes the polymer degradation (Figure 17). Analysis by gel permeation chromatography (GPC) showed a decrease in the polymer peak besides an increase in low molecular weight region corresponding to the cleavage of side groups. ^{31}P NMR analysis confirmed also backbone degradation with a visible sharp peak arising with the time due to phosphate formation. Meanwhile, the polymer showed no degradation signal in absence of H_2O_2 in the same time-frame as it could be confirmed by GPC and ^{31}P NMR spectroscopy. In order to prove the exclusive degradation of the polymer due to the presence of the arylboronate pinacol ester caging group, poly(glycine ethyl ester)phosphazene was exposed even to a higher concentration of H_2O_2 (100 mM) with no sign of degradation according to ^{31}P NMR spectroscopy.

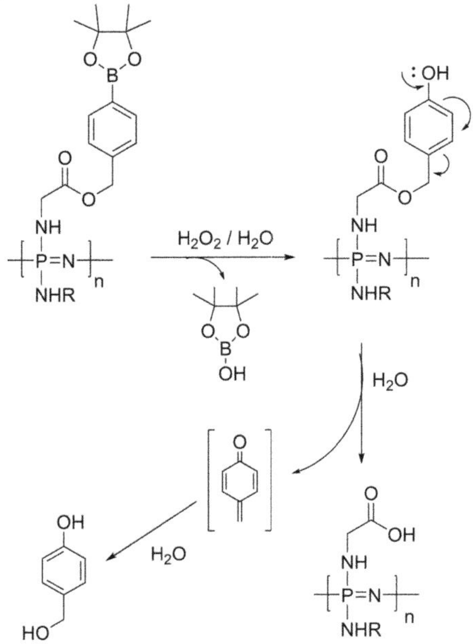

Figure 17. Proposed self-immolation mechanism of the caged poly(glycine)phosphazene upon H_2O_2 exposure. (Adapted with permission from reference (99). Copyright 2017 American Chemical Society.)

In a similar manner, coumarin-based photocages have also been incorporated into a water-soluble poly(organo)phosphazene that cleaves upon exposure to visible light (*102*). The polymer is shown to be stable in aqueous solution in the dark while fast degradation signs could be obserevd after irradiation. The sensitiveness to degradation under visible light irradiation opens the opportunity to its safe use in medical applications with longer wavelengths for deeper tissue penetration.

Summary and Outlook

Water-soluble poly(organo)phosphazenes have attracted much attention in a wide variety of applications, especially in medical applications. The facile functionalization and more importantly easily tunable degradation rates allow the synthesis of a vast number of poly(organo)phosphazenes that fully degrade to small molecules. Due to the relative lack of water soluble synthetic organic polymers with suitable and easily tunable degradation rates, this is a highly promising niche market for such polyphosphazenes.

However, while the degradation of some non-water soluble poly(organo)phosphazenes has been systematic studied in solid state (e.g. amino acid-based polymers), investigations in solution state have been much less common. Furthermore, the experimental conditions used have varied widely, hence a more systematic study is required to fully understand the relationship between the chemical structure and the hydrolytic stability in solution.

Particularly promising applications for such water soluble and degradable polymers included in medical applications are drug (*103*, *104*), gene (*52*) and vaccine delivery (*105*). Water soluble poly(organo)phosphazenes have been demonstrated to be highly convenient for such applications due to the facile functionalization with a wide variety of nucleophiles, along with the tunable degradability and fully degradation to small molecules, forming pH-buffered system. Moreover, the recent development of polyphosphazenes with responsive degradation that can be switched-on in reply to certain stimuli promises to give further momentum to the field.

Abbreviations

ATRP	atom transfer radical polymerization
BMP	bone morphogenetic protein
CD	α-cyclodextrin
CMC	critical micelle concentration
DMAA	*N,N*-dimethylacrylamide
D-NP	dual interaction of polymeric nanoparticles
DOX	doxorubicin
DPA	*N,N*-diisopropylethylenediamine
EO	ethylene oxide
GFLG	glycine-phenylalanine-leucine-glycine (Gly-Phe-Leu-Gly) sequence
GPC	gel permeation chromatography
Im-PPZ-cyclen	poly(imidazole/1,4,7,10-tetraazyclodocane)phosphazene
LCST	lower critical solution temperature
MEEP	poly[bis(2-(2-methoxyethoxy)ethoxy)phosphazene]
NIPAm	*N*-isopropylacrylamide
NMR	nuclear magnetic resonance
PCEP	poly[di(sodiumcarboxylatoethylphenoxy)phosphazene]
PCL	polycaprolactone
PCPP	poly[di(carboxylatophenoxy)phosphazene]
pDMAEA-ppz	poly[(2-dimethylaminoethylamino)phosphazene]
PEG	polyethylene glycol
PEO	poly(ethylene oxide)
PET	polyethylene terephthalate
PHB	polyhydroxybutyrate
PIDP	poly(imidazole/dimethylaminoethylamino)phosphazene
PLA	polylactic acid
PLGA	poly(lactic-co-glycolic) acid
PmPEG2000	poly(ethylene glycol) methyl ether-based poly(organo)phosphazene
PNDGPs	poly(NIPAm-co-DMAA)/Ethylglycinate-polyphosphazene

PO	propylene oxide
poly(HPMA)	poly[N-(2-hydroxypropyl)methacrylamide]
PVA	poly(vinyl alcohol)
PVP	polyvinylpyrrolidone
PYRP	poly(di[2-(2-oxo-1-pyrrolidinyl)ethoxy]phosphazene)
ROP	ring-opening polymerization
ROS	reactive oxygen species
UV	ultraviolet

References

1. Halake, K.; Birajdar, M.; Kim, B. S.; Bae, H.; Lee, C.; Kim, Y. J.; Kim, S.; Kim, H. J.; Ahn, S.; An, S. Y.; Lee, J. Recent application developments of water-soluble synthetic polymers. *J. Indust. Eng. Chem.* **2014**, *20*, 3913–3918.
2. Treiser, M.; Abramson, S.; Langer, R.; Kohn, J., Degradable and Resorbable Biomaterials. In *Biomaterials Science*, 3rd ed.; Ratner, B. D.; Hoffman, A. S.; Schoen, F. J.; Lemons, J. E., Eds.; Academic Press: San Diego, CA, 2013; Chapter I.2.6, pp 179−195.
3. Gross, R. A.; Kalra, B. Biodegradable polymers for the environment. *Science* **2002**, *297*, 803–807.
4. Lambert, S.; Wagner, M. Environmental performance of bio-based and biodegradable plastics: the road ahead. *Chem. Soc. Rev.* **2017**, *46*, 6855–6871.
5. Nair, L. S.; Laurencin, C. T. Biodegradable polymers as biomaterials. *Prog. Polym. Sci.* **2007**, *32*, 762–798.
6. Ulery, B. D.; Nair, L. S.; Laurencin, C. T. Biomedical applications of biodegradable polymers. *J. Polym. Sci., Part B: Polym. Phys.* **2011**, *49*, 832–864.
7. Delplace, V.; Nicolas, J. Degradable vinyl polymers for biomedical applications. *Nat. Chem.* **2015**, *7*, 771–784.
8. Bertrand, O.; Gohy, J.-F. Photo-responsive polymers: synthesis and applications. *Polym. Chem.* **2017**, *8*, 52–73.
9. Mutlu, H.; Barner-Kowollik, C. Green chain-shattering polymers based on a self-immolative azobenzene motif. *Polym. Chem.* **2016**, *7*, 2272–2279.
10. Wilfert, S.; Iturmendi, A.; Schoefberger, W.; Kryeziu, K.; Heffeter, P.; Berger, W.; Brüggemann, O.; Teasdale, I. Water-soluble, biocompatible polyphosphazenes with controllable and pH-promoted degradation behavior. *J. Polym. Sci., Part A: Polym. Chem.* **2014**, *52*, 287–294.
11. Göpferich, A. Mechanisms of polymer degradation and erosion. *Biomaterials* **1996**, *17*, 103–114.
12. Van Cauwenberghe, L.; Vanreusel, A.; Mees, J.; Janssen, C. R. Microplastic pollution in deep-sea sediments. *Environ. Pollut.* **2013**, *182*, 495–499.
13. Wright, S. L.; Thompson, R. C.; Galloway, T. S. The physical impacts of microplastics on marine organisms: A review. *Environ. Pollut.* **2013**, *178*, 483–492.

14. Bagheri, A. R.; Laforsch, C.; Greiner, A.; Agarwal, S. Fate of So-Called Biodegradable Polymers in Seawater and Freshwater. *Global Challenges* **2017**, *1*, 1700048.
15. Shastri, V. P. Non-degradable biocompatible polymers in medicine: Past, present and future. *Curr. Pharm. Biotechnol.* **2003**, *4*, 331–337.
16. Pan, H.; Yang, J.; Kopečková, P.; Kopeček, J. Backbone degradable multiblock N-(2-hydroxypropyl)methacrylamide copolymer conjugates via reversible addition−fragmentation chain transfer polymerization and thiol−ene coupling reaction. *Biomacromolecules* **2011**, *12*, 247–252.
17. Sponchioni, M.; Morosi, L.; Lupi, M.; Capasso Palmiero, U. Poly(HPMA)-based copolymers with biodegradable side chains able to self assemble into nanoparticles. *RSC Adv.* **2017**, *7*, 50981–50992.
18. Tappertzhofen, K.; Beck, S.; Montermann, E.; Huesmann, D.; Barz, M.; Koynov, K.; Bros, M.; Zentel, R. Bioreducible poly-L-lysine–poly HPMA block copolymers obtained by RAFT-polymerization as efficient polyplex-transfection reagents. *Macromol. Biosci.* **2016**, *16*, 106–120.
19. Teasdale, I.; Brüggemann, O. *Polyphosphazenes for Medical Applications*; Smithers RAPRA: Shrewsbury, U.K., 2014.
20. Kadajji, V. G.; Betageri, G. V. Water soluble polymers for pharmaceutical applications. *Polymers* **2011**, *3* (4), 1972–2009.
21. Haag, R.; Kratz, F. Polymer therapeutics: Concepts and applications. *Angew. Chem. Int. Ed.* **2006**, *45*, 1198–1215.
22. Duncan, R. Polymer therapeutics: Top 10 selling pharmaceuticals—What next? *J. Controlled Release* **2014**, *190*, 371–380.
23. Kopeček, J. Polymer–drug conjugates: Origins, progress to date and future directions. *Adv. Drug Deliv. Rev.* **2013**, *65*, 49–59.
24. Duncan, R. Polymer therapeutics as nanomedicines: New perspectives. *Curr. Opin. Biotechnol.* **2011**, *22*, 492–501.
25. Markovsky, E.; Baabur-Cohen, H.; Eldar-Boock, A.; Omer, L.; Tiram, G.; Ferber, S.; Ofek, P.; Polyak, D.; Scomparin, A.; Satchi-Fainaro, R. Administration, distribution, metabolism and elimination of polymer therapeutics. *J. Controlled Release* **2012**, *161*, 446–460.
26. Henke, H.; Brüggemann, O.; Teasdale, I. Branched macromolecular architectures for degradable, multifunctional phosphorus-based polymers. *Macromol. Rapid Commun.* **2017**, *38*, 1600644.
27. Allcock, H. R., *Chemistry and applications of polyphosphazenes.* Wiley: Hoboken, NJ, 2003.
28. Tian, Z.; Hess, A.; Fellin, C. R.; Nulwala, H.; Allcock, H. R. Phosphazene high polymers and models with cyclic aliphatic side groups: New structure–property relationships. *Macromolecules* **2015**, *48*, 4301–4311.
29. Allcock, H. R. The expanding field of polyphosphazene high polymers. *Dalton Trans.* **2016**, *45*, 1856–1862.
30. Allcock, H. R. A Perspective of polyphosphazene research. *J. Inorg. Organomet. Polym. Mater.* **2007**, *16*, 277–294.
31. Rothemund, S.; Teasdale, I. Preparation of polyphosphazenes: A tutorial review. *Chem. Soc. Rev.* **2016**, *45*, 5200–5215.

32. Wang, Z.; Manners, I. Synthesis and characterization of water-soluble cationic and anionic polythionylphosphazene polyelectrolytes. *Macromolecules* **2005**, *38*, 5047–5054.
33. Honeyman, C. H.; Manners, I. Ambient temperature synthesis of poly(dichlorophosphazene) with molecular weight control. *J. Am. Chem. Soc* **1995**, *117*, 7035–7036.
34. Henke, H.; Wilfert, S.; Iturmendi, A.; Brüggemann, O.; Teasdale, I. Branched polyphosphazenes with controlled dimensions. *J. Polym. Sci., Part A: Polym. Chem.* **2013**, *51*, 4467–4473.
35. Blackstone, V.; Lough, A. J.; Murray, M.; Manners, I. Probing the mechanism of the PCl5−initiated living cationic polymerization of the phosphoranimine Cl3P-NSiMe3 using model compound chemistry. *J. Am. Chem. Soc.* **2009**, *131*, 3658–3667.
36. Soto, A. P.; Manners, I. Poly(ferrocenylsilane-b-polyphosphazene) (PFS-b-PP): A new class of organometallic−inorganic block copolymers. *Macromolecules* **2009**, *42*, 40–42.
37. Wilfert, S.; Henke, H.; Schoefberger, W.; Brüggemann, O.; Teasdale, I. Chain-end-functionalized polyphosphazenes via a one-pot phosphine-mediated living polymerization. *Macromol. Rapid Commun.* **2014**, *35*, 1135–1141.
38. Suárez-Suárez, S.; Carriedo, G. A.; Soto, A. P. Tuning the chirality of block copolymers: From twisted morphologies to nanospheres by self-assembly. *Chem. Eur. J.* **2015**, *21*, 14129–14139.
39. Morozowich, N. L.; Weikel, A. L.; Nichol, J. L.; Chen, C.; Nair, L. S.; Laurencin, C. T.; Allcock, H. R. Polyphosphazenes containing vitamin substituents: Synthesis, characterization, and hydrolytic sensitivity. *Macromolecules* **2011**, *44*, 1355–1364.
40. Allcock, H. R.; Connolly, M. S.; Sisko, J. T.; Al-Shali, S. Effects of organic side group structures on the properties of poly(organophosphazenes). *Macromolecules* **1988**, *21*, 323–334.
41. Allcock, H. R. Polyphosphazene elastomers, gels, and other soft materials. *Soft Matter* **2012**, *8*, 7521–7532.
42. Allcock, H. R.; Kwon, S. An ionically crosslinkable polyphosphazene: poly[bis(carboxylatophenoxy)phosphazene] and its hydrogels and membranes. *Macromolecules* **1989**, *22*, 75–79.
43. Payne, L.; Jenkins, S.; Andrianov, A.; Roberts, B. Water-Soluble Phosphazene Polymers for Parenteral and Mucosal Vaccine Delivery. In *Vaccine Design*, Powell, M.; Newman, M., Eds.; Springer: New York, 1995; Vol. 6, pp 473−493.
44. Andrianov, A. K.; DeCollibus, D. P.; Gillis, H. A.; Kha, H. H.; Marin, A.; Prausnitz, M. R.; Babiuk, L. A.; Townsend, H.; Mutwiri, G. Poly[di(carboxylatophenoxy)phosphazene] is a potent adjuvant for intradermal immunization. *Proc. Natl. Acad. Sci.* **2009**, *106*, 18936–18941.
45. Andrianov, A. K.; Svirkin, Y. Y.; LeGolvan, M. P. Synthesis and biologically relevant properties of polyphosphazene polyacids. *Biomacromolecules* **2004**, *5*, 1999–2006.

46. Eng, N. F.; Garlapati, S.; Gerdts, V.; Potter, A.; Babiuk, L. A.; Mutwiri, G. K. The potential of polyphosphazenes for delivery of vaccine antigens and immunotherapeutic agents. *Curr. Drug Delivery* **2010**, *7* (1), 13–20.
47. Mutwiri, G.; Benjamin, P.; Soita, H.; Townsend, H.; Yost, R.; Roberts, B.; Andrianov, A. K.; Babiuk, L. A. Poly[di(sodium carboxylatoethylphenoxy)phosphazene] (PCEP) is a potent enhancer of mixed Th1/Th2 immune responses in mice immunized with influenza virus antigens. *Vaccine* **2007**, *25*, 1204–1213.
48. Andrianov, A. K.; Marin, A.; Chen, J. Synthesis, properties, and biological activity of poly[di(sodium carboxylatoethylphenoxy)phosphazene]. *Biomacromolecules* **2006**, *7*, 394–399.
49. Wycisk, R.; Pintauro, P. N. Sulfonated polyphosphazene ion-exchange membranes. *J. Memb. Sci.* **1996**, *119*, 155–160.
50. Paulsdorf, J.; Burjanadze, M.; Hagelschur, K.; Wiemhöfer, H. D. Ionic conductivity in polyphosphazene polymer electrolytes prepared by the living cationic polymerization. *Solid State Ion.* **2004**, *169*, 25–33.
51. Allcock, H. R.; Fitzpatrick, R. J.; Salvati, L. Sulfonation of (aryloxy)- and (arylamino)phosphazenes: Small-molecule compounds, polymers, and surfaces. *Chem. Mater.* **1991**, *3*, 1120–1132.
52. Luten, J.; van Steenis, J. H.; van Someren, R.; Kemmink, J.; Schuurmans-Nieuwenbroek, N. M. E.; Koning, G. A.; Crommelin, D. J. A.; van Nostrum, C. F.; Hennink, W. E. Water-soluble biodegradable cationic polyphosphazenes for gene delivery. *J. Controlled Release* **2003**, *89*, 483–497.
53. Yang, Y.; Xu, Z.; Jiang, J.; Gao, Y.; Gu, W.; Chen, L.; Tang, X.; Li, Y. Poly(imidazole/DMAEA)phosphazene/DNA self-assembled nanoparticles for gene delivery: Synthesis and in vitro transfection. *J. Controlled Release* **2008**, *127*, 273–279.
54. Ma, C.; Zhang, X.; Du, C.; Zhao, B.; He, C.; Li, C.; Qiao, R. Water-soluble cationic polyphosphazenes grafted with cyclic polyamine and imidazole as an effective gene delivery vector. *Bioconjugate Chem.* **2016**, *27*, 1005–1012.
55. Liu, X.; Zhang, H.; Tian, Z.; Sen, A.; Allcock, H. R. Preparation of quaternized organic-inorganic hybrid brush polyphosphazene-co-poly[2-(dimethylamino)ethyl methacrylate] electrospun fibers and their antibacterial properties. *Polym. Chem.* **2012**, *3*, 2082–2091.
56. Qian, Y.-C.; Chen, P.-C.; Zhu, X.-Y.; Huang, X.-J. Click synthesis of ionic strength-responsive polyphosphazene hydrogel for reversible binding of enzymes. *RSC Adv.* **2015**, *5*, 44031–44040.
57. Morozowich, N. L.; Mondschein, R. J.; Allcock, H. R. Comparison of the synthesis and bioerodible properties of N-linked versus O-linked amino acid substituted polyphosphazenes. *J. Inorg. Organomet. Polym. Mater.* **2014**, *24*, 164–172.
58. Allcock, H. R.; Kwon, S. Glyceryl polyphosphazenes: Synthesis, properties, and hydrolysis. *Macromolecules* **1988**, *21*, 1980–1985.
59. Steinhilber, D.; Sisson, A. L.; Mangoldt, D.; Welker, P.; Licha, K.; Haag, R. Synthesis, reductive cleavage, and cellular interaction studies

of biodegradable, polyglycerol nanogels. *Adv. Funct. Mater.* **2010**, *20*, 4133–4138.
60. Abbina, S.; Vappala, S.; Kumar, P.; Siren, E. M. J.; La, C. C.; Abbasi, U.; Brooks, D. E.; Kizhakkedathu, J. N. Hyperbranched polyglycerols: Recent advances in synthesis, biocompatibility and biomedical applications. *J. Mater. Chem. B* **2017**, *5*, 9249–9277.
61. Calderón, M.; Quadir, M. A.; Sharma, S. K.; Haag, R. Dendritic polyglycerols for biomedical applications. *Adv. Mater.* **2010**, *22*, 190–218.
62. Henke, H.; Posch, S.; Brüggemann, O.; Teasdale, I. Polyphosphazene based star-branched and dendritic molecular brushes. *Macromol. Rapid Commun.* **2016**, *37*, 769–774.
63. Linhardt, A.; König, M.; Iturmendi, A.; Henke, H.; Brueggemann, O.; Teasdale, I. Degradable, dendritic polyols on a branched polyphosphazene backbone. *Ind. Eng. Chem. Res.* **2018**, *57*, 3602–3609.
64. Andrianov, A. K.; Marin, A. Degradation of polyaminophosphazenes: Effects of hydrolytic environment and polymer processing. *Biomacromolecules* **2006**, *7*, 1581–1586.
65. Andrianov, A. K.; Marin, A.; Peterson, P. Water-soluble biodegradable polyphosphazenes containing N-ethylpyrrolidone groups. *Macromolecules* **2005**, *38*, 7972–7976.
66. Martinez, A. P.; Qamar, B.; Fuerst, T. R.; Muro, S.; Andrianov, A. K. Biodegradable "smart" polyphosphazenes with intrinsic multifunctionality as intracellular protein delivery vehicles. *Biomacromolecules* **2017**, *18*, 2000–2011.
67. Feinweber, D.; Verwanger, T.; Brueggemann, O.; Teasdale, I.; Krammer, B. Applicability of new degradable hypericin-polymer-conjugates as photosensitizers: principal mode of action demonstrated by in vitro models. *Photochem. Photobiol. Sci.* **2014**, *13*, 1607–1620.
68. Blonsky, P. M.; Shriver, D. F.; Austin, P.; Allcock, H. R. Polyphosphazene solid electrolytes. *J. Am. Chem. Soc.* **1984**, *106*, 6854–6855.
69. Allcock, H. R.; O'Connor, S. J. M.; Olmeijer, D. L.; Napierala, M. E.; Cameron, C. G. Polyphosphazenes bearing branched and linear oligoethyleneoxy side groups as solid solvents for ionic conduction. *Macromolecules* **1996**, *29*, 7544–7552.
70. Allcock, H. R.; Sunderland, N. J.; Ravikiran, R.; Nelson, J. M. Polyphosphazenes with novel architectures: Influence on physical properties and behavior as solid polymer electrolytes. *Macromolecules* **1998**, *31*, 8026–8035.
71. Allcock, H. R.; Dudley, G. K. Lower critical solubility temperature study of alkyl ether based polyphosphazenes. *Macromolecules* **1996**, *29*, 1313–1319.
72. Allcock, H. R.; Morozowich, N. L. Bioerodible polyphosphazenes and their medical potential. *Polym. Chem.* **2012**, *3*, 578–590.
73. Yang, Y.; Zhang, Z.; Chen, L.; Gu, W.; Li, Y. Galactosylated poly(2-(2-aminoethyoxy)ethoxy)phosphazene/DNA complex nanoparticles: In vitro and in vivo evaluation for gene delivery. *Biomacromolecules* **2010**, *11*, 927–933.

74. Chen, C.; Xu, H.; Qian, Y.-C.; Huang, X.-J. Glycosylation of polyphosphazenes by thiol-yne click chemistry for lectin recognition. *RSC Adv.* **2015**, *5*, 15909–15915.
75. Aichhorn, S.; Linhardt, A.; Halfmann, A.; Nadlinger, M.; Kirchberger, S.; Stadler, M.; Dillinger, B.; Distel, M.; Dohnal, A.; Teasdale, I.; Schöfberger, W. A pH-sensitive macromolecular prodrug as TLR7/8 targeting immune response modifier. *Chem. Eur. J.* **2017**, *23*, 17721–17726.
76. Linhardt, A.; König, M.; Schöfberger, W.; Brüggemann, O.; Andrianov, A.; Teasdale, I. Biodegradable polyphosphazene based peptide-polymer hybrids. *Polymers* **2016**, *8*, 161–176.
77. Qiu, L. Y.; Wu, X. L.; Jin, Y. Doxorubicin-loaded polymeric micelles based on amphiphilic polyphosphazenes with poly(N-isopropylacrylamide-co-N,N-dimethylacrylamide) and ethyl glycinate as side groups: Synthesis, preparation and in vitro evaluation. *Pharm. Res.* **2009**, *26*, 946–957.
78. Zheng, C.; Qiu, L.; Zhu, K. Novel polymersomes based on amphiphilic graft polyphosphazenes and their encapsulation of water-soluble anti-cancer drug. *Polymer* **2009**, *50*, 1173–1177.
79. Zheng, C.; Yao, X.; Qiu, L. Novel polymeric vesicles with pH-induced deformation character for advanced drug delivery. *Macromol. Biosci.* **2011**, *11*, 338–343.
80. Wilfert, S.; Iturmendi, A.; Henke, H.; Brüggemann, O.; Teasdale, I. Thermoresponsive polyphosphazene-based molecular brushes by living cationic polymerization. *Macromol. Symp.* **2014**, *337*, 116–123.
81. Song, S.-C.; Lee, S. B.; Jin, J.-I.; Sohn, Y. S. A new class of biodegradable thermosensitive polymers. I. Synthesis and characterization of poly(organophosphazenes) with methoxy-poly(ethylene glycol) and amino acid esters as side groups. *Macromolecules* **1999**, *32*, 2188–2193.
82. Lee, B. H.; Lee, Y. M.; Sohn, Y. S.; Song, S.-C. A Thermosensitive poly(organophosphazene) gel. *Macromolecules* **2002**, *35*, 3876–3879.
83. Lee, B. H.; Song, S.-C. Synthesis and characterization of biodegradable thermosensitive poly(organophosphazene) gels. *Macromolecules* **2004**, *37*, 4533–4537.
84. Park, M.-R.; Cho, C.-S.; Song, S.-C. In vitro and in vivo degradation behaviors of thermosensitive poly(organophosphazene) hydrogels. *Polym. Degrad. Stab.* **2010**, *95*, 935–944.
85. Seo, B.-B.; Choi, H.; Koh, J.-T.; Song, S.-C. Sustained BMP-2 delivery and injectable bone regeneration using thermosensitive polymeric nanoparticle hydrogel bearing dual interactions with BMP-2. *J. Controlled Release* **2015**, *209*, 67–76.
86. Seo, B.-B.; Chang, H.-I.; Choi, H.; Koh, J.-T.; Yun, K.-D.; Lee, J.-Y.; Song, S.-C. New approach for vertical bone regeneration using in situ gelling and sustained BMP-2 releasing poly(phosphazene) hydrogel system on peri-implant site with critical defect in a canine model. *J. Biomed. Mater. Res., Part B* **2018**, *106*, 751–759.
87. Seo, B.-B.; Koh, J.-T.; Song, S.-C. Tuning physical properties and BMP-2 release rates of injectable hydrogel systems for an optimal bone regeneration effect. *Biomaterials* **2017**, *122*, 91–104.

88. Tian, Z.; Chen, C.; Allcock, H. R. Injectable and biodegradable supramolecular hydrogels by inclusion complexation between poly(organophosphazenes) and α-cyclodextrin. *Macromolecules* **2013**, *46*, 2715–2724.
89. Cohen, S.; Bano, M. C.; Visscher, K. B.; Chow, M.; Allcock, H. R.; Langer, R. Ionically crosslinkable polyphosphazene: a novel polymer for microencapsulation. *J. Am. Chem. Soc.* **1990**, *112*, 7832–7833.
90. Potta, T.; Chun, C.; Song, S.-C. Dual cross-linking systems of functionally photo-cross-linkable and thermoresponsive polyphosphazene hydrogels for biomedical applications. *Biomacromolecules* **2010**, *11*, 1741–1753.
91. Kim, Y.-M.; Potta, T.; Park, K.-H.; Song, S.-C. Temperature responsive chemical crosslinkable UV pretreated hydrogel for application to injectable tissue regeneration system via differentiations of encapsulated hMSCs. *Biomaterials* **2017**, *112*, 248–256.
92. Allcock, H. R.; Pucher, S. R.; Scopelianos, A. G. Poly[(amino acid ester)phosphazenes]: Synthesis, crystallinity, and hydrolytic sensitivity in solution and the solid state. *Macromolecules* **1994**, *27*, 1071–1075.
93. Allcock, H. R.; Fuller, T. J.; Mack, D. P.; Matsumura, K.; Smeltz, K. M. Synthesis of poly[(amino acid alkyl ester)phosphazenes]. *Macromolecules* **1977**, *10*, 824–830.
94. Crommen, J. H. L.; Schacht, E. H.; Mense, E. H. G. Biodegradable polymers: II. Degradation characteristics of hydrolysis-sensitive poly[(organo)phosphazenes]. *Biomaterials* **1992**, *13*, 601–611.
95. DeCollibus, D. P.; Marin, A.; Andrianov, A. K. Effect of environmental factors on hydrolytic degradation of water-soluble polyphosphazene polyelectrolyte in aqueous solutions. *Biomacromolecules* **2010**, *11*, 2033–2038.
96. Carriedo, G. A.; García Alonso, F. J.; García Álvarez, J. L.; Presa Soto, A.; Tarazona, M. P.; Laguna, M. T. R.; Marcelo, G.; Mendicuti, F.; Saiz, E. Experimental and theoretical study of the acidic degradation of poly(2,2′-dioxy-1,1′-biphenyl)phosphazene. *Macromolecules* **2008**, *41*, 8483–8490.
97. Binauld, S.; Stenzel, M. H. Acid-degradable polymers for drug delivery: a decade of innovation. *Chem. Commun.* **2013**, *49*, 2082–2102.
98. Allcock, H. R.; Fuller, T. J.; Matsumura, K. Hydrolysis pathways for aminophosphazenes. *Inorg. Chem.* **1982**, *21*, 515–521.
99. Iturmendi, A.; Monkowius, U.; Teasdale, I. Oxidation responsive polymers with a triggered degradation via arylboronate self-immolative motifs on a polyphosphazene backbone. *ACS Macro Lett.* **2017**, *6*, 150–154.
100. Song, C.-C.; Du, F.-S.; Li, Z.-C. Oxidation-responsive polymers for biomedical applications. *J. Mater. Chem. B* **2014**, *2*, 3413–3426.
101. Broaders, K. E.; Grandhe, S.; Fréchet, J. M. J. A biocompatible oxidation-triggered carrier polymer with potential in therapeutics. *J. Am. Chem. Soc.* **2011**, *133*, 756–758.
102. Iturmendi, A.; Theis, S.; Maderegger, D.; Monkowius, U.; Teasdale, I. Unpublished.

103. Teasdale, I.; Wilfert, S.; Nischang, I.; Brüggemann, O. Multifunctional and biodegradable polyphosphazenes for use as macromolecular anti-cancer drug carriers. *Polym. Chem.* **2011**, *2*, 828–834.
104. Henke, H.; Kryeziu, K.; Banfić, J.; Theiner, S.; Körner, W.; Brüggemann, O.; Berger, W.; Keppler, B. K.; Heffeter, P.; Teasdale, I. Macromolecular Pt(IV) prodrugs from poly(organo)phosphazenes. *Macromol. Biosci.* **2016**, *16*, 1239–1249.
105. Cayatte, C.; Marin, A.; Rajani, G. M.; Schneider-Ohrum, K.; Snell Bennett, A.; Marshall, J. D.; Andrianov, A. K. PCPP-adjuvanted respiratory syncytial virus (RSV) sF subunit vaccine: Self-assembled supramolecular complexes enable enhanced immunogenicity and protection. *Mol. Pharm.* **2017**, *14*, 2285–2293.

Chapter 10

Designed Synthesis of Polyphosphazene Block Copolymers for Self-Assembly

Gabino A. Carriedo, Raquel de la Campa, and Alejandro Presa Soto*

Departamento de Química Orgánica e Inorgánica (IUQOEM), Facultad de Química, Universidad de Oviedo, Julián Clavería 8, 33006 Oviedo, Spain
***E-mail: presaalejandro@uniovi.es.**

This Chapter outlines recent advances on the designed synthesis of polyphosphazene two-block copolymers having a predetermined property specifically located in the phosphazene block. These materials represent a class of versatile polymers able to generate a great variety of different well-defined nanomorphologies by self-assembly in solution and in thin films. Particular focus is placed in the synthesis of linear block copolymers composed by two polyphosphazene chains (PP-*b*-PP), or hybrid polymers combining one PP chain and polystyrene segment (PP-*b*-PS). We also survey the effects of crystalline, rigid or chiral PP core-forming blocks on the self-assembly of these macromolecules to show the morphological diversity and structural complexity of the nanostructures generated in thin films or in selective solvents to one of the blocks.

Introduction: Polyphosphazene Block Copolymers in Self-Assembly

Polyphosphazenes (*1*, *2*) (PP) are macromolecules consisting of $[N{=}P(R_1,R_2)]_n$ repeating units with a pentavalent phosphorus bearing two substituents R (equal or different), and a trivalent nitrogen. The alternating –P=N-P=N- displays no electron delocalization. Because of their unique properties, such as biodegradability (*3–7*), high fire resistance (*8*, *9*) and elasticity (*10–12*), that can be easily modified by choosing the appropriate R substituents, they are among the most fascinating class of inorganic (non-carbon based main-chains) polymers and have already found a variety of potential and real applications (*1*, *2*, *13*, *14*), specially in biomedicine (e.g. bio-stable or bio-erodible materials, drug delivery systems, biomaterials and nanofabrication) (*15*, *16*). Polyphosphazene two block copolymers are an special class of polyphosphazenes that consist of two linear phosphazene chains of different chemical composition linked by a chemical bond $[(R_1,R_2)P{=}N]_n$-*b*-$[(R_3,R_4)P{=}N]_m$ ([PP-*b*-PP]). There are also hybrid inorganic-organic polymers having a phosphazene chain joined to a polymeric segment of a different nature, [PP-*b*-(Polym)]; Polym = non-phosphazenic polymer, such as, for example, a polyolefin. As most block copolymers (BCPs), block copolyphosphazenes have the ability of self-assembly in a variety of different nanostructured architectures. However, and despite the great interest aroused in the last 15 years by the self-assembly of BCPs both in thin films (*17–21*) or in solvents selective for one of the segments (*22–30*), block copolyphosphazenes have been scarcely employed in self-assembly studies. Thus, before 2012, the literature registered a sole example of the self-assembly of PP-*b*-PP, in which Allcock and coworkers reported the formation of spherical micelles by the self-assembly of the amphiphilic block copolymer, $[N{=}P(O(CH_2CH_2O)_2CH_3)_2]_n$-*b*-$[N{=}PPh(O(CH_2CH_2O)_2CH_3)]_m$ (**15** in Scheme 3), in aqueous media (*31*). This process was not studied in detail, and no experimental evidence on the nature of the spherical aggregates (*i.e.* if those aggregates are star-like micelles or hollow vesicles) was given (*31*). The self-assembly of linear hybrid BCPs having a PP block and other segment of different chemical nature has been also insufficiently explored, and most of the works have been made with PP-*b*-PS (PS = polystyrene) and PP-*b*-PEO [PEO = poly(ethylene oxide)]. Thus, H. R. Allcock *et al.* studied the micellar formation of different PP-*b*-PS having aqueous soluble PP blocks like $[N{=}P(O(CH_2CH_2O)_2CH_3)_2]_n$ (MEET, **52** in Figure 6) (*32*), $[N{=}P(OC_6H_4\text{-}CO_2Na)_2]_n$ (KPCPP, **53** in Figure 6) (*33*), or PP blocks having [adamantly/β-Cyclodextrin] complexes linked to phosphorus atoms (**54** in Figure 6) (*34*). In all those examples, the self-assembly produced spherical vesicles in which, presumably, the PS is located at the capsule wall, being the PP water-soluble block at the inner and outer coronas. The same authors also reported the self-aggregation in aqueous media of PP-*b*-PEO (PP = $[N{=}P(NHCH_2CO_2Et)_2]_n$ (*35*); or $[N{=}P(OCH_2CF_3)_2]_n$ (*36*)), and PP-*b*-poly[(dimethylamino)ethyl methacrylate] with PP = $[N{=}P(OCH_2CF_3)_2]_n$) (*37*) generating spherical aggregates (vesicle, star-like micelle, compound micelle, etc.) that were not structurally characterized. More detailed self-assembly studies were recently reported with PP-*b*-PFS, where PP = $[N{=}P(OCH_2CF_3)_2]_n$ and

PFS = poly(ferrocenyl(dimethyl)silane) (**46a-d** in Scheme 8) (*38*, *39*). In thin films, these materials gave different morphologies as a function of the volume fraction of the PFS block (φ_{PFS}). Thus, cylinders (φ_{PFS} = 0.40, **46c**) or spheres (φ_{PFS} = 0.21, **46d**) of PFS blocks were observed in the in the matrix formed by $[N{=}P(OCH_2CF_3)_2]_n$ chains (Figure 1) (*38*). On the other hand, pointed-oval shaped micelles of uniform shape and size were obtained from the self-assembly of PFS54-*b*-$[N{=}P(OCH_2CF_3)_2]290$ (**46d** in Scheme 8) in propan-2-ol (selective solvent for the $[N{=}P(OCH_2CF_3)_2]n$ blocks) (*39*).

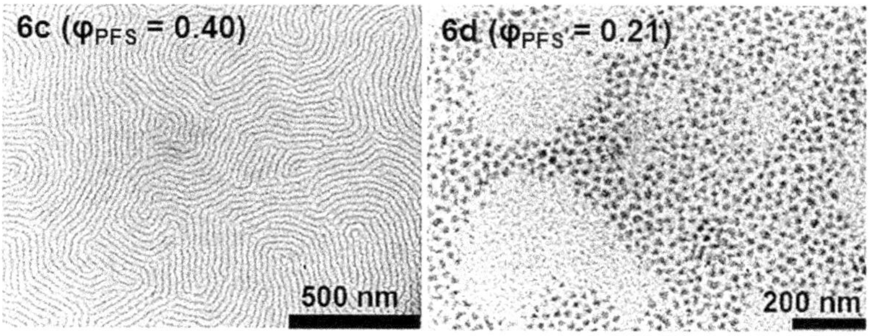

*Figure 1. Bright field TEM micrographs of drop-cast thin films PFS-b-PP block copolymers **6c** and **6d** showing cylinders (left image) and spheres (right image). Adapted with permission from ref (38). Copyright 2009, American Chemical Society.*

Considering their achievable chemical compositions, it is clear that block copolyphosphazenes allow the location of an extensive variety of specific properties (e.g., degree of flexibility, solubility in different solvents, hydrophilicity, etc.) into the PP segments. Therefore, they should be especially useful to investigate the effects of those properties on the nanostructures resulting by their self-organization. However, for those studies, the synthesis of block copolymers with well-defined chemical composition and with narrow polydispersity indexes (PDI < 1.2) is strictly necessary, as well as the fine control of the molecular weights. Unquestionably, it has been the lack of efficient synthetic routes to polyphosphazene block copolymers fulfilling those requirements, the main factor hampering the use of those copolymers in self-assembly. In 2012, our group undertook the systematic study of linear two block copolymers focusing on: (i) the development of new synthetic methodologies leading to well-defined PP-*b*-PP and PP-*b*-polystyrene (PP-*b*-PS) BCP´s; and (ii) the study of the self-assembly of these materials incorporating *crystalline*, *rigid* or *chiral PP core-forming blocks*. In this chapter we will review the synthesis of well defined (in terms of chemical composition, structure, distribution of molecular weights and low polydispersity indexes) phosphazene block copolymers, rationally designed to investigate their self-assembly as conducted by different properties, such as crystallinity, rigidity and chirality.

Synthesis of Linear Polyphosphazene-b-polyphosphazene Block Copolymers (PP-b-PP)

A great variety of polyphosphazenes (more than 700 examples) have been prepared starting from the very important precursor poly(dichloro)phosphazene) $[N{=}PCl_2]_n$ (1) by means of the very versatile and efficient macromolecular nucleophilic substitution reaction, that can be carried out sequentially (*1*, *2*, *40*) (See Chart 1). The possibilities of this macromolecular substitution for the preparation of homo and random copolymers with very different chemical compositions has been reviewed elsewhere (*1*, *2*, *16*).

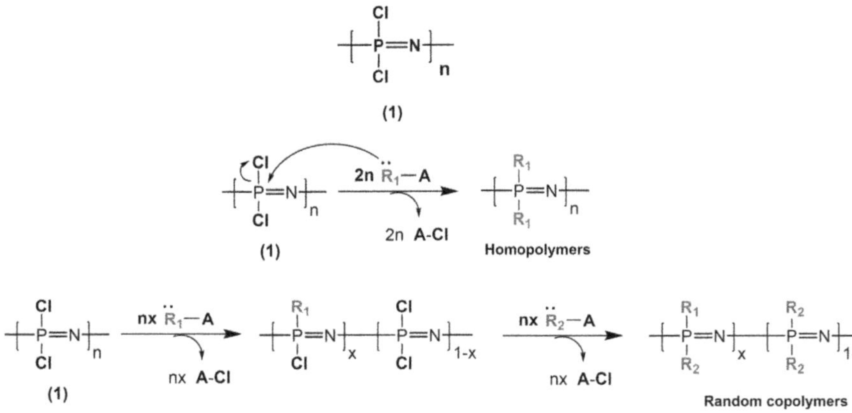

*Chart 1. The precursor Poly(dichloro)phosphazene $[N{=}PCl_2]_n$ (**1**) and the general synthesis to polyphosphazenes, homopolymers and random copolymers, by nucleophilic macromolecular substitution.*

This route, however, leads to homo and random copolymers which have very high molecular weights and PDIs (PDI $\geq$ 2). The synthesis of linear two blocks polyphosphazenes with narrow PDI was not possible until 1995 when the living cationic chain–growth condensation polymerization of the phosphoranimine $Cl_3P{=}N\text{-}SiMe_3$ (**2**) promoted by PCl_5 at ambient temperature was discovered (Scheme 1) (*41*), permitting the preparation of living poly(dichloro)phosphazene with controlled molecular weights (It should be noted that *living* and *controlled* polymerization are not necessary the same concepts (*30*).

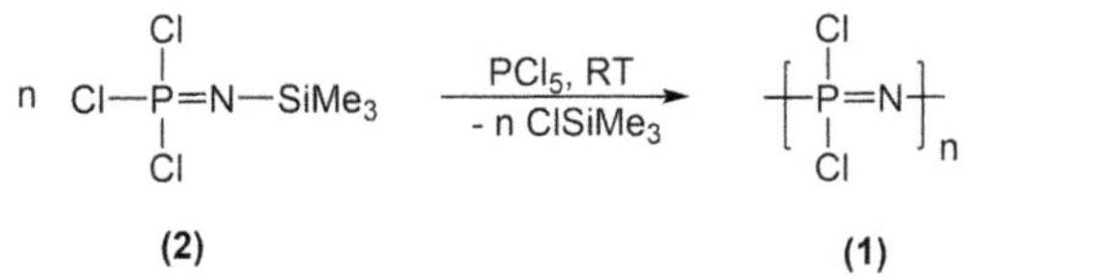

M_w = 10^4-10^5; IPD = 1.03-1.30

*Scheme 1. Living cationic polymerization of $Cl_3P{=}N\text{-}SiMe_3$ (**2**) promoted by PCl_5.*

According to the accepted mechanism of this reaction, the phosphoranimine $Cl_3P{=}N\text{-}SiMe_3$ (**2**) reacts with two equivalents of PCl_5 forming the reactive cationic initiator $[Cl_3P{=}N{=}PCl_3][PCl_6]$ ([**3**]PCl_6 in Scheme 2) (*41–43*). Subsequent condensation reactions between initiator [**3**]PCl_6 and $Cl_3P{=}N\text{-}SiMe_3$ (**2**) with elimination of $Cl\text{-}SiMe_3$ yield the poly(dichloro)phosphazene, $Cl_2P{=}N\text{-}[Cl_2P{=}N]_{(n+1)}\text{-}PCl_3^+$ $[PCl_6]^-$ (**1**[PCl_6]), having available and reactive $-PCl_3^+$ end-groups (see *Propagation Step* in Scheme 2).

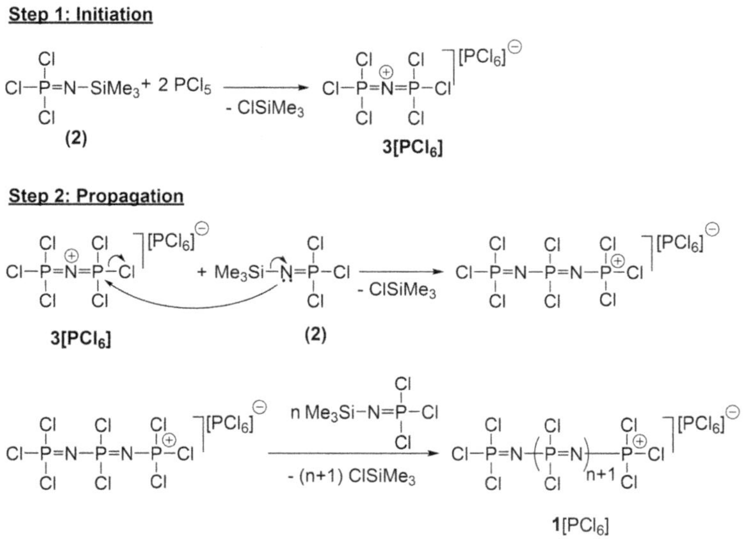

*Scheme 2. Proposed mechanism for the living cationic polymerization of $Cl_3P{=}N\text{-}SiMe_3$ (**2**) in the presence of catalytic amount of PCl_5.*

The living nature of the poly(dichloro)phosphazene (**1**[PCl_6]), provided access to a number of polyphosphazene block copolymers (PP-*b*-PP). Thus, when a phosphoranimine bearing alkyl or aryl groups on phosphorus is added to (**1**[PCl_6]), it undergoes polymerization from the reactive $-PCl_3^+$ end-group forming a block copolymer with a PP chain and a $[N{=}PCl_2]_n$ block (sequential living polymerization). The macromolecular substitution of the chlorine atoms in the latter with an appropriate nucleophile [$NaOCH_2CF_3$ or $NaO(CH_2CH_2O)_2CH_3$] allows the preparation of a variety of block copolymers (see Scheme 3) (*31, 44, 45*).

Besides $Cl_3P{=}N\text{-}SiMe_3$ (**2**), several other phosphoranimines of the type $(X)(R_1)(R_2)P{=}N\text{-}SiMe_3$, having alkyl, aryl, alkoxy and halogen substituents on the phosphorus, also undergo living cationic polymerization (see Scheme 4) (*46–48*), leading to phosphazene homopolymers (**25-32**) with narrow molecular weight distributions. The reaction, however is not general and not all phosphoranimines give well-defined materials. Moreover, as in the case of the living poly(dichloro)phosphazene (**1**[PCl_6]), the living nature of the resulting homopolymers (**25-32**), should allow a sequential monomer polymerization with a second phosphoranimine giving PP-*b*-PP block copolymers. (Scheme 5).

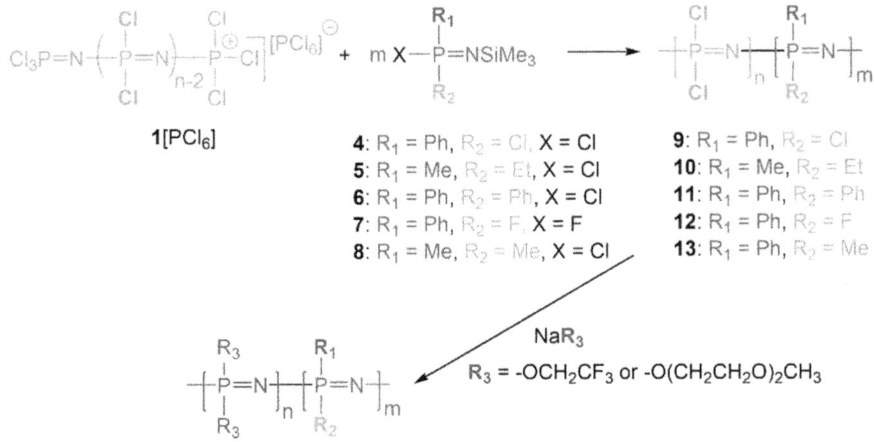

4: R_1 = Ph, R_2 = Cl, X = Cl
5: R_1 = Me, R_2 = Et, X = Cl
6: R_1 = Ph, R_2 = Ph, X = Cl
7: R_1 = Ph, R_2 = F, X = F
8: R_1 = Me, R_2 = Me, X = Cl

9: R_1 = Ph, R_2 = Cl
10: R_1 = Me, R_2 = Et
11: R_1 = Ph, R_2 = Ph
12: R_1 = Ph, R_2 = F
13: R_1 = Ph, R_2 = Me

14: R_1 = Ph, R_2 = OCH_2CF_3, R_3 = OCH_2CF_3 (refs. 44,45)
15: R_1 = Ph, R_2 = $O(CH_2CH_2O)_2CH_3$, R_3 = $O(CH_2CH_2O)_2CH_3$ (refs. 45,31)
16: R_1 = Me, R_2 = Et, R_3 = OCH_2CF_3 (refs. 44,45)
17: R_1 = Ph, R_2 = Ph, R_3 = OCH_2CF_3 (ref. 45)
18: R_1 = Ph, R_2 = Ph, R_3 = $O(CH_2CH_2O)_2CH_3$ (ref. 45)
19: R_1 = Ph, R_2 = OCH_2CF_3, R_3 = $O(CH_2CH_2O)_2CH_3$ (ref. 45)
20: R_1 = Me, R_2 = Me, R_3 = OCH_2CF_3 (ref. 44,45)
21: R_1 = Me, R_2 = Me, R_3 = $O(CH_2CH_2O)_2CH_3$ (ref. 45)

Scheme 3. Synthesis of linear polyphosphazene two-block copolymers ***14-21*** *by sequential living polymerization of phosphoranimines followed by a macromolecular substitution.*

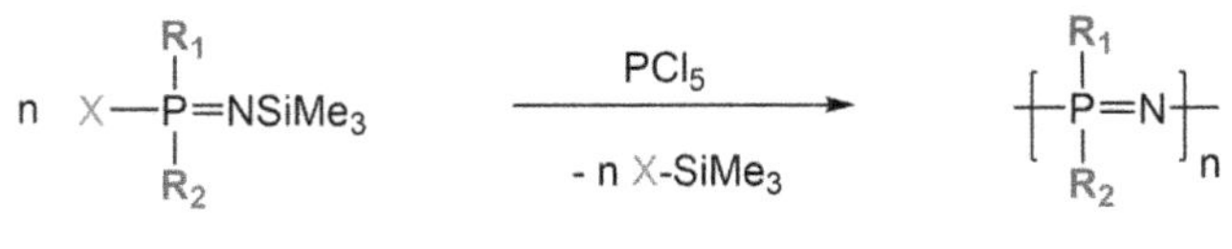

4: R_1 = Ph, R_2 = Cl, X = Cl
5: R_1 = Me, R_2 = Et, X = Cl
6: R_1 = Ph, R_2 = Ph, X = Cl
7: R_1 = Ph, R_2 = F, X = F
22: R_1 = OCH_2CF_3, R_2 = OCH_2CF_3, X = Br
23: R_1 = Ph, R_2 = CF_3, X = Br
24: R_1 = Ph, R_2 = Me, X = Br

25: R_1 = Ph, R_2 = Cl (ref. 46)
26: R_1 = Ph, R_2 = OCH_2CF_3 (ref. 46)
27: R_1 = Me, R_2 = Et (ref. 46)
28: R_1 = Ph, R_2 = Ph (ref. 46)
29: R_1 = Ph, R_2 = F (ref. 47)
30: R_1 = OCH_2CF_3, R_2 = OCH_2CF_3 (ref. 46)
31: R_1 = Ph, R_2 = CF_3 (ref. 48)
32: R_1 = Ph, R_2 = Me (ref. 48)

Scheme 4. Synthesis of polyphosphazene homopolymers ***25-32*** *by living cationic polymerization of phosphoranimines of general formula $(X)(R_1)(R_2)P=N-SiMe_3$.*

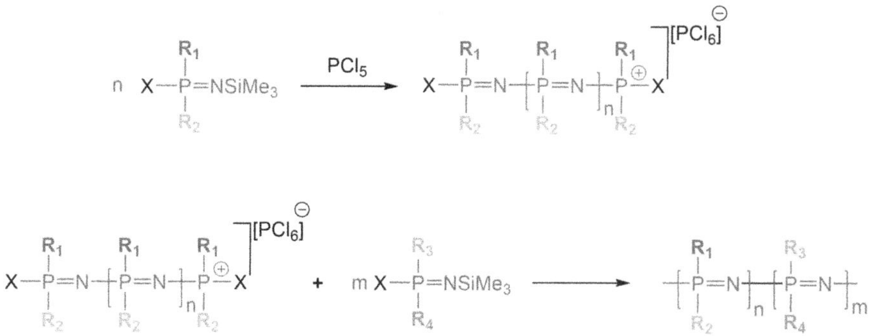

Scheme 5. General synthesis of PP-b-PP block copolymers by sequential monomer addition polymerization of two different phosphoranimines of general formula (X)(R_1)(R_2)P=N-$SiMe_3$.

In spite of its great potential, this general methodology is strongly hampered by several problems. Firstly, with most of the known organo-phosphoranimines (XR_2P=N-$SiMe_3$; X = F, Cl or Br; R = alkyl, aryl or alkoxy) the polymerizations are incomplete even after long reaction times (typically from 4 to 7 days), giving low molecular weight (M_n) oligomers and uncontrolled amounts of non-reacted monomers. Furthermore, to date, the optimal reaction conditions (time, temperature, solvent, etc.) have not been well established (*41–48*). Not surprisingly, therefore, the sequential polymerization of two different phosphoranimines is scarcely practical being limited to the examples in Scheme 4 in which a second phosphoranimine polymerizes only from living blocks of **1**[PCl_6] giving block copolymers. However, and although Allcock and coworkers suggested that the polymerization of Cl(Me)(Ph)P=N-$SiMe_3$ (**33**) can be carried out in the presence of living chains of polyphosphazene **27** ($[N=P(Me)(Et)]_n$-P^+(Me)(Et)Cl, see Scheme 6), yielding the block copolyphosphazene $[N=P(Me)(Et)]_n$-*b*-$[N=P(Me)(Et)]_m$ (**34**) (*45*), this approach to the synthesis of $[N=PR_1R_2]_n$-*b*-$[N=PR_3R_4]_m$ BCPs cannot be generalized.

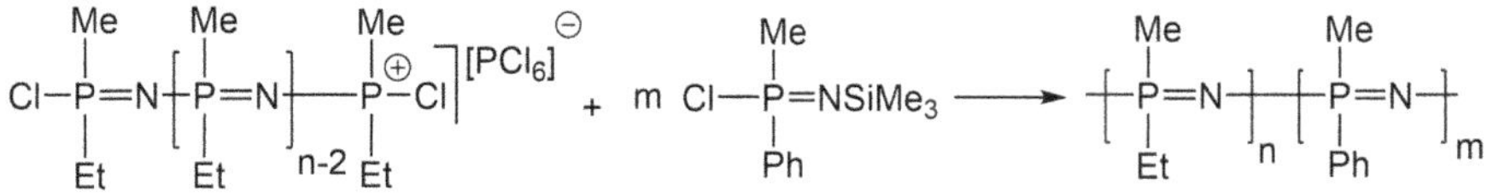

27 **33** **34**

*Scheme 6. Sequential living polymerization of phosphoranimine **33** by using active end-groups of polyphosphazene **27** as initiators. The synthesis of BCP **34** is suggested in ref. (45).*

However, the preparation of PP-*b*-PP block copolymers by sequential monomer addition of two different phosphoranimines, but using first Cl_3P=N-$SiMe_3$ (**2**), and subsequently one the type (X)(R_1)(R_2)P=N-$SiMe_3$, is potentially a very useful synthetic methodology and, therefore, further improvements are very desirable. Recently, I. Manners and co-workers demonstrated that the two $-PCl_3^+$ end-groups of the living poly(dichloro)phosphazene (**1**[PCl_6]) have different activities depending on the chain lengths of the growing polymer, leading to a *bi*-directional chain growth when monomer $Cl_3P{=}N\text{-}SiMe_3$ (**2**) was added over the living **1**[PCl_6] (*49, 50*). Based on these results, in 2012 our group re-investigated (theoretically and experimentally) this sequential phosphoranimine addition using model compounds (*51*), demonstrating that the addition of the alkyl-phosphoranimine $Br(Me)_2P{=}N\text{-}SiMe_3$ (**35**) over living chains of poly(dichloro)phosphazene, **1**[PCl_6], led to *tri*-block copolymers instead of the desired *two*-block copolyphosphazenes by a *bi*-directional chain-growth mechanism of the polyphosphazene chains (Figure 2).

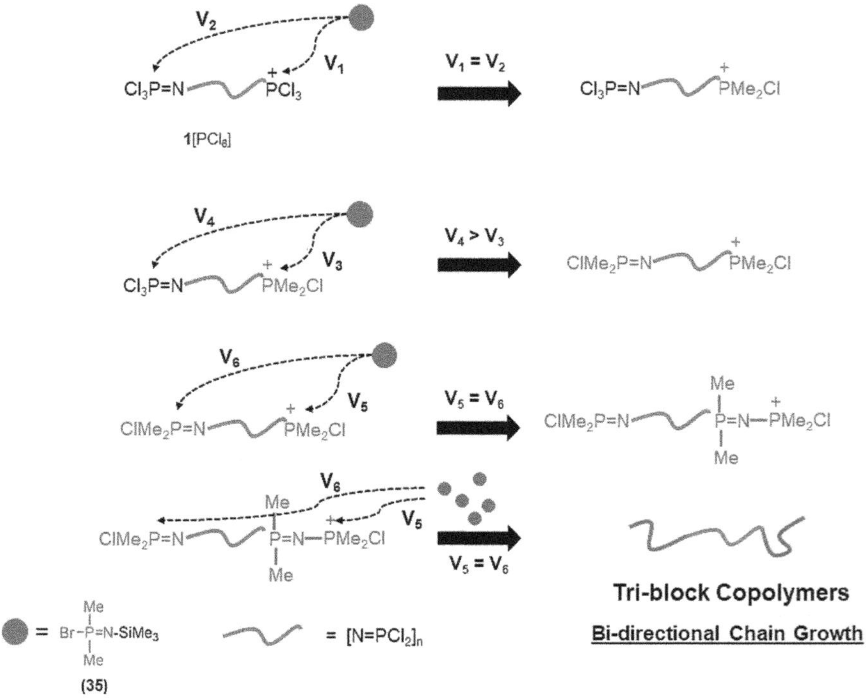

*Figure 2. Schematic representation of the reaction of phosphoranime **35** (circles) with living chains of poly(dichloro)phosphazene **1**[PF_6] leading to tri-block copolyphosphazenes by bi-directional chain growth.*

Computational studies (DFT) on model compounds demonstrated that the nature of the donor nucleophilic phosphoranimines and the electrophilic cationic end–groups of the living polyphosphazenes strongly affect the polymerization reaction. Thus, good electron donor phosphoranimines, such as those $XR_2P{=}N\text{-}SiMe_3$ bearing donor alkyl groups (R) on phosphorus and having P-N-Si angles < 180° (i.e. N atoms exhibiting imine character), react better with electron poor cationic end groups such as $=N\text{-}PCl_3^+$ (i.e., with electron withdrawing groups on the phosphorus). On the other hand, the electron deficient $Cl_3P{=}N\text{-}SiMe_3$ (**2**), having a P-N-Si angle of *ca.* 180°, did not show an imine-like nucleophilic character. The calculations also showed that the electron richer cationic end group $=N\text{-}PMe_2Cl^+$ LUMO is not sufficiently exposed to ensure an efficient attack of nucleophiles (see Figure 3), explaining why the phosphoranimines $XR_2P{=}N{-}SiMe_3$ (R = Alkyl or Aryl; X = Cl or Br) polymerize from living chains of poly(dichloro)phosphazene (i.e. from active $=PCl_3^+$ end-groups) and not the other way around (*51*).

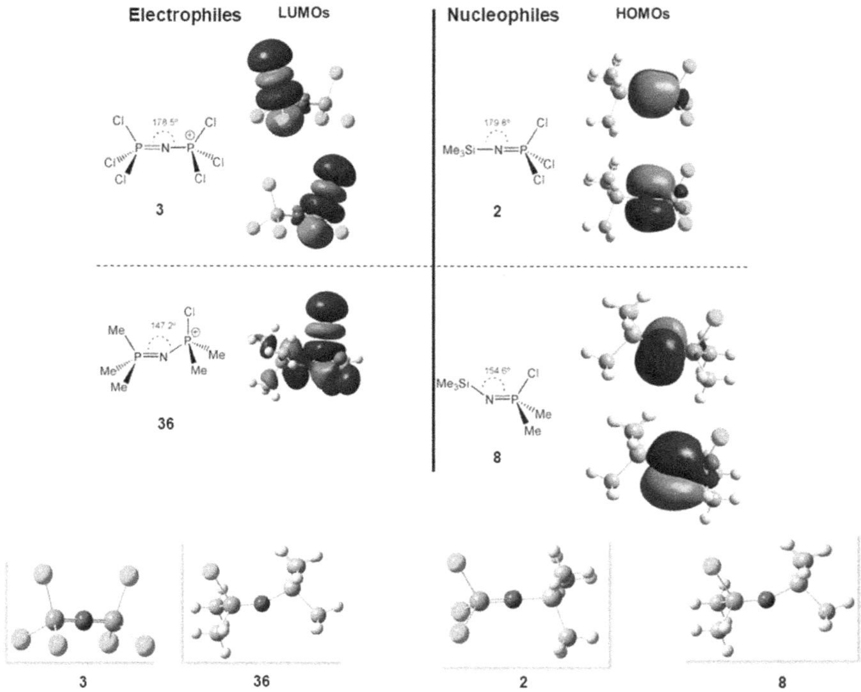

*Figure 3. DFT calculations showing P-N-Si angles, HOMOs, and LUMOs of the electrophile initiators $[Cl_3P{=}N{=}PCl_3]^+$ (**3**) and $[Me_3P{=}N{=}PMe_2Cl]^+$ (**36**); and the monomeric phosphoranimines $Cl_3P{=}N{-}SiMe_3$ (**2**) and $ClMe_2P{=}N{-}SiMe_3$ (**8**). Reprinted and adapted with permission from ref (51). Copyright 2012, American Chemical Society.*

On the other hand, the undesirable bi-directional chain growth could be avoided by using mono end-capped initiators of type $[Ph_3P{=}N{=}PCl_3][A]$ (A = Cl, **37**[Cl]; PCl_6, **37**[PCl_6]), with only one active $-PCl_3^+$ site to polymerize the phosphoranimine $Cl_3P{=}N{-}SiMe_3$ (**2**). These initiators, **37**[Cl] and **37**[PCl_6], lead to poly(dichloro)phosphazene chains with only one active end-group ($-PCl_3^+$) able to polymerize a second phosphoranimine (Scheme 7) (*51*). This synthetic approach opened the access to a variety of pure (i.e. with no impurities of three block copolymers) and well-defined (i.e. controlled molecular weights and narrow polydispersity indexes) polyphosphazene *two*-block copolymers, PP-*b*-PP, having the reactive $[N{=}PCl_2]_n$ segment, which, after a macromolecular substitution of chlorine atoms by the appropriate nucleophiles, led to phosphazene BCPs having different properties relevant in self-assembly (see Scheme 7). This approach, i.e. using end-capped initiators, was later exploited by I. Teasdale and collaborators to prepare well-defined polyphosphazenes with other architectures different than that of linear di-block copolymers (*52, 53*).

Using this methodology we prepared the synthetically very versatile block copolymer $[N{=}PCl_2]_n$-*b*-$[N{=}PMePh]_m$ (**38**) having the amorphous, flexible, and soluble in most common organic solvents $[N{=}PMePh]_m$ block (*54–58*). The corresponding macromolecular substitution of the $[N{=}PCl_2]_n$ block with the appropriately chosen nucleophiles incorporated *rigidity*, *chirality* or *crystallinity* in the final copolymers (see Figure 4) (*51, 59*). All copolymers were synthesized with different block ratios and the GPC data confirmed the very narrow (PDI < 1.2) and Gaussian-shaped distribution of molecular weights required for self-assembly studies.

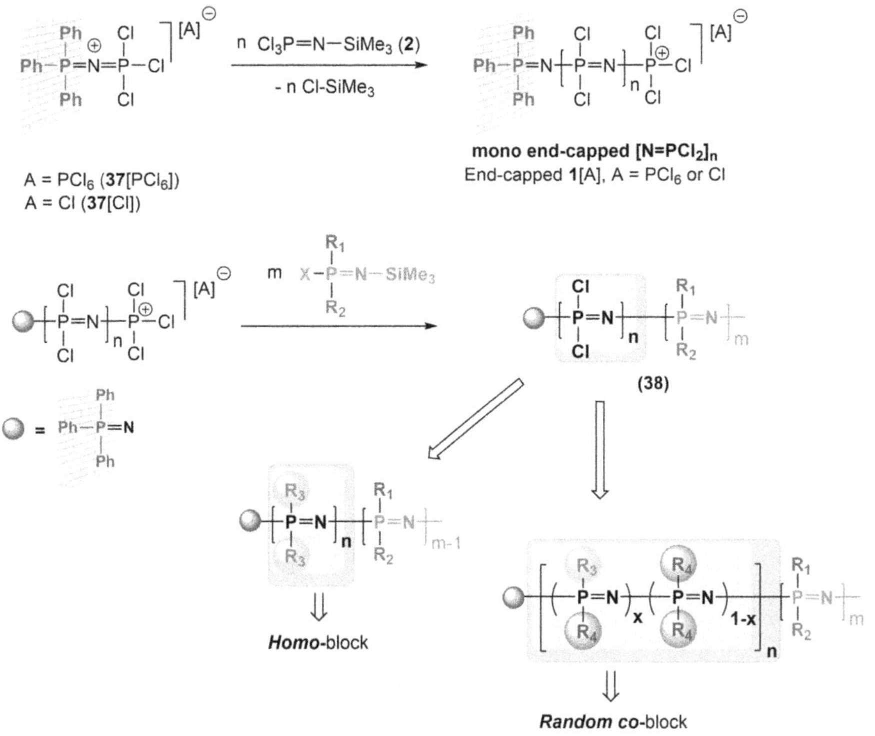

*Scheme 7. Synthesis of well-defined PP-b-PP by using mono end-capped poly(dichloro)phosphazene synthesized from end-capped initiators **37**[Cl] or **37**[PCl_6], $Cl_3P{=}N{-}SiMe_3$ (**2**), and a second chemically different phosphoranimine ($X{-}P(R_1)(R_2)N{-}SiMe_3$). The scheme also illustrates how a two sequential macromolecular substitution in the $[N{=}PCl_2]_n$ block may yield polyphosphazene block copolymers where one of the blocks has two different substituents randomly distributed along its chain.*

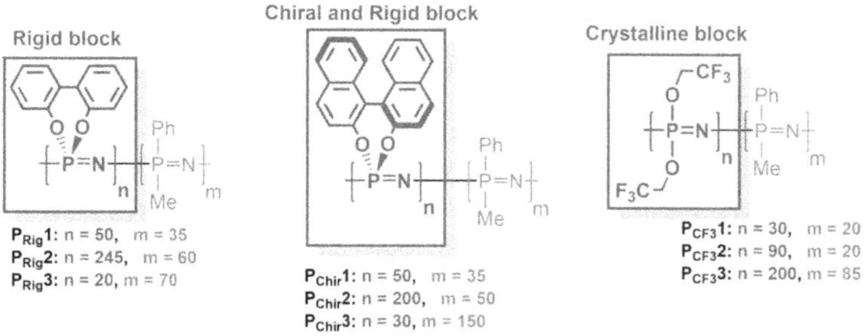

Figure 4. Synthesis of PP-b-PP combining the amorphous, flexible, and ready soluble in most common organic solvent nature of the $[N{=}PMePh]_m$, with the crystallinity ($[N{=}P(OCH_2CF_3)_2]_n$), rigidity ($[N{=}P(R\text{-}O_2C_{20}H_{12})]_n$), and chirality ($[N{=}P(R\text{-}O_2C_{12}H_8)]_n$) of the other PP block.

The chemical versatility of the polymeric intermediate $[N{=}PCl_2]_n$-*b*-$[N{=}PMePh]_m$ (**38**) also allowed the synthesis of block copolymers having functional groups randomly distributed along the chiral blocks (*60, 61*). Both, the nature (*FG* = 4-oxypyridine (**39**), morpholine (**40**) or thiomorpholine (**41**)) (*60*), and the proportion (**42a-d**) of the functional group (*61*), were conveniently modified to change the properties of the BCPs (Figure 5).

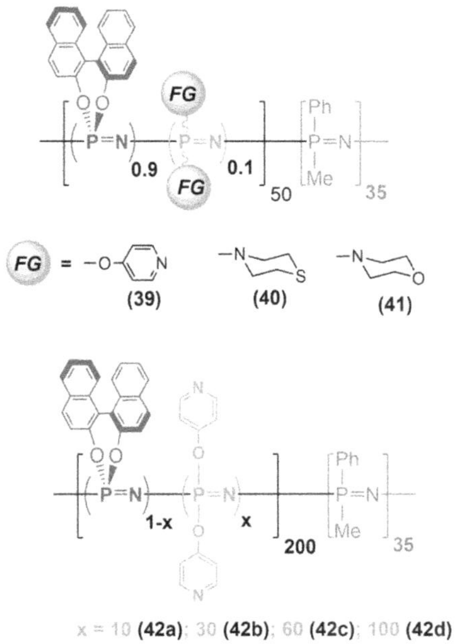

Figure 5. PP-b-PP having randomly substituted functional groups (FG) along the chirality incorporating blocks.

Synthesis of Linear Block Copolyphosphazenes with a Non-Phosphazenic Segment (PP-b-Polym; Polym = Non-Phosphazenic Polymer)

Following different methodologies polyphosphazene chains have been combined with a variety of organic segments, such as poly(dimethylsiloxane) (*62, 63*), poly(ethylene oxide) (*35, 36, 64–66*), polystyrene (*32–34, 67*), poly(dimethylaminoethyl methacrylate) (*37*), poly(propylene glycol) (*68*), poly(lactic acid) (*69*), polycaprolactone (*70*), poly(trimethylene carbonate) (*70*), and also with the organometallic block poly(ferrocenysilane) (*38*). The latter was obtained by promoting the polymerization of $Cl_3P=N\text{-}SiMe_3$ (**2**) by a macro-initiator **45**, followed by a macromolecular substitution with $NaOCH_2CF_3$ (see Scheme 8).

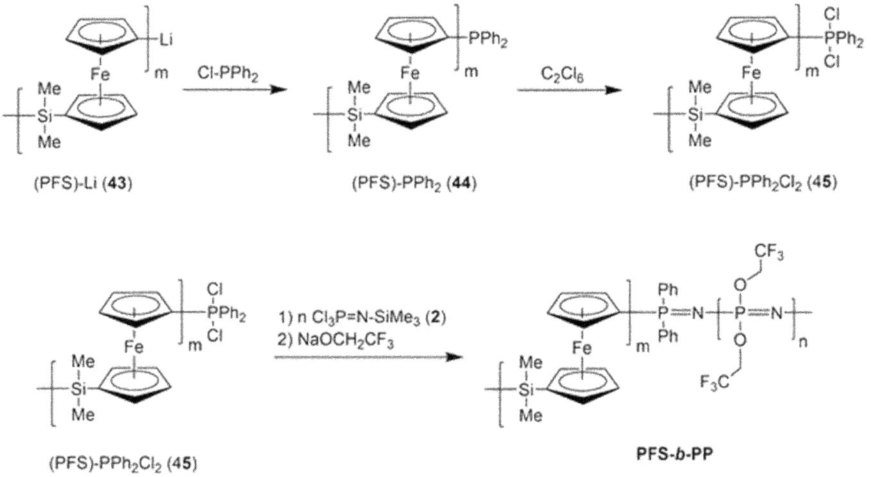

	m	n	φ (PFS)
46a	54	54	0.60
46b	54	77	0.51
46c	54	116	0.40
46d	54	290	0.21

*Scheme 8. Synthesis of well-defined polyphosphazene-b-poly(ferrocenylsilane), PP-b-PFS (**46a-d**).*

Among all the polyphosphazene-*b*-Polym BCPs, the polyphosphazene-*b*-polystyrene BCPs (PP-*b*-PS) represent a very interesting class of hybrid BCPs combining the functionality of the PP block with the chemical stability and high solubility of the PS block. These PP-*b*-PS were synthesized (see Scheme 9) by a macromolecular coupling of a phosphoranimine end-functionalized polystyrene (telechelic $(PS)_n\text{-}P(R)_2=N\text{-}SiMe_3$, **48** or **49**) and an end-capped poly(dichloro)phosphazene (**50**) (*34, 67*). The required telechelic polystyrene **48** was synthesized by quenching the anionic polymerization of styrene with

Cl-PPh_2 followed by a Staundinger reaction with trimethylsilyl azide (Scheme 9) (*67*). The polymer **49** was similarly prepared by treating the living chains of PS-Li (**47**) with a protected (chloroalkyl)amine followed by a deprotection step and the reaction of the resulting free –NH_2 group with the phosphoranimine Br-$P(OCH_3CF_3)_2$=N-$SiMe_3$ (**22**, Scheme 9) (*34*). By using this synthetic strategy BCPs of general formula [N=PCl_2]-*b*-PS (**51**) were synthesized.

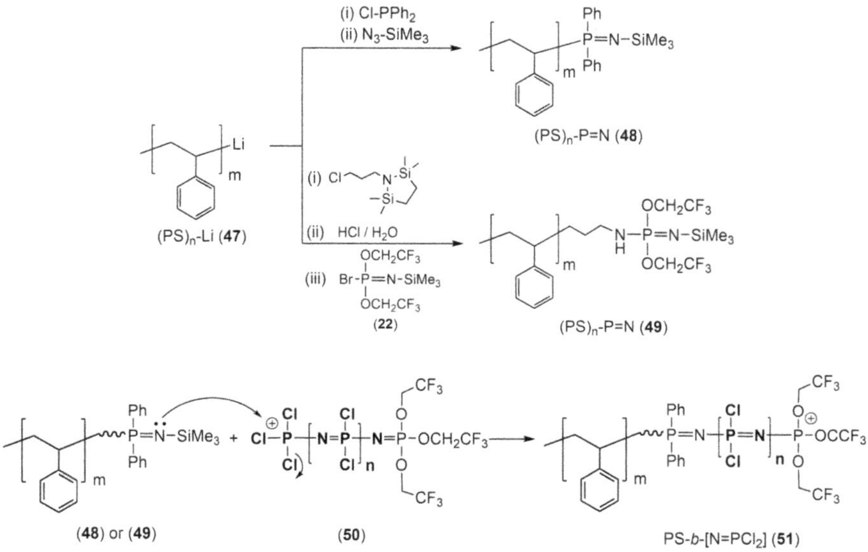

$(PS)_n$-P=N (**53, 54**)

*Scheme 9. Synthetic routes to [N=PCl_2]$_n$-b-polystyrene (**51**), via macromolecular coupling of telechelic $(PS)_n$-$P(R)_2$=N-$SiMe_3$, **48** and **49**, with an end-capped poly(dichloro)phosphazene **50**.*

[N=PCl_2]$_n$-*b*-PS (**51**), after the chlorine macromolecular substitution with the appropriate nucleophiles, yielded PP-*b*-PS BCPs having PP blocks soluble in water, such as [N=$P(O(CH_2CH_2O)_2CH_3)_2$]$_n$ (MEET, **52**) (*32*), [N=$P(OC_6H_4$-$CO_2Na)_2$]$_n$ (KPCPP, **53**) (*33*), and PP blocks having [adamantly/β-Cyclodextrin] complexes (*34*) (**54**. See Figure 6). The block copolymers **52-54** were isolated in moderately good yields (<60%) and polydispersity indexes (<1.3) (*34, 67*).

Recently, our group developed a very convenient synthesis for linear hybrid PP-*b*-PS block copolymers (Scheme 10) (*71*) using the lithiated-phosphine **55** as initiator of the anionic polymerization of styrene, which, after quenching the polymerization with Cl-$SiMe_3$, led to the telechelic polystyrene **56** (PS_m-PPh_2). The chlorination of **56** in mild conditions (25° C with C_2Cl_6), led to the polystyrene PS_m-PPh_2Cl_2 (**57**) having -PPh_2Cl_2 end-groups, able to initiate the polymerization of Cl_3P=N-$SiMe_3$ (**2**). Thus, in an one-pot synthesis, we produced the block copolymer [N=PCl_2]$_n$-PS_m (**58**) with high control of the macromolecular parameters (molecular weight and block ratio). The macromolecular substitution of the chlorine atoms by $-OCH_2CF_3$ groups yielded the block copolymer $[N{=}P(OCH_2CF_3)_2]_n$-PS_m (**59**) having crystalline PP blocks. This synthetic methodology afforded the hybrid PP-*b*-PS block copolymer in 70 % yield and with narrower polydispersity indexes (PDI = 1.2) than those of the previous reported methods.

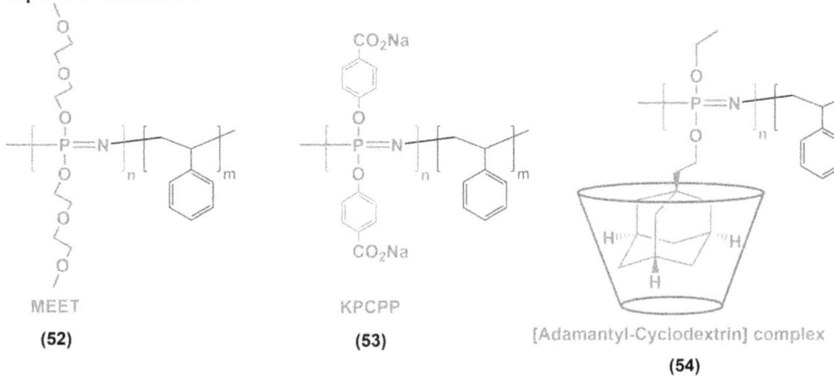

Figure 6. Molecular structure of PP-b-PS ***52-54*** *having water-soluble PP blocks.*

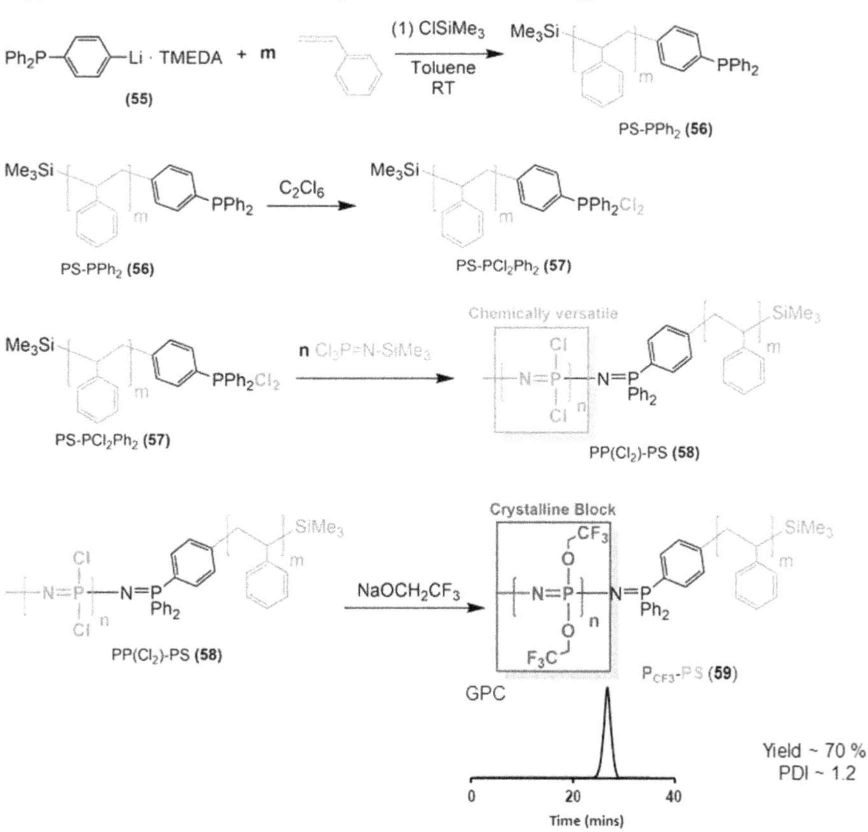

*Scheme 10. Synthetic routes to polyphosphazene-b-polystyrene, PP-b-PS (**59**), via lithiated-phosphine **55**.*

Self-Assembly of Linear Polyphosphazene Two-Block Copolymers

As stated above (see Introduction section) prior to 2012 the research on the self-assembly of phosphazene based BCP's were very scarce and was limited to a very few examples with the block copolymers: $[N{=}P(O(CH_2CH_2O)_2CH_3)_2]_n$-*b*-$[N{=}PPh(O(CH_2CH_2O)_2CH_3)]_m$ in aqueous media (*31*), PP-*b*-PS (PS = polystyrene) (*32–34*), PP-*b*-PEO [PEO = poly(ethylene oxide)] (*35–37*), and PP-*b*-PFS with PP = $[N{=}P(OCH_2CF_3)_2]_n$ and PFS = poly(ferrocenyl(dimethyl)silane) (*38*, *39*).

After resolving some of the obstructing synthetic problems, we undertook a systematically investigation of the self-assembly of polyphosphazene BCPs selecting one of the blocks to be amorphous (i.e. with not tendency to crystallize) and highly soluble in most common organic solvents (i.e. $[N{=}PMePh]_n$ for PP-*b*-PP, and polystyrene (PS) for the hybrid block copolyphosphazenes). These blocks were combined with a PP chain having a predetermined property capable of conducting the self-assembly of the resulting BCP either in thin films or in a selective solvent to $[N{=}PMePh]_n$ or PS segments.

Polyphosphazene BCPs Having a Rigid PP Block

It has been stablished that the self-assembly of linear two-block copolymers combining rigid and flexible blocks (rod-*b*-coil BCPs) originates well-defined non spherical aggregates exhibiting complex nanomorphologies (*72–77*). Thus, we found that the BCPs $[N{=}P(O_2C_{12}H_8)]_n$-*b*-$[N{=}PMePh]_m$ (n = 50, m = 35 (**P$_{Rig}$1**); n = 245, m = 60 (**P$_{Rig}$2**); n = 20, m = 70 (**P$_{Rig}$3**), see Figure 4), having $[N{=}P(O_2C_{12}H_8)]_n$ blocks with an activation energy at the glass transition (T_g *ca.* 160 °C) of ΔH_{Tg} = 500 KJ mol^{-1} (*78–80*), which is in agreement with a very rigid segment (i.e. rod-*b*-coil BCPs). In thin films created from a THF solution, those polymers spontaneously produced well-defined nano-porous films by a vesicle-to-pore morphological evolution during the solvent evaporation (Figure 7) (*81*).

The slow kinetics of the self-assembly processes allowed the characterization of most of the intermediates of this vesicle-to-pore mechanism. Thus, we demonstrated that the presence of the rigid $[N{=}P(O_2C_{12}H_8)]_n$ blocks favors the formation of vesicles in solution which, by close packing during the solvent evaporation, constructed the film. The loss of the solvent occluded in the inner parts of vesicles produced voids (free volume) inside the film. Those voids located near the top layer of the film, create the pores. The diameter of the pores could be adjusted in the interval of 120 nm to 80 nm by controlling the speed of solvent evaporation using spin-coating techniques (*81*).

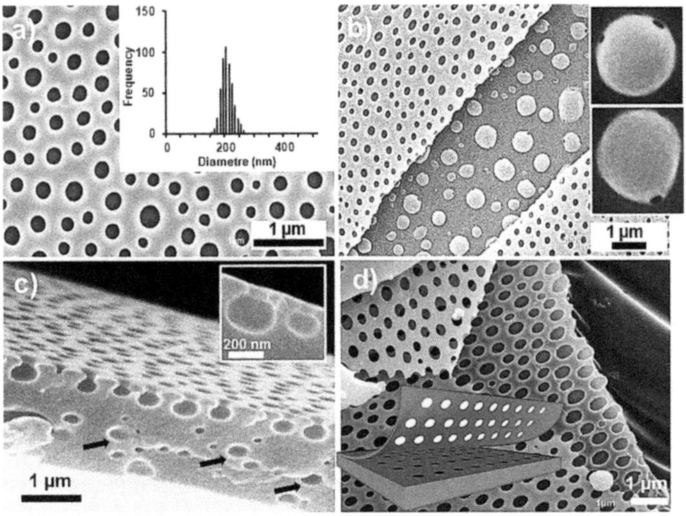

Figure 7. SEM images of porous films (a) and vesicles (b) from $\boldsymbol{P_{Rig}1}$*. (c) SEM image of a cross fracture of the film showing voids in the inner part. (d) Pilling-off the top-layer of the porous film generating a porous flexible sheet. Adapted with permission from ref (81). Copyright 2016, Royal Society of Chemistry.*

Polyphosphazene BCPs Having a Chiral and/or Helical PP Block

Helicity has been historically postulated as one of the main factors inducing homochirality in nature (*82*). Mimicking nature, researchers have used synthetic one-handed block copolymers (BCPs) as versatile building blocks to create supramolecular helical nanoarchitectures by self-assembly (*83*). In 1998 Nolte *et al.* reported the first example of one-handed helical nanostructures created by the self-assembly of BCPs (*84*). Authors applied the term "*chiral effect*" to describe the transfer of chiral information from molecules to nanostructures (*83–91*). Our group examined the "*chiral effect*" using the block copolymers ([N=P(*R*-$O_2C_{20}H_{12}$)]$_n$-*b*-[N=PMePh]$_m$ (n = 50, m = 35 ($\mathbf{P_{Chir}1}$); n = 200, m = 50 ($\mathbf{P_{Chir}2}$); n = 30, m = 150 ($\mathbf{P_{Chir}3}$)), see Figure 4), equipped with chiral [N=P(*R*-$O_2C_{20}H_{12}$)]$_n$ blocks exhibiting a secondary helical structure with preferential helical sense (*92–95*), and very high rigidity (T_g = 320 °C) (*92*). However, whilst the stiffness of the chiral [N=P(*R*-$O_2C_{20}H_{12}$)]$_n$ blocks are independent on the chain length, the degree of preference of the helical sense increases with the number of repeating units. Thus, whereas the self-assembly of $\mathbf{P_{Chir}1}$, having a short [N=P(*R*-$O_2C_{20}H_{12}$)]$_n$ block (50 repeat units), would be controlled by the rigidity, that of $\mathbf{P_{Chir}2}$ with longer [N=P(*R*-$O_2C_{20}H_{12}$)]$_n$ block (200 repeat unit), would be directed by the helicity. Consistently, like the polymers $\mathbf{P_{rig}1\text{-}3}$ (with a rigid [N=P($O_2C_{12}H_8$)]$_n$ block), the self-assembly of $\mathbf{P_{Chir}1}$ in thin films led to chiral nano-porous films (Figure 8) (*59*). Studies with polarized-light microscopy demonstrated that the chiral [N=P(*R*-$O_2C_{20}H_{12}$)]$_n$ blocks were preferentially located at the edge of the pores. This fact was exploited in the self-assembly of [N=P(*R*-$O_2C_{20}H_{12}$)$_{0.9}$(OC_5H_4N)$_{0.1}$]$_{50}$-*b*-[N=PMePh]$_{35}$ (**39**, see Figure 5), having the same degree of polymerization of both blocks than that of **P_{Chir}1**, but with 10 % of 4-pyridinoxy groups randomly substituted along the chains bearing the chiral units. Under the same experimental conditions than those for **P_{Chir}1**, the self-assembly of **39**, produced nano-porous films endowed with coordinative active pyridine ligands placed at the edge of the pore. Those functionalities were exploited to stabilize gold nanoparticles (AuNPs) originating specifically decorated nano-porous films (Figure 8c-e) (*60*).

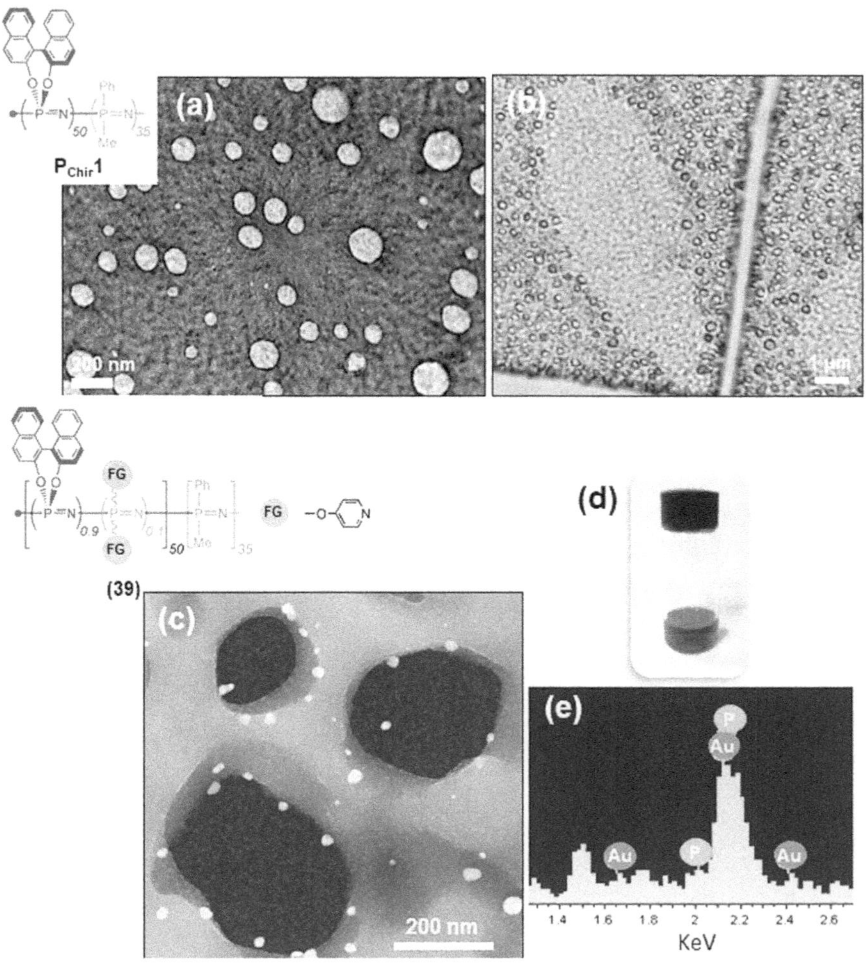

*Figure 8. (a) Bright-TEM image of the nano-porous film prepared by the self-assembly of **P_{Chir}1**. (b) High resolution polarized light microscopy of drop-casting films of **P_{Chir}1**. (c) Dark-field-TEM image of AuNPs decorated porous films of **39** (note that, in the dark-field TEM micrograph, AuNPs appear as white dots and pores as dark circles). (d) Stabilized solution of AuNPs and BCP **39** (THF). (e) EDX analysis. Adapted with permission from ref (59) and (60) Copyright 2013, Wiley-VCH Verlag GmbH & Co.*

The self-assembly of **P$_{Chir}$2**, having longer isotactic [N=P(*R*-$O_2C_{20}H_{12}$)]$_n$ segments than **P$_{Chir}$1**, afforded supramolecular twisted-morphologies of 200-300 nm long, with a pitch and projected diameter of 80 and 72 nm (on average), respectively (Figure 9) (*59*). Importantly, these twisted-morphologies were not observed in the self-assembly of the non-chiral analogue [N=P(***R/S***-$O_2C_{20}H_{12}$)]$_{200}$-*b*-[N=PMePh]$_{50}$ (**P$_{R/S}$2**), which has 50 % of [N=P(***R***-$O_2C_{20}H_{12}$)] and 50 % [N=P(***S***-$O_2C_{20}H_{12}$)] units randomly substituted within the block, and therefore lacks of preferential helical sense conformations (Figure 9). The slow kinetics of the self-assembly of the chiral **P$_{Chir}$2** allowed the characterization of some intermediates and to propose a mechanism showing that the formation of the twisted-aggregates is due to the transfer of the chiral information from the initial spherical micelles in solution to helical pearl necklaces and, finally, to twisted worm-like textures (Figure 9) (*59*).

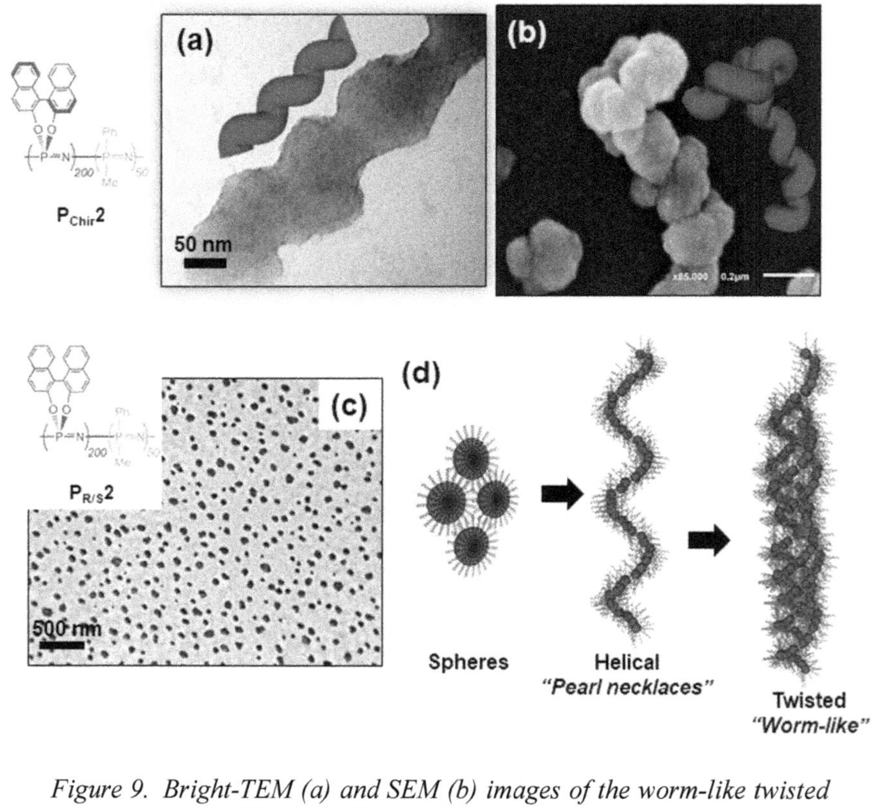

Figure 9. Bright-TEM (a) and SEM (b) images of the worm-like twisted aggregates created by the self-assembly of ***P$_{Chir}$2*** *in THF. (c) Bright-TEM image of the ill-defined spherical aggregates created by the self-assembly of non-isotactic analogous* ***P$_{R/S}$2*** *in THF. (d) Schematic representation of the proposed mechanism to explain the formation of twisted-morphologies. Adapted with permission from ref (59) Copyright 2013, Wiley-VCH Verlag GmbH & Co.*

In order to probe this mechanism, we varied the helicity degree of the chiral core-forming PP blocks by synthesizing the BCPs $\{[N{=}P(R\text{-}O_2C_{20}H_{12})]_{200\text{-}x}[N{=}P(OC_5H_4N)_2]_x\}$-*b*-$[N{=}PMePh]_{50}$ [x = 10 (**42a**), y = 30 (**42b**), y = 60 (**42c**), y = 100 (**42d**), see Figure 5] with the same degrees of polymerization than $\mathbf{P_{Chir}2}$, but having an increasing proportion (x) of the non-chiral $[N{=}PR_2]$ units (R = OC_5H_4N) randomly substituted within the binaphtoxy-phosphazene segment (Figure 10) (*61*). The chiroptical properties ($[\alpha]_D$ *vs* T and CD) of those polymers confirmed the expected decrease of the helicity degree as the proportion of non-chiral $[N{=}PR_2]$ units (x) increases. Thus, whereas **42a**, with only 5% of $[N{=}P(OC_5H_4N)_2]$ units, exhibited a preferential helical sense, **42d** with 50 % of those units, adopted non preferred helical conformations. The influence of the helicity on the self-assembly of these materials was very significant and produced a variety of morphologies extending from helical nanostructures (**42a**), to "*pearl necklace*" aggregates of different lengths (**42b-c**), and, finally, to nano-spheres (**42d**) (Figure 10) (*61*).

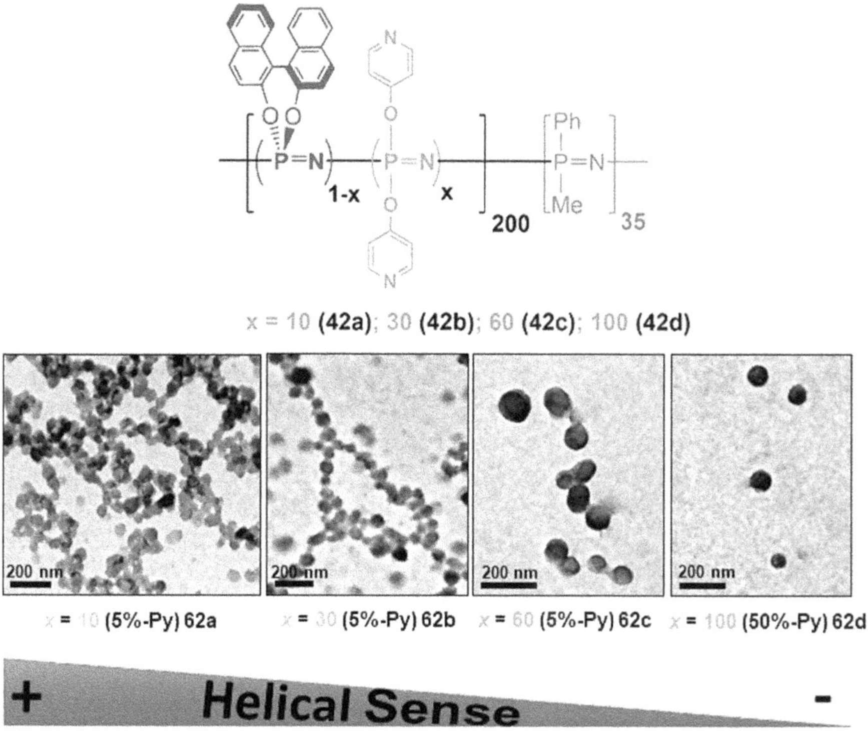

*Figure 10. Bright TEM images of the thin films prepared by drop-casting of the 0.5 mg/mL solutions of block copolymers **42a**, **42b**, **42c**, and **42d**. Adapted with permission from ref (61) Copyright 2015, Wiley-VCH Verlag GmbH & Co.*

Polyphosphazene BCPs Having a Crystalline PP Block

The self-assembly of BCPs containing a crystallisable core-forming block has been exploited to create complex nanomorphologies by crystallization-driven living self-assembly (CDSA) (*96–104*). The presence of crystalline blocks facilitates, under an appropriate experimental conditions, the creation of nanomorphologies with curvatures, at the core-corona interface, lower than the corresponding to an sphere (*105–110*). In order to investigate the effects of the crystallinity (i.e. the tendency to give highly ordered solid phases) on the self-assembly of PP-*b*-PP polymers, the BCPs $[N{=}P(OCH_2CF_3)_2]_n$-*b*-$[N{=}PMePh]_m$ (**P_{CF3}1**: n = 30, m = 20; **P_{CF3}2**: n = 90, m = 20; **P_{CF3}3**: n = 200, m = 85), were synthesized having crystalline $[N{=}P(OCH_2CF_3)_2]_n$ segments (see Figure 4). Their self-assembly in THF produced stable giant unilamellar vesicles (GUVs, diameter ≥ 1000 nm, **P_{CF3}1**) and big-sized vesicles (diameter ≥ 500 nm, **P_{CF3}2-3**, see (Figure 11) (*111*). The formation and stability of these vesicles are provided by the crystallization of the $[N{=}P(OCH_2CF_3)_2]_n$ blocks at the capsule wall (the $[N{=}PMePh]_m$ blocks are located at the inner and outer coronas). Very interestingly, the GUVs proved to be responsive to strong acids (HOTf). Thus, after the selective protonation of the coronal $[N{=}PMePh]_m$ block, they undergo a morphological evolution to smaller spherical star-like micelles in which the core and corona roles have been inverted. This morphological evolution is totally reversible by neutralization with a base (NEt_3), which reproduces the original GUVs. The monitoring of this process by DLS allowed us to propose a mechanism where **P_{CF3}1** and its protonated form **$P_{CF3}1^+$** are the intermediates as free solvated block copolymers in solution and it is their different self-assembly behavior in THF which originates the GUVs or the spherical star-like aggregates respectively (Figure 11) (*111*).

Very recently, our group studied the self-assembly of hybrid block copolymers combining a crystalline PP block ($[N{=}P(OCH_2CF_3)_2]_n$) with a polystyrene (PS). Thus, crystalline-*b*-coil $[N{=}P(OCH_2CF_3)_2]_{35}$-*b*-$PS_{35}$ (**59** in Scheme 10) block copolymer self-assembled in THF leading to bicontinuous nanospheres or toroidal micelles, two very rare nanomorphologies (Figure 12) (*71*). The processes can be directed to one morphology or the another by simply adjusting the initial BCP concentration. Thus, whereas bicontinuous nanospheres were obtained at 0.33 mg/mL, toroidal micelles resulted at 2 mg/mL. Moreover, the bicontinuous nanospheres and the toroidal micelles can be reversibly transformed one into the other in solution, establishing an unprecedented "*bicontinuous nanosphere-to-toroidal micelles*" morphological transition (Figure 12). WAXS analysis on dried films containing bicontinuous or toroidal aggregates demonstrated that whereas the $[N{=}P(O_2CH_2CF_3)_2]_n$ blocks are essentially amorphous in the structure of the bicontinuous nanospheres, the crystallization of those blocks in the core of the toroidal micelles is crucial to stabilize these morphologies. Thus, the crystallinity of the core forming $[N{=}P(O_2CH_2CF_3)_2]_n$ blocks is the main guide to create bicontinuous nanospheres and toroidal micelles (*71*).

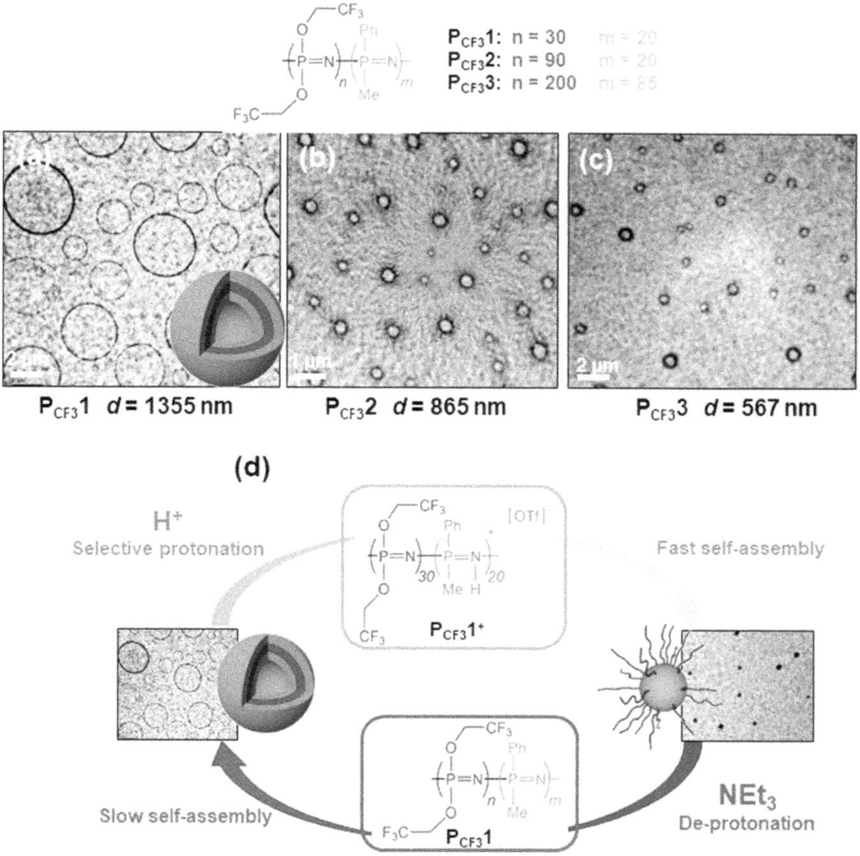

Figure 11. Bright-field TEM images of GUVs (a), and large vesicles (b and c) prepared by the self-assembly of BCPs ***$P_{CF3}1$, $P_{CF3}2$,*** *and* ***$P_{CF3}3$*** *(d = diameter of the vesicle by TEM). (d) Schematic representation of the reversible morphological evolution from GUVs to spherical star-like vesicles promoted by a selective acid-base reaction. Adapted with permission from ref (111) Copyright 2016, Wiley-VCH Verlag GmbH & Co.*

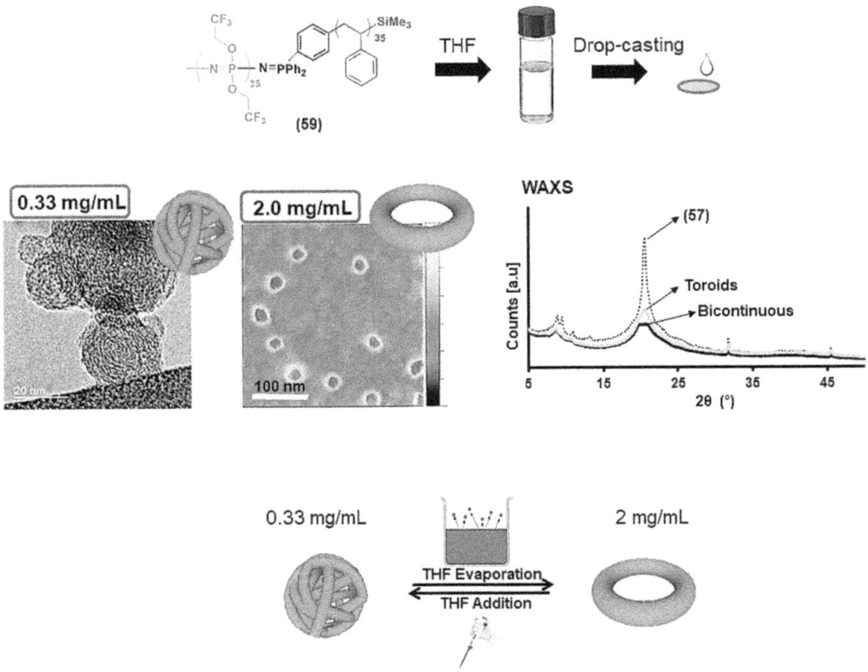

*Figure 12. Self-assembly of $[N{=}P(OCH_2CF_3)_2]_{35}$-b-$PS_{35}$ (**59**) in THF. Formation of bicontinuous nanospheres or toroidal micelles. Schematic representation of the "bicontinuous-to-toroidal micelle" is shown at the bottom of the figure. Adapted with permission from ref (71) Copyright 2016, Wiley-VCH Verlag GmbH & Co.*

Conclusions

Polyphosphazene block copolymers are outstanding materials with high potential for self-assembly studies. However, block copolyphosphazenes had been scarcely employed due principally to the lack of synthetic routes leading to well-defined block copolyphosphazenes, i.e. with controllable molecular weight, block ratios, and importantly, with very narrow polydispersity indexes (PDI < 1.2). Recent works have resolved most of these problems providing with new synthetic protocols to achieve well-defined two-block copolymers combining two polyphosphazene blocks (PP-*b*-PP), or a PP chain with a non phosphazenic segment. The high chemical adaptability (almost "*á la carte*") of phosphazene polymers enables the synthesis of polyphosphazene block copolymers with PP chains incorporating a variety of properties, such as crystallinity, rigidity or chirality. The study of the self-assembly of these polymers facilitates the investigation of the factors controlling the creation of nanostructured materials. As revealed in this Chapter, a diversity of different well-defined nanomorphologies can be achieved from BCPs having crystalline, rigid or chiral core-forming PP blocks. Moreover, the slow self-assembly observed by PP-based block copolymers allows the characterization of intermediate kinetically stabilized nanostructures helping to clarify the mechanisms of the aggregation processes. All the work described herein, strongly suggest that the self-assembly of polyphosphazene-based BCPs will be of increasing interest from both fundamental and applied scientific perspectives.

Acknowledgments

The authors are grateful to FICYT (Projects SV-PA-13-ECOEMP-83 and FC-15-GRUPIN14-106), Universidad de Oviedo (Project UNOV-13-EMERG-GIJON-08) and the MINECO (Project CTQ2014-56345-P and CTQ2017-88357-P) for the funding. We are also grateful to the COST action Smart Inorganic Polymers (SIPs-CM1302 - http://www.sips-cost.org/home/index.html). APS would like to thank the MEC for the Juan de la Cierva and Ramón y Cajal programs.

References

1. Allcock, H. R. *Chemistry and Applications of Polyphosphazenes*; John Wiley & Sons: New York, 2003.
2. De Jaeger, R.; Gleria, M. *Phosphazenes: A Worldwide Insight*; NOVA Science Publishers: New York, 2005.
3. Krogman, N. R.; Weidel, A. L.; Kristhart, K. A.; Nukavarapu, S. P.; Deng, M.; Nair, L. S.; Laurencin, C. T.; Allcock, H. R. The influence of side group modification in polyphosphazenes on hydrolysis and cell adhesion of blends with PLGA. *Biomaterials* **2009**, *30*, 3035–3041.
4. Deng, M.; Narir, L. S.; Nukayarapu, S. P.; Jiang, T.; Kanner, W. A.; Li, X.; Kumbar, S. G.; Weikel, A. L.; Krogman, N. R.; Allcock, H. R.; Laurencin, C. T. Dipeptide-based polyphosphazene and polyester blends for bone tissue engineering. *Biomaterials* **2010**, *31*, 4898–4908.
5. Linhardt, A.; König, M.; Schöfberger, W.; Brüggemann, O.; Andrianov, A. K.; Teasdale, I. Biodegradable polyphosphazene based peptide-polymer Hybrids. *Polymers* **2016**, *8*, 161–178.
6. Rothemund, S.; Aigner, T. B.; Iturmendi, A.; Rigau, M.; Husár, B.; Hildner, F.; Oberbauer, E.; Prambauer, M.; Olawale, G.; Forstner, R.; Liska, R.; Schröder, K. R.; Brüggemann, O.; Teasdale, I. Degradable glycine-based photo-polymerizable polyphosphazenes for use as scaffolds for tissue regeneration. *Macromol. Biosci.* **2015**, *15*, 351–363.
7. Wilfert, S.; Iturmendi, A.; Schoefberger, W.; Kryeziu, K.; Heffeter, P.; Berger, W.; Brüggemann, O.; Teasdale, I. Water-soluble, biocompatible polyphosphazenes with controllable and pH-promoted degradation behavior. *J. Polym. Sci., Part A: Polym. Chem.* **2014**, *52*, 287–294.
8. Mayer-Gall, T.; Knittel, D.; Gutmann, J. S.; Opwis, K. Permanent flame retardant finishing of textiles by allyl-functionalized polyphosphazenes. *ACS Appl. Mater. Interfaces* **2015**, *7*, 9349–9363.
9. Amarnath, N.; Appavoo, D.; Lochab, B. Eco-friendly halogen-free flame retardant cardanol polyphosphazene polybenzoxazine networks. *ACS Sustainable Chem. Eng.* **2018**, *6*, 389–402.

10. Modzelewski, T.; Wilts, E.; Allcock, H. R. Elastomeric polyphosphazenes with phenoxy–cyclotriphosphazene side groups. *Macromolecules* **2015**, *48*, 7543–7549.
11. Modzelewski, T.; Wonderling, N. M.; Allcock, H. R. Polyphosphazene elastomers containing interdigitated oligo-p-phenyleneoxy side groups: Synthesis, mechanical properties, and X-ray scattering studies. *Macromolecules* **2015**, *48*, 4882–4890.
12. Modzelewski, T.; Allcock, H. R. An unusual polymer rrchitecture for the generation of elastomeric properties in fluorinated polyphosphazenes. *Macromolecules* **2014**, *47*, 6776–6782.
13. Nair, L. S.; Bhattacharyya, S.; Bender, J. D.; Greish, Y. E.; Brown, P. W.; Allcock, H. R.; Laurencin, C. T. Fabrication and optimization of methylphenoxy substituted polyphosphazene nanofibers for biomedical applications. *Biomacromolecules* **2004**, *5*, 2212–2220.
14. Allcock, H. R.; Wood, R. M. Design and synthesis of ion-conductive polyphosphazenes for fuel cell applications: Review. *J. Polym. Sci.: Part B: Polym. Phys.* **2006**, *44*, 2358–2368.
15. Akram, M.; Wang, L.; Yu, H.; Amer, W. A.; Khalid, H.; Abbasi, N. M.; Cheng, Y.; Zain-ul-Abdin; Saleem, M.; Tong, R. Polyphophazenes as anti-cancer drug carriers: From synthesis to application. *Prog. Polym. Sci.* **2014**, *39*, 1987–2009.
16. Andrianov, A. K. *Polyphosphazenes for Biomedical Applications*; Wiley: Hoboken, NJ, 2009.
17. Bates, F. S.; Fredrickson, G. H. Block copolymers-designers soft materials. *Phys. Today* **1999**, *52*, 32–38.
18. Alexandridis, P.; Lindman, B. *Amphiphilic Block Copolymers: Self-Assembly and Applications*; Elsevier, Amsterdam, 2000.
19. Förster, S.; Plantenberg, T. From self-organizing polymers to nanohybrid and biomaterials. *Angew. Chem., Int. Ed.* **2002**, *41*, 688–714.
20. Kim, J. K.; Yang, S. Y.; Lee, Y.; Kim, Y. Functional nanomaterials based on block copolymer self-assembly. *Prog. Polym. Sci.* **2010**, *35*, 1325–1349.
21. Orilall, M. C.; Wiesner, U. Block copolymer based composition and morphology control in nanostructured hybrid materials for energy conversion and storage: solar cells, batteries, and fuel cells. *Chem. Soc. Rev.* **2011**, *40*, 520–535.
22. Blanazs, A.; Armes, S. P.; Ryan, A. J. Self-assembled block copolymer aggregates: From micelles to vesicles and their biological applications. *Macromol. Rapid Commun.* **2009**, *30*, 267–277.
23. Hamley, I. *Block Copolymers in Solution: Fundamentals and Applications*; Wiley, Chichester, UK, 2005.
24. Gohy, J. F. Block Copolymer Micelles. In *Block Copolymers II. Advances in Polymer Science*; Abetz, V., Ed.; Springer: Berlin, Heidelberg, 2005; Vol. 190, pp 65−136.
25. Cui, H.; Chen, Z.; Wooley, K. L.; Pochan, D. J. Origins of toroidal micelle formation through charged triblock copolymer self-assembly. *Soft Matter* **2009**, *5*, 1269–1278.

26. Riess, G. Micellization of block copolymers. *Prog. Polym. Sci.* **2003**, *28*, 1107–1170.
27. Jain, S.; Bates, F. S. On the origins of morphological complexity in block copolymer surfactants. *Science* **2003**, *300*, 460–464.
28. Discher, D. E.; Eisenberg, A. Polymer vesicles. *Science* **2002**, *297*, 967–973.
29. Won, Y. Y.; Davis, H. T.; Bates, F. S. Giant wormlike rubber micelles. *Science* **1999**, *283*, 960–963.
30. M. Lazzari, G. Liu, S. Lecommandoux *Block Copolymers in Nanocience*; Wiley-VCH: Weinheim, Germany, 2006.
31. Chang, Y.; Lee, S. C.; Kim, K. T.; Kim, C.; Reeves, S. D.; Allcock, H. R. Synthesis and micellar characterization of an amphiphilic diblock copolyphosphazene. *Macromolecules* **2001**, *34*, 269–274.
32. Allcock, H. R.; Powell, E. S.; Chang, Y. Synthesis and micellar behavior of amphiphilic polystyrene-poly[bis(methoxyethoxyethoxy)phosphazene] block copolymers. *Macromolecules* **2004**, *37*, 7163–7167.
33. Chang, Y.; Powell, E. S.; Allcock, H. R. Environmentally responsive micelles from polystyrene–poly[bis(potassium carboxylatophenoxy)phosphazene] block copolymers. *J. Polym. Sci., Part A: Polym. Chem.* **2005**, *43*, 2912–2920.
34. Cho, S. Y.; Allcock, H. R. Synthesis of adamantyl polyphosphazene-polystyrene block copolymers, and β-cyclodextrin-adamantyl side group complexation. *Macromolecules* **2009**, *42*, 4484–4490.
35. Chang, Y.; Bender, J. D.; Phelps, M. V. B.; Allcock, H. R. Synthesis and self-Association behavior of biodegradable amphiphilic poly[bis(ethyl glycinat-N-yl)phosphazene]-poly(ethylene oxide) block copolymers. *Biomacromolecules* **2002**, *3*, 1364–1369.
36. Chang, Y.; Prange, R.; Allcock, H. R. Amphiphilic poly[bis(trifluoroethoxy)phosphazene]-poly(ethylene oxide) block copolymers: synthesis and micellar characteristics. *Macromolecules* **2002**, *35*, 8556–8559.
37. Tian, Z.; Liu, X.; Chen, C.; Allcock, H. R. Synthesis and micellar behavior of novel amphiphilic poly[bis(trifluoroethoxy)phosphazene]-copoly[(dimethylamino)ethyl methacrylate] block copolymers. *Macromolecules* **2012**, *45*, 2502–2508.
38. Soto, A. P.; Manners, I. Poly(ferrocenylsilane-b-polyphosphazene) (PFS-b-PP): A new class of organometallic-inorganic block copolymers. *Macromolecules* **2009**, *42*, 40–42.
39. Soto, A. P.; Gilroy, J. B.; Winnik, M. A.; Manners, I. Pointed-oval-shaped micelles from crystalline-coil block copolymers by crystallization-driven living self-assembly. *Angew. Chem., Int. Ed.* **2010**, *49*, 8220–8223.
40. Rothemund, S.; Teasdale, I. Preparation of polyphosphazenes: A tutorial review. *Chem. Soc. Rev.* **2016**, *45*, 5200–5215.
41. Honeyman, C. H.; Manners, I.; Morrissey, C. T.; Allcock, H. R. Ambient temperature synthesis of poly(dich1orophosphaene) with molecular weight control. *J. Am. Chem. Soc.* **1995**, *117*, 7035–7036.
42. Allcock, H. R.; Crane, C. A.; Morrissey, C. T.; Nelson, J. M.; Reeves, S. D.; Honeyman, C. H.; Manners "Living" cationic polymerization of

phosphoranimines as an ambient temperature route to polyphosphazenes with controlled molecular weights. *Macromolecules* **1996**, *29*, 7740–7747.
43. Allcock, H. R.; Reeves, S. D.; de Denus, C. R.; Crane, C. A. Influence of reaction parameters on the living cationic polymerization of phosphoranimines to polyphosphazenes. *Macromolecules* **2001**, *34*, 748–754.
44. Allcock, H. R.; Reeves, S. D.; Nelson, J. M.; Crane, C. A.; Manners, I. Polyphosphazene block copolymers via the controlled cationic, ambient temperature polymerization of phosphoranimines. *Macromolecules* **1997**, *30*, 2213–2215.
45. Allcock, H. R.; Reeves, S. D.; Nelson, J. M.; Manners, I. Synthesis and characterization of phosphazene di- and triblock copolymers via the controlled cationic, ambient temperature polymerization of phosphoranimines. *Macromolecules* **2000**, *33*, 3999–4007.
46. Allcock, H. R.; Nelson, J. M.; Reeves, S. D.; Honeyman, C. H.; Manners, I. Ambient-temperature direct synthesis of poly(organophosphazenes) via the "living" cationic polymerization of organo-substituted phosphoranimines. *Macromolecules* **1997**, *30*, 50–56.
47. Nelson, J. M.; Allcock, H. R.; Manners, I. Cationic homo- and copolymerization of fluorophosphoranimines as an ambient temperature synthetic route to poly(fluorophosphazenes), [N=PF(R)]n, with controlled architectures. *Macromolecules* **1997**, *30*, 3191–3196.
48. Allcock, H. R.; de Denus, C. R.; Prange, R.; Nelson, J. M. Synthesis of trifluoromethyl- and methylphosphazene polymers: differences between polymerization and initiator/terminator properties. *Macromolecules* **1999**, *32*, 7999–8004.
49. Blackstone, V.; Lough, A. J.; Murray, M.; Manners, I. Probing the mechanism of the PCl_5-initiated living cationic polymerization of the phosphoranimine $Cl_3P{=}NSiMe_3$ using model compound chemistry. *J. Am. Chem. Soc.* **2009**, *131*, 3658–3667.
50. Blackstone, V.; Pfirrmann, S.; Helten, H.; Staubitz, A.; Soto, A. P.; Whittell, G. R.; Manners, I. A cooperative role for the counteranion in the PCl_5-initiated living, cationic chain growth polycondensation of the phosphoranimine $Cl_3P{=}NSiMe_3$. *J. Am. Chem. Soc.* **2012**, *134*, 15293–15296.
51. Suárez-Suárez, S.; Presa-Soto, D.; Carriedo, G. A.; Soto, A. P.; Staubitz, A. Experimental and theoretical study of the living polymerization of N-silylphosphoranimines. synthesis of new block copolyphosphazenes. *Organometallics* **2012**, *31*, 2571–2581.
52. Wilfert, S.; Henke, H.; Schoefberger, W.; Brüggemann, O.; Teasdale, I. Chain-end-functionalized polyphosphazenes via a one-pot phosphine-mediated living polymerization. *Macromol. Rapid Commun.* **2014**, *35*, 1135–1141.
53. Henke, H.; Brüggemann, O.; Teasdale, I. Branched macromolecular architectures for degradable, multifunctional phosphorus-based polymers. *Macromol. Rapid Commun.* **2016**, *38*, 1600644.

54. Wisian–Neilson, P.; Neilson, R. H. Poly(dimethylphosphazene), (Me_2PN). *J. Am. Chem. Soc.* **1980**, *102*, 2848–2849.
55. Neilson, R. H.; Hani, R.; Wisian–Neilson, P.; Meister, J. J.; Roy, A. K.; Hagnauer, G. L. Synthesis and characterization of poly(alkyl/arylphosphazenes). *Macromolecules* **1987**, *20*, 910–916.
56. Wisian–Neilson, P.; Allcock, H. R.; Wynne K. J. *Inorganic and Organometallic Polymers II*; American Chemical Society: Washington, DC, 1994.
57. Walker, C. H.; St. John, J. V.; Wisian–Neilson, P. Synthesis and size control of gold nanoparticles stabilized by poly(methylphenylphosphazene). *J. Am. Chem. Soc.* **2001**, *123*, 3846–3847.
58. Neilson, R. H.; Wisian-Neilson, P. Poly(alkyl/arylphosphazenes) and their precursors. *Chem. Rev.* **1988**, *88*, 541–562.
59. Suárez-Suárez, S.; Carriedo, G. A.; Tarazona, M. P.; Soto, A. Twisted morphologies and novel chiral macroporous films from the self-assembly of optically active helical polyphosphazene block copolymers. *Chem. Eur. J.* **2013**, *19*, 5644–5653.
60. Suárez-Suárez, S.; Carriedo, G. A.; Soto, A. P. Gold-decorated chiral macroporous films by the self-assembly of functionalised block copolymers. *Chem. Eur. J.* **2013**, *19*, 15933–15940.
61. Suárez-Suárez, S.; Carriedo, G. A.; Soto, A. P. Tuning the chirality of block copolymers: from twisted morphologies to nanospheres by self-assembly. *Chem. Eur. J.* **2015**, *21*, 14129–14139.
62. Prange, R.; Allcock, H. R. Telechelic syntheses of the first phosphazene siloxane block copolymers. *Macromolecules* **1999**, *32*, 6390–6392.
63. Allcock, H. R.; Prange, R. Properties of poly(phosphazene-siloxane) block copolymers synthesized via telechelic polyphosphazenes and polysiloxane phosphoranimines. *Macromolecules* **2001**, *34*, 6858–6865.
64. Nelson, J. M.; Primrose, A. P.; Hartle, T. J.; Allcock, H. R.; Manners, I. Synthesis of the first organic polymer/ polyphosphazene block copolymers: ambient temperature synthesis of triblock poly(phosphazene-ethylene oxide) copolymers. *Macromolecules* **1998**, *31*, 947–949.
65. Allcock, H. R.; Prange, R.; Hartle, T. J. Poly(phosphazene-ethylene oxide) di- and triblock copolymers as solid polymer electrolytes. *Macromolecules* **2001**, *34*, 5463–5470.
66. Chang, Y.; Powell, E. S.; Allcock, H. R.; Park, S. M.; Kim, C. Thermosensitive behavior of poly(ethylene oxide)-poly[bis(methoxyethoxyethoxy)-phosphazene] block copolymers. *Macromolecules* **2003**, *36*, 2568–2570.
67. Prange, R.; Reeves, S. D.; Allcock, H. R. Polyphosphazene-polystyrene copolymers: block and graft copolymers from polyphosphazene and polystyrene macromonomers. *Macromolecules* **2000**, *33*, 5763–5765.
68. Linear *Tri*-block copolymer PP-PPG-PP (PPG = poly(propylene glycol): Allcock, H. R.; Cho, S. Y.; Steely, L. B. New amphiphilic poly[bis(2,2,2-trifluoroethoxy)phosphazene]/ poly(propylene glycol) triblock copolymers: synthesis and micellar characteristics. *Macromolecules* **2006**, *39*, 8334–8338.

69. Weikel, A. L.; Cho, S. Y.; Morozowich, N. L.; Nair, L. S.; Laurencin, C. T.; Allcock, H. R. Hydrolysable polylactide–polyphosphazene block copolymers for biomedical applications: synthesis, characterization, and composites with poly(lactic-co-glycolic acid). *Polym. Chem.* **2010**, *1*, 1459–1466.
70. Krogman, N. R.; Steely, L.; Hindenlang, M. D.; Nair, L. S.; Laurencin, C. T.; Allcock, H. R. Synthesis and characterization of polyphosphazene-block-polyester and polyphosphazene-block-polycarbonate macromolecules. *Macromolecules* **2008**, *41*, 1126–1130.
71. Presa-Soto, D.; de la Campa, R.; Carriedo, G. A.; Soto, A. P. Formation and reversible morphological transition of bicontinuous nanospheres and toroidal micelles by the self-assembly of a crystalline-b-coil diblock copolymer. *Angew. Chem., Int. Ed.* **2016**, *55*, 10102–10107.
72. Lim, Y.-B.; Moon, K.-S.; Lee, M. Rod–coil block molecules: Their aqueous self-assembly and biomaterials applications. *J. Mater. Chem.* **2008**, *18*, 2909–2918.
73. Olsen, B. D.; Segalman, R. A. Self-assembly of rod–coil block copolymers. *Mat. Sci. Eng.* **2008**, *R 62*, 37–66.
74. Wu, D.; Huang, Y.; Xu, F.; Mai, Y.; Yan, D. Recent advances in the solution self-assembly of amphiphilic "rod-coil" copolymers. *J. Polym. Sci. Part A: Polym. Chem.* **2017**, *55*, 1459–1477.
75. He, M.; Qiu, F.; Lin, Z. Conjugated rod–coil and rod–rod block copolymers for photovoltaic applications. *J. Mater. Chem.* **2011**, *21*, 17039–17048.
76. de Cuendias, A.; Hiorns, R. C.; Cloutet, E.; Vignauc, L.; Cramail, H. Conjugated rod–coil block copolymers and optoelectronic applications. *Polym. Int.* **2010**, *59*, 1452–1476.
77. Chen, X.-F.; Shen, Z.; Wan, X.-H.; Fan, X.-H.; Chen, E.-Q.; Ma, Y.; Zhou, Q.-F. Mesogen-jacketed liquid crystalline polymers. *Chem. Soc. Rev.* **2010**, *39*, 3072–3101.
78. Carriedo, G. A.; Gómez-Elipe, P.; García-Alonso, F. J.; Lizaso, E.; Muño, M. E.; Santamaría, A. Rheological features of a novel poly(spirophosphazene). *Macromol. Chem. Phys.* **2001**, *202*, 3437–3443.
79. Carriedo, G. A.; Fernández-Catuxo, L.; García-Alonso, F. J.; Gómez-Elipe, P.; González, P. A. Preparation of a new type of phosphazene high polymers containing 2,2′-dioxybiphenyl groups. *Macromolecules* **1996**, *29*, 5320–5325.
80. Carriedo, G. A.; García-Alonso, F. J.; González, P. A.; García Álvarez, J. L.; Gómez-Elipe, P. Direct synthesis of new types of phosphazene polymers using alkaline carbonates. *Phosphorus Sulfur* **1999**, *73*, 144–146.
81. Suárez-Suárez, S.; Carriedo, G. A.; Soto, A. P. Porous films by the self-assembly of inorganic rod-b-coil block copolymers: mechanistic insights into the vesicle-to-pore morphological evolution. *Soft Matter* **2016**, *12*, 3084–3092.
82. Green, M. M.; Jain, V. Homochirality in life: two equal runners, one tripped. *Origins Life Evol. Biospheres* **2010**, *40*, 111–118.
83. See an excellent recent revision of the topic in: Yashima, E.; Ousaka, N.; Taura, D.; Shimomura, K.; Ikai, T.; Maeda, K. Supramolecular helical

systems: helical assemblies of small molecules, foldamers, and polymers with chiral amplification and their functions. *Chem. Rev.* **2016**, *116*, 13752–13990.

84. Cornelissen, J. J. L. M.; Fischer, M.; Sommerdijk, N. A. J. M.; Nolte, R. J. M. Helical superstructures from charged Poly(styrene)-poly(isocyanodipeptide) block copolymers. *Science* **1998**, *280*, 1427–1430.
85. Wen, T.; Wang, H. F.; Li, M. C.; Ho, R. M. Homochiral evolution in self-assembled chiral polymers and block copolymers. *Acc. Chem. Res.* **2017**, *50*, 1011–1021.
86. Grason, G. M. Chirality transfer in block copolymer melts: emerging concepts. *ACS Macro Lett.* **2015**, *4*, 526–532.
87. Ho, R. M.; Chen, C. K.; Chiang, Y. W. Novel nanostructures from self-assembly of chiral block copolymers. *Macromol. Rapid Commun.* **2009**, *30*, 1439–1456.
88. de Bruin, A. G.; Barbour, M. E.; Briscoe, W. H. Macromolecular and supramolecular chirality: a twist in the polymer tales. *Polym. Int.* **2014**, *63*, 165–171.
89. Chiang, Y. W.; Ho, R. M.; Burger, C.; Hasegawa, H. Helical assemblies from chiral block copolymers. *Soft Matter* **2011**, *7*, 9797–9803.
90. Ho, R. M.; Chiang, Y. W.; Lin, S. C.; Chen, C. K. Helical architectures from self-assembly of chiral polymers and block copolymers. *Prog. Polym. Sci.* **2011**, *36*, 376–453.
91. Yashima, E.; Maeda, K.; Iida, H.; Furusho, Y.; Nagai, K. Helical polymers: synthesis, structures, and functions. *Chem. Rev.* **2009**, *109*, 6102–6211.
92. Carriedo, G. A.; Garcia Alonso, F. J.; Gonzalez, P. A.; Garcia-Alvarez, J. L. Synthesis of new phosphazene high molecular weight polymers containing functionalized and optically active spirocyclic groups. *Macromolecules* **1998**, *31*, 3189–3196.
93. Carriedo, G. A.; Garcia Alonso, F. J.; Gómez-Elipe, P.; Garcia-Alvarez, J. L.; Tarazona, M. P.; Rodriguez, M. T.; Saiz, E.; Vazquez, J. T.; Padron, J. I. Spectroscopic and solution properties of phenoxyphosphazene random copolymers containing optically active binaphthoxy groups. *Macromolecules* **2000**, *33*, 3671–3679.
94. Carriedo, G. A.; García-Alonso, F. J.; García Álvarez, J. L.; Soto, A. P.; Tarazona, M. P.; Mendicuti, F.; Marcelo, G. Thermal atropisomerization and photoluminiscence spectra of very high glass transition temperature chiral poly(binaphthoxyphosphazenes) with a secondary helicoidal structure. *Macromolecules* **2004**, *37*, 5437–5443.
95. Carriedo, G. A.; García-Alonso, F. J.; García Álvarez, J. L.; Lombardo, G.; Pappalardo, G. C.; Punzo, F. Molecular dynamics (MD) simulations and large-angle X-ray ccattering (LAXS) studies of the solid-state structure and assembly of isotactic (R)-poly(2,2′-dioxy-1,1′-binaphthyl-)phosphazene in the bulk state and in the cast film. *Chem. Eur. J.* **2004**, *10*, 3775–3782.
96. Gilroy, J. B.; Gädt, T.; Whittell, G. R.; Chabanne, L.; Mitchels, J. A.; Richardson, R. M.; Winnik, M. A.; Manners, I. Monodisperse cylindrical micelles by crystallization-driven living self-assembly. *Nat. Chem.* **2010**, *2*, 566–570.

97. Gädt, T.; Ieong, N. S.; Cambridge, G.; Winnik, M. A.; Manners, I. Complex and hierarchical micelle architectures from diblock copolymers using living, crystallization-driven polymerizations. *Nat. Mater.* **2009**, *8*, 144–150.
98. Massey, J. A.; Temple, K.; Cao, L.; Rharbi, Y.; Raez, L.; Winnik, M. A.; Manners, I. Self-assembly of organometallic block copolymers: the role of crystallinity of the core-forming polyferrocene block in the micellar morphologies formed by poly(ferrocenylsilane-b-dimethylsiloxane) in n-alkane solvents. *J. Am. Chem. Soc.* **2000**, *122*, 11577–11584.
99. Wang, X.; Guerin, G.; Wang, H.; Wang, Y.; Manners, I.; Winnik, M. A. Cylindrical block copolymer micelles and co-micelles of controlled length and architecture. *Science* **2007**, *317*, 644–647.
100. Lazzari, M.; López-Quintela, M. A. Block copolymers as a tool for nanomaterial fabrication. *Adv. Mater.* **2003**, *15*, 1583–1594.
101. Hailes, R. L. N.; Oliver, A. M.; Gwyther, J.; Whittell, G. R.; Manners, I. Polyferrocenylsilanes: synthesis, properties, and applications. *Chem. Soc. Rev.* **2016**, *45*, 5358–5407.
102. Crassous, J. J.; Schurtenberger, P.; Ballauff, M.; Mihut, A. M. Design of block copolymer micelles via crystallization. *Polymer* **2015**, *62*, A1–A13.
103. Schmelz, J.; Schacher, F. H.; Schmalz, H. Cylindrical crystalline-core micelles: pushing the limits of solution self-assembly. *Soft Matter* **2013**, *9*, 2101–2107.
104. He, W. N.; Xu, J. T. Crystallization assisted self-assembly of semicrystalline block copolymers. *Prog. Polym. Sci.* **2012**, *37*, 1350–1400.
105. Wu, J.; Wang, L.; Yu, H.; Zain-ul-Abdin; Khan, R. U.; Haroon, M. Ferrocene-based redox-responsive polymer gels: Synthesis, structures and applications. *J. Organomet. Chem.* **2017**, *828*, 38–51.
106. Manners, I. Organometallic soft materials from the ring-opening polymerization of strained metallocenophanes and related species with p-hydrocarbon ligands. *J. Organomet. Chem.* **2011**, *696*, 1146–1149.
107. Bellas, V.; Rehahn, M. Polyferrocenylsilane-based polymer systems. *Angew. Chem., Int. Ed.* **2007**, *46*, 5082–5104.
108. Rider, D. A.; Manners, I. Synthesis, self-assembly, and applications of polyferrocenylsilane block copolymers. *Polym. Rev.* **2007**, *47*, 165–195.
109. Kulbaba, K.; Manners, I. Polyferrocenylsilanes: metal-containing polymers for materials science, self-assembly and nanostructure applications. *Macromol. Rapid Commun.* **2001**, *22*, 711–724.
110. Manners, I. Poly(ferrocenylsilanes): novel organometallic plastics. *Chem. Commun.* **1999**, *10*, 857–865.
111. Suárez-Suárez, S.; Cariedo, G. A.; Soto, A. P. Reversible morphological evolution of responsive giant vesicles to nanospheres by the self-assembly of crystalline-b-coil polyphosphazene block copolymers. *Chem. Eur. J.* **2016**, *22*, 4483–4491.

Chapter 11

Molecular Simulation of Polyphosphazenes

Joel R. Fried*

Department of Chemical Engineering, University of Louisville, Ernst Hall 106, Louisville, Kentucky 40292, United States
***E-mail: joel.fried@louisville.edu**

The use of molecular simulation, including molecular dynamics and Monte Carlo methods, in the study of polyphosphazenes is reviewed. The development and validation of the Class II force field COMPASS is discussed with examples on its utility for predicting physical and thermal properties such as density, permeability, and the glass transition temperature. Applications for the use of molecular simulations in the study of polyphosphazenes that are reviewed include polymer electrolytes, gas-separation membranes, and biomedical materials.

Over 700 poly(organophosphazene)s have been synthesized based on the general structure shown in Figure 1. The substituent groups, R_1 and R_2, can be any of a large number of substituent groups such as an alkoxy, alkyl, or amine group. The nature of the substituent group has significant control on the properties of the polyphosphazene including the glass transition temperature (T_g), degree of crystallinity, and gas permeability. Values of T_g can vary from 173 to 423 K (*1*). Poly-phosphazenes having aromatic-amine substituent groups typically have the highest T_g (*2*). Some semicrystalline polyphosphazenes exhibit both a crystalline melting temperature (T_m) and a sub-T_m transition, $T(l)$, that marks a transition from a crystalline to a mesophasic state.

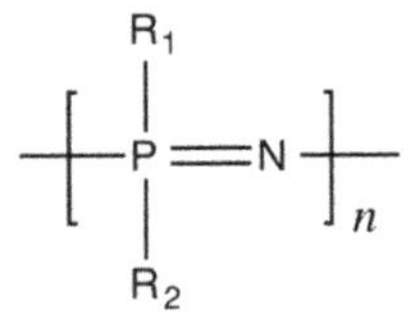

Figure 1. Generic structure of a repeating unit of a polyphosphazene where R_1 and R_2 represent substituent groups.

This chapter focuses on the use of molecular simulations to investigate properties of polyphosphazenes such as intermolecular interactions, density, diffusivity, sorption, and T_g. Applications include a wide range of important uses such as fuel cells, polymer electrolytes, membrane separations, and controlled drug release. In order to investigate such properties using atomistic molecular simulations, it is necessary to have a well-parameterized and extensively validated molecular mechanics force field. The COMPASS (*3*) (***C****ondensed-phase* ***O****ptmized* ***M****olecular* ***P****otentials for* ***A****tomistic* ***S****imulation* ***S****tudies*) force field is one that has been well-parameterized and validated for polyphosphazenes as discussed in the following section.

Development of the COMPASS Force Field for Polyphosphazenes

The first step toward the parameterization of the COMPASS force field was an ab initio study of the structure of polyphosphazenes. This was accomplished by Sun (*4*) who used density functional (DFT) calculations of trimers to study polyphosphazene structure and conformations. One conclusion from that study was that the P–N bonding of the main chain consists of an ionic *s*- and a *p*-bond induced primarily by negative hyperconjugation with charges on P and N close to +1 and –1, respectively. This type of bonding results in a very flexible polymer chain with rotational barriers comparable to that of the C–C bond in ethane (ca. 2 kcal/mol). This high mobility results in low T_g and high permeability of polyphosphazenes that approaches that of poly-siloxanes (*5*).

COMPASS is classified as a Class II force field that had been derived from the earlier force fields such as the Consistent Force Field (CFF), the Consistent Valence Force Field (CVFF), and its most important derivative, the Polymer Consistent Force Field (pcff). The functional forms of COMPASS are the same as CFF93 (*6*, *7*) given as

$$E_{\text{total}} = E_{\text{bonded}} + E_{\text{non-bonded}} = \left(E_b + E_\theta + E_\phi + E_\chi + E_{bb'} + E_{b\theta} + E_{b\phi} + E_{\theta\theta'} + E_{\theta\theta'\phi}\right) + \left(E_{\text{elec}} + E_{\text{LJ}}\right) \quad (1)$$

where the potential energy, E, has two contributions—bonded terms and non-bonded terms. The bonded terms include potential terms for bond-stretching (b), angle-bending (θ), and torsion (φ). COMPASS also includes a term for improper torsion (i.e., out-of-plane, χ). The bond-stretching and angle-bending terms in COMPASS are quartic. Charges and bonded terms had been derived from Hartree Fock (HF/6-31G*) quantum chemical calculations in the parent force fields. Non-bonded terms were initially transferred from pcff and optimized using MD simulations of condensed-phase properties. Characteristic of Class II force fields is a large number of cross terms that reflect the coupling of different potential motions such as bond stretching and angle bending (e.g., $E_{b\theta}$) on contiguous bonds. These cross-coupling terms are important for predicting vibrational frequencies and structure variations associated with conformational changes. The non-bonded terms are traditionally electrostatic (identified as *elec*) and steric, typically Lennard-Jones (LJ). The electrostatic contribution is in the form of a Coulombic term while the LJ term is a 6-9 potential compared to the more commonly used but less accurate 6-12 potential. For the electrostatic contributions, the net partial charge of an atom, q_i, is obtained as a summation of all charge bond-increments, δ_{ij}, related to the specific atom as

$$q_i = \sum_j \delta_{ij}. \qquad (2)$$

The parameterization of COMPASS for phosphazenes was accom-plished by a collaboration of Sun, Ren, and Fried (*8*). Details of the parameterization procedures are given in that publication. Validations included comparisons of structural parameters, vibrational frequencies, X-ray data, conformational energies, and densities with experimental data and DFT calculations. Of importance in regards to subsequent discussion, molecular dynamics (MD) simulations of *isomers* of poly(butoxyphosphazenes) were used to predict density and T_g for comparison with experimental data. These amorphous polyphos-phazenes were selected due to their amorphous character and the availability of good experimental data (*9*), especially for density and T_g. For poly(di-*n*-butoxyphosphazene (PnBuP) and poly(di-*sec*-butoxyphos-phazene) (PsBuP), bond lengths, bond angles, and torsional angles agreed well with X-ray data and T_gs obtained by NPT MD agreed well with results from thermal analysis (*8*). Simulated values of density and T_g for the a isomer, poly(di-*i*-butoxyphosphazene) (PiBuP), were reported to be in good agreement with experimental values in a subsequent publication by Fried and Ren (*10*). As shown in Table I, there is excellent agreement between simulated and experimental values (DSC measurements) of T_g for the three isomers. A comprehensive study of the T_g of the polybutoxyphosphazenes as well as poly[bis(2,2,2-trifluoroethoxy)phosphazene] (PTFEP) have been given by Fried and Ren (*11*).

Table I. Structures and Values of Density and T_g of Three Isomers of Poly(butoxy phosphazenes) and Poly[bis(2,2,2-trifluoroethoxy)-phosphazene] (PTFEP)

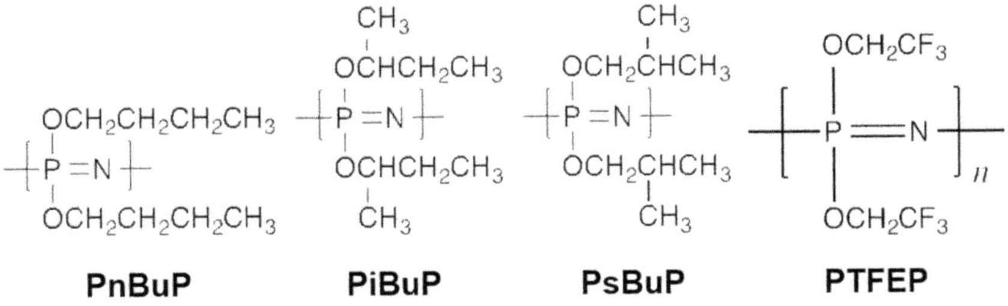

Polymer	T_g (K)	Density (g/cc)
PnBuP	162 (exp 165, 166)[a]	1.051±0.005[a] (exp 1.047)[a]
PiBuP	178 (exp 178)[a]	1.059±0.013[a] (exp 1.059)[a]
PsBuP	159 (exp 182)[a]	1.087±0.005[b] (exp 1.104)[a]
PTFEP	212 (exp 198, 203, 207)[b]	1.633±0.0375 (exp 1.71)[b]

a. Fried and Ren[10]; b. Fried and Ren[11]

Many of the simulation studies reported in this chapter used the COMPASS force field. A review of COMPASS as well as several other force fields that have been used in the simulation of polyphosphazenes and other polymers has been given by Fried (*12*). COMPASS has also been used successfully by our group for the molecular simulation of phosphoric acid (*13*). An updated version of COMPASS (COMPASS II) extends coverage to include additional polymers and drug-like molecules while initial parameterization is conserved (*14*). Other force fields that have been used in the simulation of polyphosphazenes are CHARMm (*15*), AMBER (*16*). DREIDING (*17*), and UFF (*18*) as discussed in subsequent sections. It is noted that CHARMm and AMBER are the most widely used force fields for the molecular simulation of biological materials including proteins, carbohydrates, and nucleic acids. The form of these force fields have been reviewed in the previously cited publication (*12*). Since these force fields were not parameterized for polymers, parameters in CHARMm and AMBER were separately developed for each polyphosphazene. Different force fields that have been used in the molecular simulation studies of polyphosphazenes (*10*, *19–34*) are identified in Table II.

Table II. Molecular Simulation Studies of Polyphosphazenes

Polymer	*Force Field/ Method*	*Comments*	*Ref*
PnBuP[a], PsBuP[b], PiBuP[c]	COMPASS	MD simulation of gas permeability	(*10*)
PTFEP[d]	COMPASS	Gas permeability simulation Specific CO_2 interactions	(*19–21*)
PDCP[e]	ADMP COMPASS AMBER CHARMm	*Ab-initio* MD of hydrolysis Conformational model Crystal structure Parameterization of CHARMm	(*22*) (*23*) (*24*) (*25*)
oligomers[f]	Custom FF	Conformational analysis	(*26*)
PMPP[g]	CHARMm	Combined X-ray/MD study	(*27*)
P-DPNP[h]	CHARMm	Combined X-ray/MD study	(*28*)
PDOBPN[I]	AMBER	Thermal degradation, solution properties	(*29*)
Copolymer[j]	AMBER	Analysis of secondary structure	(*30*)
MEEP[k]	COMPASS	Li^+ transport	(*31*)
PBMP[l]	COMPASS	Direct methanol fuel cell	(*20*)
MP[m]	DREIDING	Ion diffusion	(*32*)
MEEP-like	AMBER	Ionic conduction	(*33*)
MEEP-like	UFF DREIDING	Li+ distribution in a polyphosphazene mixture	(*34*)

[a] PnBuP, poly(di-*n*-butoxyphosphazene. [b] PsBuP, poly(di-*sec*-butoxy-phosphazene). [c] PiBuP, poly(di-*i*-butoxyphosphazene). [d] PTFEP, poly-[bis(2,2,2)-trifluoroethoxyphosphazene]. [e] PDCP, poly[bis(chloro)phos-phazene]. [f] chloro- and difluorophosphazenes]. [g] PMPP, poly[di(4-methylphenoxy)phosphazene]. [h] P-DBNP, poly(2.2′-dioxy-1,1′-binaph-thyl)phosphazene). [I] PDOBPN, poly(2,2′-dioxybiphenylphosphazene). [j] copolymer of phenoxy and binaphthoxy chromophoric groups. [k] MEEP, poly[bis(2-(2-methoxyethoxy)ethoxy)phosphazenes]. [l] PBMP, poly[bis(3-methylphenoxy)phosphazene]. [m] MP, poly[bis(2-(2′-methoxyl)phospha-zene].

Molecular Simulations of Polyphosphazenes

There are many applications for polyphosphazenes. For example, their high permeability makes polyphosphazenes attractive as mem-branes for gas, liquid, and vapor separations. Other important uses for polyphosphazenes include controlled drug delivery, micro-encapsula-tion, polymer electrolytes, non-linear optical and electro-optic materials, and proton-exchange membranes for direct methanol fuel cells. For many of these applications, molecular simulations have provided impor-tant insight into polyphosphazene function as discussed in the following sections. Detailed reviews that address molecular simulations of poly-phosphazenes are provided in two other sources (*35*, *36*).

Polymer Electrolytes

Following the discovery of ion conduction in poly(ethylene oxide) in 1973 by Wright (*37*), there has been interest in looking at polymers with similar chemical structure that could provide attractive performance. One polyphosphazene structure that has attracted interest is poly[bis(2-(2-methoxyethoxy)ethoxy)phosphazene] (MEEP) whose structure is shown in Figure 2. There have been a number of simulation studies using different force fields to look at ion transport and ion coordination in polyphosphazenes (*31–34*). Several of these studies were cited in Table II. A more review of transport in polymer electrolytes is given by Fried (*35*). One study by Luther et al. (*31*) utilized both spectroscopic and MD simulation (COMPASS force field). This simulation employed 20-mer chains and focused on the coordination of Li^+ with MEEP. A mechanism generally proposed for ionic transport in MEEP is that Li^+ association is limited to the oxygen nuclei on the substituent group. The study by Luther et al. (*31*) showed that lithium association also occurs with the nitrogen nuclei of the polymer backbone. They suggested that the association is probably a coordination between the nitrogen nuclei and the oxygen nuclei of the substituent groups as illustrated in Figure 3.

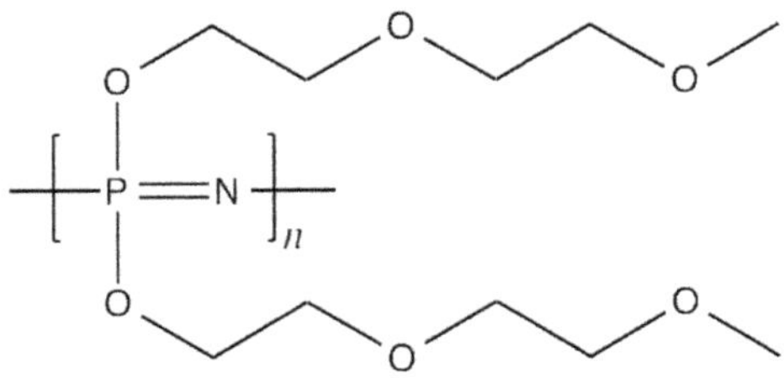

Figure 2. Chemical structure of MEEP.

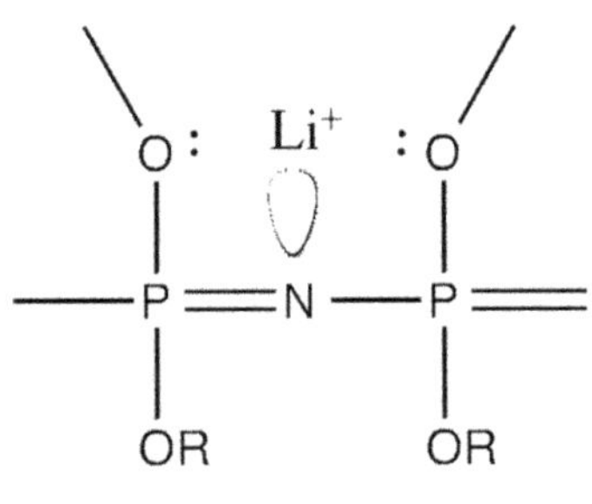

Figure 3. Lithium association with MEEP.

Gas-Separation Membranes

A comprehensive review of the molecular simulation of gas and vapor transport in highly permeable polymers, including polyphosphazenes, has been given by Fried (*36*). A review of the simulation of gas diffusion and solubility specifically for poly(organophosphazenes) appears in another review (*38*). Due to the high flexibility of the polyphosphazene chain and low T_g, the permeability of polyphosphazenes is among the highest of all major classes of polymers with the exception of polysiloxanes and poly[1-(trimethylsilyl)-1-propyne] (PTMSP) which is a glassy nanoporous polymer. In terms of simulation of gas transport properties, the most extensively studied polyphosphazenes have been the isomers of polybutoxyphosphazenes, mentioned earlier in reference to force field validation, and poly[bis(2,2,2-trifluoroethoxy)phosphazene] or PTFEP whose structure is shown in Table I.

Poly(dibutoxyphosphazenes)

A detailed molecular simulation study of gas permeability of all three amorphous poly(dibutoxyphosphazenes), PnBuP, PsBuP, and PiBuP (see Table I for structures) using the COMPASS force field parameter-ized for phosphazenes (*8*) was made by Fried and Ren (*10*). The permeability coefficient is a product of the solubility (S) and diffusivity (D) coefficients of the penetrant ($P = DS$). Both S and D can be obtained from molecular simulation.

In the study by Fried and Ren, the gas diffusion coefficient (D) of six gases (He, Ne, O_2. N_2, CO_2, and CH_4) were obtained from NVT molecular dynamics by use of the Einstein relationship written in the form

$$D_\alpha = \frac{1}{6N_\alpha} \lim_{t\to\infty} \frac{\mathrm{d}}{\mathrm{d}t} \sum_{i=1}^{N_\alpha} \left\langle \left| r_i(t) - r_i(0) \right|^2 \right\rangle \tag{3}$$

where N_α is the number of diffusing particles of type α, $r_i(0)$ and $r(\mathrm{t})$ are the initial and final positions of the center of mass of particle i over the time interval t, and $\left\langle \left| r_i(t) - r_i(0) \right|^2 \right\rangle$ is the mean-square displacement (MSD) averaged over the ensemble. The diffusion coefficients followed the correlation given by Toi (*39*) as

$$\log D = K_1 - K_2 d_{\mathrm{eff}}^2 \tag{4}$$

where K_1 and K_2 are constants and d_{eff} is the effective diameter of the gas molecule. The solubility coefficients were obtained by using the Grand Canonical Monte Carlo (GCMC) method that measures the concentration of sorbed gas (C) and S was obtained from the limiting slope of the sorption isotherm at zero pressure as

$$S = \lim_{p\to 0} \left(C/p \right) \tag{5}$$

where C is in units of cm^3(STP)/cm^3 polymer and p is pressure. The solubility coefficients obtained in this way followed the correlation proposed by Teplyakov and Meares (*40*) given as

$$\log S = \log S^{o} + m\left(\varepsilon/k\right) \quad (6)$$

where S^{o} is a constant and the ratio ε/k is the Lennard-Jones potential well depth of the gas.

Poly[bis(2,2,2-trifluoroethoxy)phosphazene]

Following the earlier simulation study of gas diffusion and solubility in poly(dibutoxyphosphazenes), Hu and Fried (*19*) reported results of molecu-lar simulation of diffusion and solubility of seven gases (He, H_2, O_2. N_2, CH_4, CO_2, and Xe) in PTFEP. Since PTFEP is semicrystalline, gas diffusion and solubility were obtained in an amorphous cell and in all α-orthorhombic crystalline cell of PTFEP. The COMPASS force field was also used in this study. Diffusion and solubility coefficient were obtained by use of eqs. (3) and (5), respectively. It was shown that only He showed any solubility in the α-orthorhombic crystalline cells of PTFEP although diffusion was unrestricted.

It was noted in this study that the solubility coefficient for CO_2 was elevated compared to the prediction expected by applying the correlation given by eq (6) to the other six gas molecules. Similar observations have been made for sorbed CO_2 in other fluorine-containing polymers. As shown by Fried and Hu (*21*), the cause of this elevated CO_2 solubility is a moderately strong dipole–quadrupole interaction (-11.5 kJ mol^{-1}) involving the polar fluoroethoxy group and a strong CO_2 quadrupole (*21*). The nature of this interaction was supported by both MD simulation (RDF) and high-level ab-initio calculations MP2/6-311++G**) in that study.

Polyphosphazenes for Biomedical Applications

Kroger and Fried (*23*) have reported a detailed molecular simulation of poly(dichlorophosphazne) (PDCP), both crystalline and amorphous state, poly(glycinethylesterphosphazene) (PGEEP), poly[bis(carbonylato-phenoxy)phosphazene) (PCPP), and poly[di(imidoazole)phosphazene] (PDIP). Chemical structures, T_gs, densities, and solubility parameters are given in Table III. The COMPASS force field was used to obtain density, T_g, and the solubility parameter (obtained from the cohesive energy density, CED), as well as time-correction functions that provided information concerning the flexibility of the P–N backbones and side chains.

Table III. Structures and Properties of Biomedically Relevant Polyphosphazenes

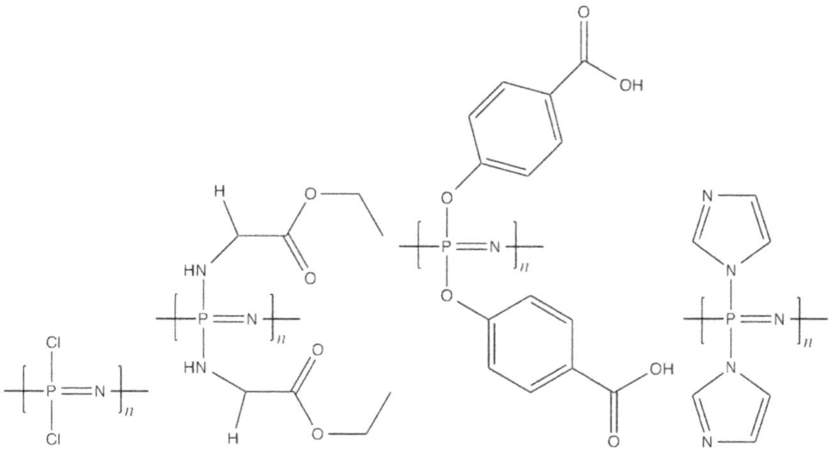

PDCP **PGEEP** **PCOO** **PDIP**

Polymer	T_g (K)	Density[a] (g/cc)	Solubility Parameter[a] $(J/cm^3)^{1/2}$
PDCP	NA	1.93±0.012 (exp 1.98) 2.174±0.006 (exp 2.17)	14.437±0.003
PGEEP	255 K (exp 233-255 K)	1.269±0.006	13.928±0.007
PCPP	NA	1.454±0.008	19.082±0.006
PDIP	NA	1.373±0.005	14.647±0.006

a. MD simulation results taken from Kroger and Fried[23]; error bounds represent a 95% confidence interval.

References

1. Allcock, H. R. *Chemistry and Applications of Polyphosphazenes*; John Wiley & Sons: Hoboken, NJ, 2003.
2. Potin, P.; Delaeger, R. Polyphosphazenes: synthesis, structure, properties, applications. *Eur. Polym. J.* **1991**, *27* (4-5), 341–348.
3. Sun, H. COMPASS: An ab initio force-field optimized for condensed-phase applications—overview with details on alkane and benzene compounds. *J. Phys. Chem. B* **1998**, *102* (38), 7338–7364.
4. Sun, H. Molecular structures and conformations of polyphosphazenes — A study based on density functional calculations of oligomers. *J. Am. Chem. Soc.* **1997**, *119* (15), 3611–3618.

5. Mark, J. E.; Schaefer, D. W.; Lin, G. *The Polysiloxanes*; Oxford University Press: Oxford, 2015.
6. Hwang, M. J.; Stockfisch, T. P.; Hagler, A. T. Derivation of class II force fields. 2. Derivation and characterization of a class II force field, CFF93, for the alkyl functional group and alkane molecules. *J. Am. Chem. Soc.* **1994,** *116*, 2515–2525.
7. Maple, J. R.; Hwang, M.-J.; Stockfisch, T. P.; Dinur, U.; Waldman, M.; Ewig, C. S.; Hagler, A. T. Derivation of class II force fields. I. Methodology and quantum force field for the alkyl functional groups and alkane molecules. *J. Comput. Chem.* **1994**, *15*, 162–182.
8. Sun, H.; Ren, P.; Fried, J. R. The COMPASS force field: parameterization and validation for phosphazenes. *Comput. Theor. Polym. Sci.* **1998**, *8* (1/2), 229–243.
9. Hirose, T.; Mizoguchi, K. Gas transport in poly(alkoxyphosphazenes). *J. Appl. Polym. Sci.* **1991**, *43*, 891–900.
10. Fried, J. R.; Ren, P. The atomistic simulation of the gas permeability of poly(organophosphazenes). Part 1. Poly(dibutoxyphosphazenes). *Comput. Theor. Polym. Sci.* **2000**, *10* (5), 447–463.
11. Fried, J. R.; Ren, P. Molecular simulation of the glass transition of polyphosphazenes. *Comput. Theor. Polym. Sci.* **1999**, *9*, 111–116.
12. Fried, J. R. Computational Parameters. In *Physical Properties of Polymers Handbook*, second ed.; James, E. M., Ed.; Springer: New York, 2007.
13. Li, S.; Fried, J. R.; Sauer, J.; Colebrook, J.; Dudis, D. S. Computational chemistry and molecular simulations of phosphoric acid. *Int. J. Quantum Chem.* **2011**, *111*, 3212–3229.
14. Sun, H.; Jin, Z.; Yang, C.; Akkermans, R. L. C.; Robertson, S. H.; Spenley, N. A.; Miller, S.; Todd, S. M. COMPASS II: extended coverage for polymer and drug-like molecule databases. *J. Mol. Model.* **2016**, *22*.
15. Brooks, B. R.; Bruccoleri, R. E.; Olafson, B. D.; States, D. J.; Swaminathan, S.; Karplus, M. CHARMM: A program for macromolecular energy, minimization, and dynamic calculations. *J. Comput. Chem.* **1983**, *4* (2), 187–217.
16. Cornell, W. D.; Cieplak, P.; Bayly, C. I.; Gould, I. R.; K. M. Merz, J.; Ferguson, D. M.; Spellmeyer, D. C.; Fox, T.; Caldell, J. W.; Kollman, P. A. A second generation force field for the simulation of proteins, nucleic acids, and organic molecules. *J. Am. Chem. Soc.* **1995**, *117*, 5179–5197.
17. Mayo, S. L.; Olafson, B. D.; Goddard, I. W. A. DREIDING: a generic force field for molecular simulations. *J. Phys. Chem.* **1990**, *94* (26), 8897–8909.
18. Rappe, A. K.; Casewell, C. J.; Colwell, K. S.; III, W. A. G.; Skiff, W. M. UFF, a full periodic table force field for molecular mechanics and molecular dynamics simulations. *J. Am. Chem. Soc.* **1992**, *114* (25), 10024–10035.
19. Hu, N.; Fried, J. R. The atomistic simulation of the gas permeability of poly(organophosphazenes). Part 2. Poly[bis(2,2,2-trifluoroethoxy)phosphazene]. *Polymer* **2005**, *46*, 4330–4343.
20. Hu, N. *Molecular Simulation of Polyphosphazenes As Gas Separation and Direct Methanol Fuel Cell Membranes*. Ph.D. Dissertation, University of Cincinnati, 2003.

21. Fried, J. R.; Hu, N. The molecular basis of CO_2 interaction with polymers containing fluorinated groups: computational chemistry of model compounds and molecular simulation of poly[bis(2,2,2-trifllluoroethoxy)-phosphazene]. *Polymer* **2003**, *44*, 4363–4372.
22. Kroger, J. L.; Fried, J. R.; Skelton, A. A. Computational simulations of hydrolysis of phosphazene oligomer utilizing atom-center density matrix propagation. *Int. J. Quantum Chem.* **2013**, *113*, 63–70.
23. Kroger, J. L.; Fried, J. R. Molecular simulations of polyphosphazenes for biomedical applications. *J. Inorg. Orgomet. Polym.* **2012**, *22*, 973–984.
24. Tarazona, M. P.; Saiz, E. A conformational model for pol(dichloro-phosphazene) derived from molecular dynamics simulations. *Polymer* **2000**, *41*, 3337–3347.
25. Amato, M. E.; Grassi, A.; Lipkowitz, K. B.; Lombardo, G. M.; Pappalardo, G. C.; Sadun, C. Molecular dynamics simulations of polyphosphazenes: poly[bis)chloro)phosphazene]{$NPCl_2$]. *J. Inorg. Orgomet. Polym.* **1996**, *6* (3), 237–253.
26. Boyd, R. H.; Kesmer, L. Conformational analysis of phosphazenes. A force field for the calculation of the molecular structures of halophosphazenes. *J. Am. Chem. Soc.* **1977**, *99* (13), 4248–4256.
27. Caminiti, R.; Gleria, M.; Lipkowitz, K. B.; Lombardo, G. M.; Pappalardo, G. C. Molecular dynamics simulations combined with large angle X-ray scattering technique for the determination of the structure, conformation, and conformational dynamics of polyphosphazene in amorphous phase: study of poly[di(4-methylphenoxy)phosphazene]. *J. Am. Chem. Soc.* **1997**, *119*, 2196–2204.
28. Carriedo, G. A.; Alonso, F. J. G.; Alvarez, J. L. G.; Lombardo, G. M.; Pappalardo, G. C.; Punzo, F. Molecular dynamics (MD) simulations and large-angle X-ray scattering (LAXS) studies of the solid-state structure and asssembly of isotactic (R)-poly(2,2′-dioxy-1,1″-binaphthyl)-phosphazene in the bulk state and in the cast film. *Chem. Eur. J.* **2004**, *10* (15), 3775–3782.
29. Laguna, M. T. R.; Tarazona, M. P.; Carriedo, G. A.; Alonso, F. J. G.; Fidalgo, J. I.; Saiz, E. Thermal degradation and solution properties of poly(2,2′-dioxybiphenyl phosphazene). *Macromolecules* **2002**, *35* (19), 7505–7515.
30. Marcelo, G.; Saiz, E.; Mendicuti, F.; Carriedo, G. A.; Alonso, F. J. G.; Alvarez, J. L. G. Analysis of the secondary structure of random copolymers by a combination of fluorescence and molecular dynamics methods. Application to polyphosphazenes containing phenoxy and binaphthoxy groups. *Macromolecules* **2006**, *39* (2), 877–885.
31. Luther, T. A.; Stewart, F. F.; Budzien, J. L.; LaViolette, R. A.; Bauer, W. F.; Harrup, M. K.; Allen, C. W.; Elayan, A. On the mechanism of ion transport through polyphosphazene solid polymer electrolyte: NMR, IR, and Raman spectroscopic studies and computational analysis of 15N-labeled polyphosphazenes. *J. Phys. Chem. B* **2003**, *107*, 3168–3176.
32. Balbuena, P. B.; Lamas, E. J.; Wang, Y. Molecular modeling studies of polymer electrolytes for power sources. *Electrochim. Acta* **2005**, *50*, 3788–3795.

33. Allcock, H. R.; Napierala, M. E.; Olmeijer, D. L.; Best, S. A.; K. M. Merz, J. Ionic conduction in polyphosphazene solids and gels: ^{13}C, ^{31}P, and ^{15}N NMR spectroscopy and molecular dynamics simulations. *Macromolecules* **1999**, *32*, 732–741.
34. Wang, Y.; Balbuena, P. B. Combined ab initio quantum mechanics and classical molecular dynamics studies of polyphosphazene polymer electro-lytes: Competitive solvation of Li^+ and $LiCF_3SO_3$. *J. Phys. Chem. B* **2004**, *108*, 15694–15702.
35. Fried, J. R., Transport properties of polyphosphazenes. In *Polyphosphazenes for Biomedical Applications*; Andrianov, A. K., Ed.; John Wiley & Sons: New York, 2009; pp 325−344.
36. Fried, J. R., Molecular simulation of gas and vapor transport in highly permeable polymers. In *Materials Science of Membrane for Gas and Vapor separation*; Yampolski, Y.; Pinnau, I.; Freeman, B. D., Eds.; John Wiley & Sons, Ltd.: West Sussex, England, 2006; pp 95−136.
37. Fenton, D. E.; Wright, J. M.; Wright, P. V. *Polymer* **1973**, *121*, 91–99.
38. Fried, J. R. Gas diffusion and solubility in poly(organophosphazenes): Results of molecular simulation studies. *J. Inorg. Organometallic Polym. Mater.* **2006**, *16* (4), 407–418.
39. Toi, K.; Morel, G.; Paul, D. R. Gas sorption and transport in poly(phenylene oxide) and comparisons with other polymers. *J. Appl. Polym. Sci.* **1982**, *27* (8), 2997–3005.
40. Teplyakov, V.; Meares, P. Correlation aspects of the selective gas permeabilities of polymeric materials and membranes. *Gas Sep. Purif.* **1990**, *4* (2), 66–74.

Editors' Biographies

Alexander K. Andrianov

Alexander K. Andrianov is a Research Professor at the Institute for Bioscience and Biotechnology Research, University of Maryland. Prior to this he has served in various executive and managerial roles in a number of biotechnology companies focusing on drug delivery technologies and biomaterials. Dr. Andrianov has long-standing research interests in the discovery and development of novel synthetic macromolecules for life sciences applications, including biodegradable materials, self-assembling polymers, immunostimulants, and nanoparticulate drug delivery systems. He has been involved in all aspects of technology development and commercialization, including product advancement from research laboratories to manufacturing and clinical trials. He is an author of over 70 peer-reviewed articles, editor of the *Polyphosphazenes for Biomedical Applications* book, and an inventor on over 40 issued and pending patents.

Harry R. Allcock

Harry R. Allcock was born and educated in England. His B.Sc. and Ph.D. degrees were awarded by the University of London. After a year as a postdoctoral scientist at Purdue University and two years at the Canadian National Research Council in Ottawa, he spent five years as a research scientist in the laboratories of the American Cyanamid Company in Stamford, Connecticut, before moving to the Pennsylvania State University in 1966. In the Penn State Chemistry Department, he and his coworkers have focused on polyphosphazene synthesis, properties, and possible applications. His group has published 640 research papers on polyphosphazenes and small molecule models, and he has authored or co-authored six books including *Contemporary Polymer Chemistry* (1981, 1990, and 2003), *Phosphorus-Nitrogen Compounds* (1972), *Chemistry and Applications of Polyphosphazenes* (2003), and *Introduction to Materials Chemistry* (2008; 2nd edition 2018). Professor Allcock is the recipient of three American Chemical Society National Awards in Polymers and Materials, three ACS Polymer Division Awards, and the International Award of the Polymer Society of Japan. He is a member of the American National Academy of Engineering, with membership in the Medical and Chemical Engineering sections.

Indexes

Author Index

Subject Index

R

S

W

www.ingramcontent.com/pod-product-compliance
Ingram Content Group UK Ltd.
Pitfield, Milton Keynes, MK11 3LW, UK
UKHW020251290926
15470UKWH00002B/862